ONE WEEK LOAN

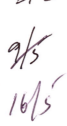

$$D + B =$$

excavate
Demolish $= GC/Works/1$

$+$

ICE Contracts

$\}$ Page.

258-

civil
engineering
works $= ICE \quad Contract \quad = \quad 279.$

Construction Management
Principles and Practice

Alan Griffith
and
Paul Watson

First published 2004 by
PALGRAVE MACMILLAN
Houndmills, Basingstoke, Hampshire RG21 6XS and
175 Fifth Avenue, New York, N.Y. 10010
Companies and representatives throughout the world

PALGRAVE MACMILLAN is the global academic imprint of the Palgrave
Macmillan division of St. Martin's Press, LLC and of Palgrave Macmillan Ltd.
Macmillan® is a registered trademark in the United States, United Kingdom
and other countries. Palgrave is a registered trademark in the European
Union and other countries.

ISBN 0–333–96878–6

This book is printed on paper suitable for recycling and made from fully
managed and sustained forest sources.

A catalogue record for this book is available from the British Library.

10 9 8 7 6 5 4 3 2 1
13 12 11 10 09 08 07 06 05 04

Printed and bound in China

Acknowledgements

The authors gratefully acknowledge *all* those individuals who have contributed and allowed the
inclusion of the supporting materials used in this book. Particular thanks are given to
Amalgamated Construction Company Ltd and to Taylor Woodrow Construction Ltd.

Contents

Preface

Construction Management: Principles and Practice focuses on the activities of the principal contracting organisation when undertaking a construction project. The aim of the book is to introduce and explain those key management functions and processes applicable throughout the siteworks of a construction project and consider the basic concepts, principles, systems and procedures which impact on the management functions and processes carried out. The fundamental tenet of this book is that the comprehensive understanding and considered undertaking of key site management functions and processes is a prerequisite to organisational effectiveness and project success. While the book focuses on site management aspects it must always be remembered that appropriate, efficient and effective company policies, strategies and systems, construction technology and production methods are essential to the undertaking of any project. Moreover, the contribution and management of the people who make construction happen lie at the heart of any construction project. All of these aspects contribute to project success.

Many long-standing management functions, concepts, principles and procedures are traditionally applied to the site management of construction works: site organisation; planning and progressing; cost control; plant and materials procurement. These and others are well understood, recognised and accepted. The effective undertaking of such aspects has successfully delivered many construction projects. In more recent times, not necessarily new but evolving and reconfigured site management functions and processes have become prominent in meeting the increasing and substantial demands which have emerged from both within and outside the construction industry. Over the past twenty-five years quality assurance management, for example, has evolved to become a fundamental, accepted and significant part of construction management, with recent developments seeing the introduction of the quality management standard BS EN ISO 9000: 2000 and the European Foundation for Quality Management Business Excellence Model (EFQM). Moreover, quality management has served as a driver for a more systematic approach to other management functions such as health and safety and environmental safeguard. Site management needs to be aware of the many both traditional and evolving concepts and practices within construction which impinge on the company and project site organisations with which they are involved. The book seeks to assist in meeting this need.

ALAN GRIFFITH
PAUL WATSON

List of Abbreviations

AA%RR	Average Annual Percentage Rate of Return
AAR	Average Annual Return
ADR	Alternative Dispute Resolution
BEP	Break-even point
BPR	Business Process Re-engineering
BS	British Standard
BSI	British Standards Institution
CDM	Construction (Design and Management) Regulations 1994
CEO	Chief Executive Officer
CHSWR	Construction (Health, Safety and Welfare) Regulations 1996
CII	Construction Industry Institute
CIOB	Chartered Institute of Building
CIRIA	Construction Industry Research and Information Association
COSHH	Control of Substances Hazardous to Health Regulations 1999
CPR	Construction Process Re-engineering
CSF	Critical Success Factor
CSM	Construction Site Manager
EA	Environment Agency
EC	European Community
EFQM	European Foundation for Quality Management
EIA	Environmental Impact Assessment
EMP	Environmental Management Plan
EMS	Environmental Management System
EPA	Environmental Protection Act 1990
EU	European Union
FC	Fixed Costs
H&SMS	Health and Safety Management System
HSE	Health and Safety Executive
HSWA	Health and Safety at Work, etc. Act 1974
IiP	Investors in People Standard
IRR	Internal Rate of Return
ISO	International Organisation for Standardisation
IT	Information Technology
JCT	Joint Contracts Tribunal
JIT	Just in Time
KF	Key Factor
MD	Managing Director
MHSWR	Management of Health and Safety at Work Regulations 1999
MRC	Market Requirements Curve
NEDO	National Economic Development Council
NPV	Net Present Value
PDCA	Plan, Do, Check, Act Control Cycle

PEST	Political/Legal, Economic, Social/Cultural and Technological
PPE	Personal Protective Equipment at Work Regulations 1992
PV	Present Value
QA	Quality Assurance
QMS	Quality Management System
RADAR	Results, Analysis, Deployment, Assessment and Review
RIDDOR	Reporting of Injuries, Diseases and Dangerous Occurrences Regulations 1995
SCM	Supply Chain Management
SMEs	Small and Medium Enterprises
SWOT	Strengths, Weaknesses, Opportunities, Threats
TC	Total Costs
TQM	Total Quality Management
TR	Total Revenue
VC	Variable Costs
VM	Value Management

1 Introduction

The purpose of this chapter is to introduce the topics to be discussed in subsequent chapters of the book. Within each chapter the aim is to provide an introduction to the relevant management concept, principles, systems, procedures and practices that might be considered by a principal contracting organisation when managing that aspect. Each management aspect presented has been the subject of many textbooks over the years. For a comprehensive consideration and understanding of particular management concepts the reader is directed, as a starting point, to the Bibliography at the end of this book.

Although this book focuses on the management activities of the principal contracting organisation, reference is made, where required, to the roles, responsibilities and practices of other parties working alongside the principal contractor in managing a construction project. As a principal contracting organisation often supports a large company infrastructure in addition to its many temporary project site infrastructures, corporate as well as project site activities are presented where appropriate. Examples from recently completed construction projects are presented to support the text and references are provided at the end of chapters to assist further reading. The national, European and international standards referred to in this introductory chapter are referenced within the main chapter in which they appear. It will be evident that some topics overlap, and this is deliberate and essential in outlining the linkages that exist between particular management concepts.

It is taken as axiomatic that any construction project is undertaken within the framework of the form of contract being administered. It is not an objective of this book to define and describe potential forms of contract such as the JCT 80 Standard Form of Building Contract and its derivatives or the New Engineering Contract (NEC) branded suite of contracts. Although specific duties and responsibilities are placed upon the principal contractor under various forms of contract, the management systems and procedures used in meeting the site management of construction works are applicable generally to all construction contracts. By entering into a contract for construction works the principal contractor is charged with delivering the project to the client's requirements for time, cost and quality while ensuring that welfare, health and safety and environmental aspects are safeguarded.

It is emphasised that it is not this book's intention to be prescriptive nor to lay down the best practice for the activities of any principal contracting organisation. Any contracting company will have its own best ways of approaching each management aspect of site construction works within the context of its individual, and possibly unique, organisational circumstances and the particular nature of its construction projects.

Corporate and Project Management

It will become evident from the subsequent chapters of this book that the principal contracting organisation will utilise management concepts and principles which require application within both the company, or corporate, organisation and the project, or site, organisation of the construction contracts that it undertakes. A clear example of this is the holistic approach to health and safety management. The company organisation will provide strategic vision, ethos, policies, objectives and senior management commitment while the project organisation will provide safe working procedures in delivering the

1

construction product on site. To deliver successful projects, and to enable the business of the company to flourish, the principal contractor needs to put strong and effective links in place between the company organisation and the construction project organisation.

Although having many elements and functions common to all projects, construction also differs from project to project. The use of decentralised management structures across wide geographic locations, delegation to virtually autonomous project sites, the temporary nature of project site organisation and the involvement of a transient workforce all serve to illustrate that the management of construction can be far from simple and straightforward.

For any principal contracting organisation to be managed effectively there must be a well-conceived and recognised company structure. This is necessary to provide the foundation upon which the formal organisation of the business activities and assignment of management responsibilities are established. It is also a prerequisite to developing the clear business aims, objectives and policies which guide and stabilise the business.

It must always be remembered that a principal contracting organisation, like any company, involves people. Individuals and groups lie at the heart of organisational activities. They are central to shaping the organisation, to getting the work done and to driving and maintaining the company. Within the many management and operational processes and resources that combine to deliver a construction project it is easy to forget the contribution of those people that really make it happen. Effective company organisation will establish the formal structure that is necessary to ensure that people can work at their best and support the business of the company.

Also necessary are methods, or systems, by which the principal contracting organisation will deliver and drive its business. Systems include management procedures which guide the undertaking of the organisation's business and working instructions which guide the processes, operations and tasks that are needed to deliver the business, be it a product or a service. A system may be simple, reflected in a schedule of tasks needed to complete a brief and routine process, or may encompass the whole functional management aspect within the organisation, for example an environmental management system. A system can be tailor-made to meet the requirements of part or all of a management function and be recognised only by the intra-organisation, or it may be standards-based and have external certification such as a quality management system meeting the international standard BS EN ISO 9000: 2000.

Whatever the management systems used by the company it is essential that they are interactive with and mutually supportive of the project organisation on site. To facilitate the translation and application of the management systems to its construction projects, the company will need to establish robust site organisation. An effective project, or site, structure will provide the means for clear lines of authority, responsibility, communication and feedback to be established within and around the principal contracting organisation and its activities – all fundamentally necessary to managing the project and achieving success.

Effective and Efficient Site Management

Both efficiency and effectiveness are vital components of the management of any construction project. The key aspects contributing to these two components are: the role and functions of the site manager; the management styles adopted; organisational structure and the corporate planning processes; anthropocentric production systems; and motivation of staff and operatives. All of these can be encapsulated under the umbrella of achieving 'best practice' and, in so doing, the attainment of a 'sustainable competitive advantage' for the principal contracting organisation.

It is important that the holistic nature of an organisation is not forgotton, and therefore although the key aspects can be appreciated individually they are not mutually exclusive. For example, leadership will have a considerable impact upon motivation and hence upon the efficiency and effectiveness of the construction project. A further and important dimension to these key components is that they exist to support the critical function of providing client satisfaction. In practice, this needs to be delivered with due consideration to the utilisation of finite organisational resources. Therefore, these aspects of the principal contracting organisation's operation are far from simple and straightforward.

Some of these key organisational components have traditionally been problematic in practical application by principal contracting organisations. Nevertheless, generic approaches are recognised and advocated as a positive way forward in accommodating some of the difficulties which prevail. The thrust of appreciating the issues more comprehensively involves having a good understanding of what organisational resources are available. Furthermore, there must be a full appreciation of what strategies and management techniques need to be deployed in order to maximise resource utilisation, or efficiency, and to attain set organisational goals, or effectiveness.

Planning and Programmes

For any construction project there is the fundamental requirement to plan, monitor and control all activities and all resources. Planning, the process of determining, analysing, devising and organising the resources required for a construction project, and programming, or sequencing, of those resources using any one or more of a number of programme types are two of the most traditional of all construction management functions. On almost all construction projects planning is undertaken at three key phases within the total construction process, at the pre-tender, pre-contract and contract phase. Each phase is an important management consideration to the principal contracting organisation, providing the tender price, outlining the strategic resourcing for the project and providing detailed sequences for the sitework operations.

Large construction projects are complex and interdisciplinary in nature. They embrace many independent yet interdependent items of work, many operational tasks and involve the contribution of many participants. Effective management requires the implementation of accurate and robust planning and programming. Moreover, the principal contracting organisation must not only ensure the efficient and effective management of any one construction project but may need to manage many projects simultaneously. Effective construction planning lies at the heart of good project performance and hence company success; it can make the difference between project profit or loss for the principal contractor. The primary objective of planning and programming is to allow site management to obtain the necessary volume, speed and quality of work with the best use of resources and with the greatest economy to the contractor. In this way, value for money is delivered to the client and the principal contracting organisation secures a financial profit on the project.

The most important aspect of construction planning is the project's construction technology and the determination of the construction methods to be used for the siteworks. As a consequence of this, a key element of planning is the preparation of method statements. These consider and depict the crucial methods of undertaking the works and only when these are determined can sequences be considered and programmes drawn up. Planning and programmes will be of little value unless they accurately reflect the construction methods to be used.

While other management functions such as quality assurance, environmental safeguard and health and safety are characterised by their standards-based, certificated systems, management approach to construction planning is structured and systematised predominently by those particular programming methods and tools used. The most commonly utilised methods of programming can be broadly grouped into Gantt, or bar, charts and network analysis, although a number of other methods can be applied in specific situations. Tools exist to aid the application of these methods and these are computer-based and utilise a variety of software packages. These assist with producing method statements, sequences of operational elements and also with the presentation of planning outputs in the forms of many available construction programmes. A principal contracting organisation will probably favour particular programming methods and tools and these will form the basis of the company approach to managing the planning of their construction projects.

Operations, Cost Planning and Monitoring Expenditure

A fundamental activity for any principal contracting organisation is to ensure that operational functions provide the means for achieving a sustainable competitive advantage. Therefore, strategic decisions should not be taken without due consideration being given to their impact upon operations. If this is not ensured, the resulting outcome could be that, far from providing a competitive advantage, they become a corporate millstone. The core function for success is an appropriate operations strategy, together with the implementation of key management tools and techniques such as business process re-engineering (BPR), partnering, and financial control and decision-making aids.

It is important to recognise the integrated rather than the separate and often disparate nature of tools and techniques in application. Project success will depend on understanding how the various key concepts, principles and practices are understood holistically and are translated into operational strategies which through implementation underpin company strategies.

The decision-making processes used are crucial to making timely and effective determinations of what needs to be done, when things have to be done and how things will be done. Effective decision-making relies upon the collection, analysis and evaluation of information. In the construction siteworks phase the focus is invariably upon the assessment of quantitative data although it should be remembered that thorough and transparent decision-making also requires the assessment of qualitative data.

The key function of financial planning lies at the core of understanding data. It is vital to the operation of the whole organisation and therefore involves both operational and corporate levels. Cash flow provides the life-blood for any corporate enterprise and so a comprehensive understanding of cumulative cash-flow analysis is required.

Site Establishment

Running in parallel with the pre-tender, pre-contract and contract planning stages of any large construction project is the planning of site establishment. Almost all construction projects require the principal contractor to operate remotely from its corporate organisation, providing a temporary project infrastructure and organisation to support the siteworks. While site establishment may be minimal on small building works, it may be costly and extensive on large construction engineering projects. Site establishment on a large project could include project preliminary items, management and supervisory organisation, site layout and temporary facilities, and personnel health and welfare provision as key elements. The careful and thorough consideration of site establishment can have a considerable influence on the efficient and effective undertaking of the siteworks. Inexhaustive organisational and project specific variables that alter from one project to another preclude the postulation of any particular approach, although recognised common principles and practice can be considered.

Aspects of the site establishment will be intrinsic components of the contract between the client and the principal contractor and be costed within the tender price. There are also formal responsibilities placed upon the principal contractor by current legislation such as that on health and welfare. In addition, many aspects are voluntary undertakings by the principal contractor and follow the sound construction management practice of establishing sensible and workable site layout and clear and simple management organisation. A clear management and supervisory structure is essential to putting in place the necessary channels of communication, responsibility, authority and control among the members of the site team. Good site layout is essential if a safe and healthy working environment for all persons working on and coming into contact with the operation of the site is to be provided.

Within a raft of comprehensive construction legislation, the Construction (Design and Management) Regulations 1994 require that the principal contractor provides a planned safe working environment while the Construction (Health, Safety and Welfare) Regulations 1996 require, among many needs, that good welfare facilities are available to employees on site. Good site establishment will not only satisfy such legislation but provide a comfortable and rewarding environment in which management and operatives feel able to work positively and productively. As such, effective site establishment makes a valuable contribution to the holistic success of any construction project.

Plant and Materials

A principal contracting organisation will only remain a profitable business if it maintains a sustainable competitive advantage over other contractors in undertaking its projects, and this relies upon the efficient and effective procurement and use of plant and materials.

Today, with the greater propensity for subcontracting, the principal contractor may not have to carry a large stock of plant and hiring-in can also provide considerable financial advantage. Nevertheless, some large principal contractors have significant plant holdings and these must be deployed to site through careful and comprehensive planning. Likewise, the careful planning of materials is essential to the successful running of any project. Plant and materials planning for construction projects impinges upon the planning processes at both corporate and project levels. The critical link between strategic planning and operational planning must be recognised and embraced within the management techniques and practices adopted.

Terotechnology, Just-in-Time procurement (JiT) and Supply Chain Management (SCM) are techniques which provide the principal contracting organisation with the basis for efficient and effective management of plant and materials on site while also forging the planning links between the corporate and project organisations. Although terotechnology appreciates economic life-cycle costs of plant acquisition, ownership and use, JiT focuses on the precision delivery of materials and components to the project as and when they are required to maintain production continuity, while SCM addresses the sequences of ordering, manufacture and supply to site. All focus on providing positive, continuous and synergistic improvements to the processes involved. This chapter focuses on these important advanced management concepts, embracing both the advantages of application and some of the key difficulties. The content does not address the technological aspects of materials choice and use nor the selection, maintenance and utilisation of plant *per se*, as many texts on both subjects already exist.

Progress and Control

Maintaining progress throughout the siteworks and taking decisions leading to timely and constructive action to keep the works on track with the planned programme are essential to the effective management of any construction project. Accurate short-term planning is key to maintaining progress and control. Short-term planning facilitates the detailed programming of specific sections of the contract programme, allowing focused progress monitoring and closer control. Feedback from these aspects into the contract, or master, programme allows the progress and control of the total siteworks to be presented and reviewed as required. Effective control of progress ensures that resources are allocated in their appropriate configuration and sequence and at the right time to conduct operational tasks in line with the programme. Both programming and progressing are therefore fundamental and significant to the efficient and effective management of the project.

Progress and control mechanisms are only useful where timely action is taken following any departure from the works that have been planned as reflected in the contract programme. Effective monitoring and review together with active and on-going progress charting are necessary. These should be commensurate with the reporting and evaluation mechanisms used on the project, which are usually presented at the client's formal site progress meetings. Progress monitoring, charting, evaluation and review at appropriate and regular meetings will, as far as practicable, provide the most effective control of progress for the project.

Quality Assurance

To any principal contracting organisation the quality of service provision to its clients is paramount. Quality is a key performance indicator by which principal contracting organisations differentiate themselves from their competitors, and in today's highly competitive marketplace it is fundamental to the organisation's continued operation of the enterprise and in the drive for increased profits. Construction organisations simply cannot operate without corporate policies and strategies for quality management and mechanisms to translate these into effective operational practice on the construction projects they undertake.

There has been a great proliferation of principal contracting organisations who are certificated under the BS EN ISO 9000 series quality management systems. The increase in the number of certificated construction firms reinforces the realisation and importance of quality in marketplace positioning within the construction industry. It is therefore crucial to appreciate that the BS EN ISO 9000 series has, in recent times, undergone considerable change, evolving into the BS EN ISO 9000: 2000 series. Principal contracting organisations need to be aware of such changes together with the suitable frameworks and approaches for accommodating them.

Many organisations have implemented, or are considering at this time, the principles and practices of total quality management (TQM) although this has proved to be far more involved and problematic than application of quality assurance under the BS EN ISO 9000 series. An alternative approach for the implementation of TQM has been established by the European Foundation for Quality Management (EFQM) through its quality 'Business Excellence Model'. This model is a dynamic approach for engaging in the corporate drive for continued improvement based upon detailed self-assessment against challenging performance indicators, including external benchmarking. Principal contracting organisations will find that a good understanding of EFQM may assist them to extend the boundaries of quality management achievement within their organisations. As with any corporate-based management system the concepts and principles must be translated into practical management procedures and working instructions for application on site. This is achieved through the implementation of project quality plans which incorporate the many generic and project specific facets of systematic quality management.

The role and commitment of people are essential in all construction site activities together with the contributions of those individuals and groups who support construction projects from the company, or corporate, organisation. Investors in People (IiP) has in recent years become prominent as organisations seek to optimise the skills, knowledge and commitment of their employees. Within the BS EN ISO 9000: 2000 series and the EFQM model the contribution of people is clearly evident and therefore there are strong links with the concept and principles of IiP. Investing in people and creating a learning-organisation can assist construction companies to get more from their resources, develop a culture of improvement and become better placed to meet the increasing demands of construction both now and in the future.

Health and Safety

In the late 1980s the Health and Safety Executive (HSE) published *Blackspot Construction*, a report on the occurrence of accidents within the construction industry (HSE, 1988). Considered to be the definitive report, it showed in stark and clear terms the potential hazards present on all construction sites and the possible dangers to the health, welfare and safety of construction workers. In highlighting the principal reasons for fatal accidents within the construction processes the overriding conclusion of the report was that accidents within construction are avoidable. The Egan report *Rethinking Construction* (DETR, 1998) stated that "*the health and safety record of construction is the second worst of any industry.*" Although a declining trend in the number of fatal accidents within construction was evident between the late 1980s and the late 1990s, future accident profiles will always be uncertain.

Over the last twenty-five years the welcome implementation of European Directives and associated national regulations of European Community (EC) member states has led to significant developments in UK health and safety legislation. Among many revised and new regulations augmenting the Health and Safety at Work, etc. Act 1974, The Construction (Design and Management) Regulations 1994 are perhaps the most significant and well recognised (HSE, 1974; 1994). Their introduction saw a departure from the traditional and long-standing expectation for health and safety management on a construction project to the meeting of legislative requirements by duty holders throughout a construction project.

Construction work is by its very nature hazardous. Health and safety is a paramount consideration for all those who are involved with or come into contact with a construction project. The CDM Regulations 1994 places the principal contracting organisation at the forefront of health and safety management on the construction site. On almost all construction projects, following the provision of a pre-tender health and safety plan procured by the client organisation, the principal contractor is required to develop a construction-phase health and safety plan and moreover to ensure that health and safety aspects are effectively managed throughout the production stage on site.

Many principal contracting organisations recognise that an appropriate way to meet their onerous responsibilities for health and safety is to deliver the establishment of a safe and healthy working environment and to ensure that health and safety aspects of a construction project are clearly recognised, risk-assessed, planned, organised, controlled, monitored, recorded, audited and reviewed in a systematic and robust way. One approach to achieving this is to implement a health and safety management system (H&SMS). Moreover, it is essential that effective health and safety management on site is well supported by an effective corporate organisation which provides a strong and positive commitment to health and safety reinforced by clear company policies, plans, procedures and practice. A health and safety management system may be configured around existing organisational management systems or be a dedicated standards-based system following BSI-OHSAS 18001, the specification for occupational health and safety management systems. Whatever form a management approach may take, the focus of the principal contracting organisation is to plan, establish and maintain a safe and healthy working environment on site throughout the duration of the construction project.

Environment

The construction industry plays a key role in shaping and developing the built environment. It also has an undisputed and significant impact upon the environment. The industry has for some time been seriously challenged to become more environmentally friendly and promote sustainability. This challenge requires the delivery of change within the industry, with new forms of relationships being configured, new and revised methods of working being utilised and more demanding levels of performance being met. Along with all participants to the total construction process the principal contracting organisation has a key and vital role to perform as a driver of change. Principal contractors must deliver timely, cost-conscious, safe and quality performance on its projects, all within a context of effective environmental safeguard.

Compared to industries in the manufacturing, chemicals and process engineering sectors over the last twenty-five years, construction has perhaps been slow to respond to environmental demands. Environmental pressure is growing however, with increasingly stringent environmental legislation being introduced, the requirement placed by clients upon contractors for environmental prequalification being more widely demanded and, indeed, greater interest, cogniscance and expectations being held within the business and commercial world and also by the general public.

Within the construction industry, strong parallels have emerged over the last decade in the development of environmental management concepts and the established concepts of quality management and quality assurance. Quality management has given rise to performance benchmarking, contract prequalification and management-systems certification widely throughout the service and supply sectors to construction. Since the introduction of BS 7570: Specification for Environmental Management Systems in 1992, and its international counterpart ISO 14001 in 1996, now integrated with the European standard and known as BS EN ISO 14001, major client organisations in both the public and private sectors have considered environmental management as an important aspect of project evaluation and development.

In meeting the requirements of current legislation, and to provide active environmental safeguard on construction projects, many principal contracting organisations have established formal environmental management systems (EMS) within both their corporate and project site organisations. As with health and safety management systems, an organisation can implement an EMS to meet any desired degree of inclusiveness. A comprehensive standards-based system requires that the principal contracting organisation should develop, implement and maintain appropriate management procedures and working instructions to ensure that its activities conform to the company environmental policy, strategy and objectives. Furthermore, the approach must meet all the current environmental legislation that regulates its business activities. On site, effective management of the project environment through the utilisation of a project environmental plan (EMP) will focus on ensuring that the siteworks are considered, planned, organised, monitored and controlled with comprehensive awareness of the environmental impacts that those works may create.

Post-Contract Review

For reasons of both intra-organisational and inter-organisational reflection, post-contract review is a concluding yet important part of managing any construction project. Feedback to the principal contractor on project performance is likely to come from the client, as many carry out a formal post-contract evaluation of their project consultants and contractors. Contract performance, measured against pre-tender benchmarks, allows the client organisation to evaluate the effectiveness of the contractor in satisfying the requirements of the project. The procedure may be used to consider the principal contractor's suitability for future work and in determining preferred-contractor status.

Intra-organisational post-contract review is an important final stage in the undertaking of any construction project by the principal contractor. It affords the opportunity for the project team, together with the corporate organisation who provided tender and pre-contract inputs to the contract, to summarise and reflect on their activities and the success, or otherwise, of the project. With so many facets of site management having to integrate and function efficiently and effectively to make a construction project successful, post-contract review allows the principal contractor to take stock of the systems and procedures implemented in managing the project. For example, quality assurance procedures may need to be improved in the light of workmanship indicators while environmental management systems might be modified in the light of difficulties experienced so that they are more effective on the next construction project.

A further dimension is the requirement within specific construction legislation, for example, the CDM Regulations, to provide information upon contract completion for the project's construction health and safety file. With regular and active auditing becoming a prominent part of systems management implementation, it is likely that inspection audits will have taken place throughout the project's duration. Post-contract review is a suitable way of collating and synthesising the on-going findings from periodic auditing to identify aspects of good practice and less effective practice with a view to systems improvement. With the generic elements of project documentation, such as health and safety plans and environmental management plans, being used from project to project it is essential that procedures are reviewed upon project completion and updated for subsequent application.

Summary

This chapter has presented a general introduction to the book and outlined the content of each subsequent chapter. It will be evident that the management of any construction project may be far from simple and straightforward, with a great many management concepts, principles and practices to be incorporated. Likewise, the interaction of individuals and contractual organisations, the technology of the works and the characteristics of the project present many variables which the principal contracting organisation must consider in successfully delivering the project. Focusing on management explicitly, each subsequent chapter will present an introduction to key management concepts, principles, systems, procedures and practices that might be considered by a principal contracting organisation when implementing construction management.

References

Department of the Environment, Transport and the Regions (DETR) (1998) *Rethinking Construction*. Construction Task Force. HMSO, London.

Health and Safety Executive (HSE) (1974) *The Health and Safety at Work, etc. Act 1974*. HMSO, London.

Health and Safety Executive (HSE) (1988) *Blackspot Construction: A Study of Five Years Fatal Accidents in the Building and Civil Engineering Industries*. HMSO, London.

Health and Safety Executive (HSE) (1994) *The Construction (Design and Management) Regulations 1994*. HMSO, London.

Health and Safety Executive (HSE) (1996) *The Construction (Health, Safety and Welfare) Regulations 1996*. HMSO, London.

2 Corporate and Project Management

Introduction

This chapter focuses on the recognition of and the link between corporate and project management within a principal contracting organisation. Almost all such organisations are medium to large, well developed, may have an extensive range of business and utilise a decentralised structure serving sub-organisations and project sites which may be dispersed widely geographically. The prime objective of the corporate organisation is to develop company ethos, policies, aims and objectives which fuel and drive the business and allow the company to survive and prosper. The prime objective of the project organisation is to carry out efficiently and effectively those projects which the company undertakes, delivering its product on time, to budget and with high quality while making a profit. Both parts of the organisation need the support of each other, and together they must capitalise on their synergistic link to establish holistic success. Strong and robust structure and organisation are needed to provide effective management from executive level, through directive and operational levels to supervision and management on site. The encouragement of teamwork with integrated and holistic commitment is absolutely essential, as construction is and will always be a 'people' activity. Company policies, aims and objectives will emanate from executive level and these must be communicated throughout the management hierarchy. Moreover, they must be translated into actions throughout the management hierarchy, formalised and written down for clarity and direction of purpose. This will be achieved by establishing management systems – sets of procedures and working instructions – which guide and direct functional management activities throughout the organisation and work practices, together with their supervision and control at the project site. To successfully deliver any construction project, a principal contracting organisation would need to ensure that it has a sound corporate structure to provide effective organisation, carefully developed systems to provide effective planning and management mechanisms, and a robust project structure to provide effective working practices on the project site. This chapter looks at these important aspects, essential to framing the principal contracting organisation's approach to managing the project – the site management of construction works.

Management

Management is a fundamental aspect of all human activity and endeavour. The use of the word 'management' is sometimes misrepresented, often misunderstood and is sometimes perceived as an over-used cliché. Providing a robust definition of the meaning of management can be problematic, as each user of the word often has in mind a discrete application. It is best thought of as being a term with broad practical application. Management can range from an individual's handling of the most routine and simple matter to the planning and resourcing of the most complex technological projects. Management is not a single discipline but rather one which crosses the boundaries between many disciplines. It can therefore, be multi-disciplinary in nature and inter-disciplinary in application.

In the context of this book, management is explicitly used to emphasise the recognised and well-accepted application of functional management concepts, principles and practices within the scenario of the construction environment. More explicit is the focus on the activities of the principal contracting organisation and its site activities. Management is a collection of practices for handling concepts, principles and procedures, and these are transferable from organisation to organisation as organisational circumstances dictate. Such practices are generic throughout the management of organisations across industry sectors, yet specific in their application within their sectors as company, or corporate, management is translated into the management of particular types of projects.

Management Theories

Modern management theories, or those emerging after World War II, made great strides from the schools of management thought which preceded them – the *scientific management school* and the *human relations school*. Scientific management propounded by F.W. Taylor focused on universal principles of production and its organisation with a view to determining the best way of undertaking tasks by simplifying its component parts. Taylor's work was followed by a host of related contributions from Frank Gilbreth, noted for time and motion studies, Henry Gantt, famous for the Gantt, or bar, chart, Henry Ford, renowned for assembly line production, and Henri Fayol, the French industrialist and propounder of the well-accepted five processes of management – forecasting, or planning, organising, commanding, or instructing, co-ordinating, or harmonising, and controlling.

The human relations school, emerging in the 1920s, focused on productivity and its link to the welfare of workers and their motivation through positive management. The chief advocate of the human approach was Elton Mayo, famous for the Hawthorne Experiments at the Western Electric Hawthorne Plant in the USA. Essentially, workers with the greatest sense of belonging to a group had the greatest motivation and work outputs. An important dimension to the work was the finding that informal structures develop within the formal structure, and this can have immeasurable positive effects on communication, teamworking, morale and motivation, all important aspects within the construction project situation on site.

While both the scientific management school and human relations school have many positive aspects to offer to the management of construction projects, modern management theories are perhaps those which have the greatest propensity to assist. Many theories have emerged in the post-1945 period, of which two have become prominent and influential – *systems theory* and *contingency theory*.

Systems Theory

Systems theory (Checkland, 1993) is generally attributed to Hegel [1770–1831]. Concerned with creating a systems map to help explain organic biology, he focused on the various sub-systems that combine to create the whole. Hegel suggested that: the whole adds up to become greater than the sum of the parts, or the product of synergy; the whole determines the characteristics of the parts, the parts cannot be fully understood if they are seen in isolation from the whole; and the parts are interrelated and therefore interdependent, a view commonly held about most organisations today. From this understanding, systems theory presents concepts to an organisation to develop their: philosophy – way of thinking; management – design and operation of the organisation as a whole system made up of sub-systems; analysis – use of problem-solving techniques; and systematic thinking – logical, thorough and regular consideration (Hamilton, 1997). Essentially, systems theory suggests that an organisation is the whole, comprising many parts, and that the organisation exists in an environment which exerts many pressures upon its activities, indeed the very characteristics of almost all construction companies.

Contingency Theory

Contingency theory focuses upon the need for any organisation to adapt to change. It suggests that there is no one right way to manage an organisation indefinitely and that, over time, organisations evolve and must therefore (accommodate change,) both (within and outside the organisation.) Contingency theory has a clear overlap with systems theory in that the organisation's environment is always a catalyst for change and must be managed with total flexibility if it is to be successful and to prosper. Principal contracting organisations, like virtually all organisations in all business sectors, must be aware of their environment and be vigilant to its changing requirements. Perhaps more than any other sector, construction is susceptible to the volatility of the marketplace and so organisations must maintain flexibility to manoeuvre their way through changing commercial climates.

Both contingency theory and systems theory are prominent in the management approach to construction today. However, it must always be remembered that there is an assembly of different and varied approaches to managing construction. Companies will apply their own management approach and style to meet the individual operating characteristics of their business and given their specific situation and circumstances. Nevertheless almost all organisations may adopt some degree of systems approach and will recognise the need to accommodate contingency and change.

Management Practice: Systems Theory Applied to the Principal Contracting Organisation

Organisations operating within the manufacturing or process industries will probably be businesses based on a single and centralised corporate management infrastructure and location with perhaps one or a small number of production or processing sites. In such situations, the management systems will be relatively simple and straightforward. A principal contracting organisation differs through the accepted characteristics of construction, in that there is greater decentralised management through sub-organisations, each project is in many ways a unique temporary organisation, the parent company is usually managing multiple projects simultaneously, and each project can have a wide range of inter-disciplinary elements. In the construction situation, the principal contractor must establish the core corporate structure and systems for management, which are often duplicated at other regional locations, and also translate corporate systems for application at each of its project site organisations (see Figure 2.1). In this way, tiers of management organisation must be configured – the corporate and the project.

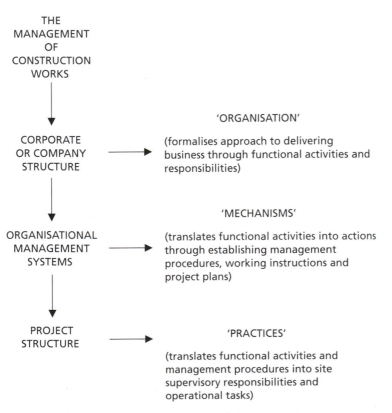

Figure 2.1 *Key elements of the management for construction works by the principal contracting organisation*

Corporate, or Company, Structure

The Need for Structure

For any organisation to be managed efficiently and effectively there must be clear business aims, objectives and policies, commensurate with the core activities of the organisation. To translate these key elements into workable procedures throughout both corporate and project management the company must establish a strong, recognised and accepted company structure. Within and around this structure the company is able to put into place the necessary formal organisations and also allow the natural development of positive informal organisations which will stabilise, drive and sustain its business.

A company must develop, implement and maintain an overarching business structure which is sufficiently rigid to provide the necessary formal structure yet remain sufficiently flexible to accommodate change and adaptability to fluctuating commercial and business environmental demands. Very few, if any, organisations would be successful if they were to remain static with an introspective disposition. All must be morphogenic, or outward looking, with a strong perspective of their holistic business and with a keen eye on the performance of their products, services or projects. A robust structure supported by well-conceived organisational management systems and project management allows the company to remain aware of its business surroundings and inter-organisational influences while also maintaining a good sense of its intra-organisational dimension (Griffith *et al.*, 2000).

All organisations require structure to give form and focus to their business pursuits, based predominantly around the company's thrust of specialisation depicted by their core business. Specialisation forms the basis of a company's structure, with vertical specialisation determining the hierarchical structure of authority while horizontal specialisation gives the differentiation of management functions within the various levels of the hierarchy. Together, these thrusts of specialisation form the basis of the structure within which the company establishes formal organisations, processes, procedures and practices. Organisations with strong vertical specialisation and control, where policies, processes and procedures are well documented and with defined divisions of labour and resources, are termed mechanistic. Organisations with a flat hierarchy but strong horizontal specialisation with few documented policies, processes and procedures and with predominantly informal control, are termed organic (Pilcher, 1992). Almost all companies are in some ways a combination of the two, formalising an overarching management structure through formal organisations while encouraging the all important informal interactions of its employees, who are central to the running of its business.

It should always be remembered that people are at the core of any organisation (Harris and McCaffer, 1995). Within construction organisations, and in particular any principal contracting organisation, people form the nucleus of resources both at corporate and project levels. In addition, whereas businesses in other commercial sectors focus on products or services, the principal contracting organisation focuses on delivering a project. Central to any construction organisation therefore is the project, or production, site (Newcombe *et al.*, 1990). The temporary and in some ways unique characteristics of construction projects with their individual needs, organisations and people inputs mean that the structure developed by the company must be robust to provide the necessary stability for business operation within not only the corporate organisation but within the many site organisations that will be established to deliver its projects.

The organisational structure must also be configured within the context of the company's constitution and surroundings. Its constitution will be influenced by: the nature of the business; the arrangement of labour and resources; the operational characteristics; and the people skills and abilities needed. Its surroundings will impart the influences of: market, natural and physical environments; competition; and commercial, public and legal constraints (see Figure 2.2). It is essential that the principal contracting organisation develops its organisational structure within the intra-organisational, inter-organisational and wider environmental context and within a holistic perspective of serving its whole business which encompasses both the corporate and project dimensions. Only in this way will it develop, survive and prosper as a successful organisation. The organisational structure must reflect the activities of corporate management which maintain and drive the business, the activities of site organisations which carries out the business through the projects they undertake, and also provide a clear and workable link between the two levels.

ASPECT OF ORGANISATION	ASPECT OF INFLUENCE
CORPORATE RESPONSIBILITIES	• ethics • ethos • policies • organisational effects • responsibilities • accountability
BUSINESS STRATEGY	• business mission • commitment to policy • market awareness • commercial positioning • customer focus
LEGAL and FINANCIAL REGULATION	• EC Directives • international standards • national legislation • investment • risk assessments • liabilities • insurances
MARKETING and CORPORATE IMAGE	• public perception • internal image • company promotion • corporate literature • media interfacing and communications
ORGANISATIONAL MANAGEMENT SYSTEMS	• organisational structure • management systems and procedures • performance indicators • operational and project control • business delivery improvement • communication, information technology and knowledge
HUMAN RESOURCES	• staff deployment • recruitment • training and development • incentives • investment in people
OPERATIONS	• time, cost and quality • environment • health, welfare and safety • resources • knowledge and skill base

Figure 2.2 *Prominent organisational influences upon organisational structure*

Type of Structure

Structure and organisation are prerequisites for a project's success. Organisation can be considered to be "*the process of arranging people and other resources to work together to accomplish a goal*" (Schermerhorn and Chappell, 2000). Structure and organisation are usually depicted in the form of an organisation chart which identifies functional management disciplines, responsibilities, relationships between groups and individuals, chains of command and communication routes. Structure and organisation are essential to facilitating the principal contracting organisation in setting and communicating its corporate policies, aims and objectives and translating these into management procedures and working practices at the project level.

A small-scale contracting organisation, of which there are many within the construction sector, is the simplest of company organisational forms. With a centralised organisation, the manager, frequently the owner, oversees and directs all business and project management activities utilising a small labour force on each construction site. As a company increases its business turnover, its workload may change, expanding in range of type, size and technical complexity. As this occurs the organisation will naturally divide into functional management elements based on specialist skills. As further expansion occurs there will probably be separation into divisions based on types of work or location, giving the traditional organisation of a medium-size principal contracting organisation.

Figures 2.3 and 2.4 show an organisation chart for a medium-sized contracting organisation and the direct link between the two company management levels – the corporate, or head office, organisation and the project, or site, organisation. It can be clearly seen that the purpose of this structure and organisation is to channel management responsibility from the company, or executive, management down to the project level through the division of responsibility to specialised functional management. Directive functional managers play a key role within this structure by translating the company's policies, aims and objectives into management systems – procedures and practices – implemented at project level to undertake these works. Application of organisation within the larger principal contracting organisation follows later in this chapter.

The structure adopted is vital also to creating teamwork. Effective teamwork with integrated and committed managers, supervisors and operatives is absolutely essential. Any construction project is only as good as the team who undertakes it. Teamwork, through the effective structuring of individuals and groups, provides the sound basis for developing synergy, harmony and commitment throughout the project. It is often said that around a half of all problems on construction projects emanates from technological matters while the other half results from human issues. People management is therefore fundamental and a prerequisite to project success. The company must foster effective human resource management through its corporate structure and organisation and through commitment to important initiatives such as Investors in People. In addition, the project organisation must foster good employee

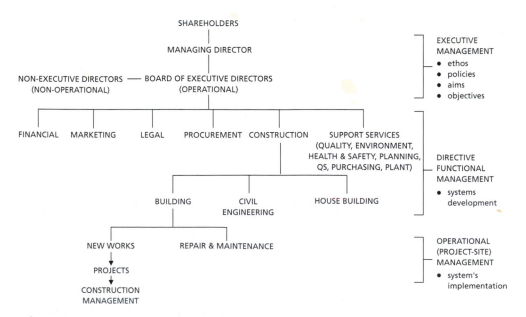

Figure 2.3 *Corporate, or head office, structure for a medium-sized contracting organisation*

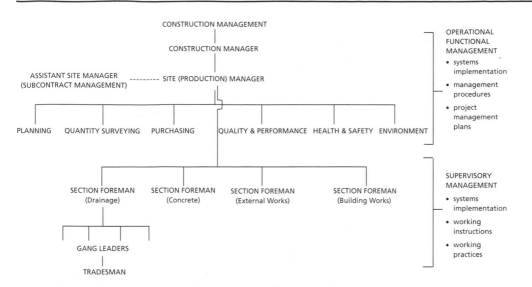

Figure 2.4 *Project, or site, structure for a medium-sized contracting organisation*

relations through structuring and organising an excellent working environment on site, underpinned by effective leadership and motivation of the project team and workforce.

Organisational Management Systems

"*A management system is, simply put, a way of doing things. Systems develop protocols and sets of procedures and instructions which bring structure, order and therefore, stability to an organisation where otherwise there might be chaos*" (Griffith, 1999). Organisational management systems describe methods of working both with the company organisation and the site organisations established to deliver its projects.

The system, or systems, used in managing the company will have key generic elements common to the corporate organisation and the project organisation supplemented by specific elements to accommodate the requirements of a particular project. In addition, an organisation may adopt standards-based systems such as those specified by international management standards, for example, a BS EN ISO 9000 quality system, or develop its own in-company approach based around processes and procedures it uses together with the influence of any computer software utilised.

In practice a system is a set of management procedures for guiding the management of the organisation's business activities and a set of working instructions to guide the processes, operations and tasks that need to be performed to deliver the service or product that forms that business. Management procedures and working instructions embrace both the corporate and project dimensions of the company's business, established through the company's systems of management and implemented through management and supervision on the project site. The need for any system to be capable of translation into easily conducted procedures and tasks is particularly important within construction where management is function orientated and work on site is task based. The essence of a management system is that it should be simple, but not be simplistic, and it should be easily understood by those persons who must implement the system and work with and around it. Systems are relatively easy to see and understand where a company's structure, organisation and procedures meet the formal requirements of national and international management standards, such as ISO 9000 for quality and accompanying certification schemes, but less easy where company procedures are systematised by the particular management methods or tools utilised. For example, planning procedures may use a particular piece of computer planning software. Nevertheless, planning procedures must still be based on solid policies, procedures and practices used throughout the company organisation and on those projects that the company undertakes.

The true focus of any management system is that it should fulfil the company's business policy. In that context, it could be argued that 'the system' is 'the business' with 'management' being the control mechanism of the parts, or sub-systems which come together to perform all the functions of management – financial, legal, progress, quality, health and safety – and more (Griffith *et al.*, 2000). One can therefore perceive the business driven by a parent core management system supported by others, or sub-systems, which plan, monitor and control particular elements and act as the interface between management of the corporate organisation and the project organisation.

The various systems through which specific functions are managed, for example planning, procurement, quality, health and safety or environment, must be interactive and intrinsically supportive. Moreover, they must link together to support the corporate organisation holistically while also supporting its given management function at project level. While respecting the specific needs to fulfil the particular management function, the systems must also have commonality and consistency of approach to other systems. In this way information may be transmitted across the boundaries to other systems. This characteristic also serves to assist an individual manager in handling a number of management functions simultaneously, for example a mechanism used to check environmental project factors could also be used to check health and safety factors. Although, traditionally, contracting organisations may have used separate and vertical structures to manage specific functions, there is now greater application of horizontal cross-functional management systems leading to combined and integrated management systems where information and practices are shared across management boundaries. This is depicted in Figures 2.5 and 2.6 (Griffith, 1999). This is reflected in standards-based certification schemes for management systems such as the British Standards Institution Integrated Management System Assessment (IMSA).

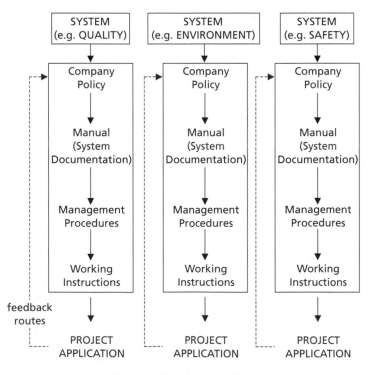

*Each system is separate giving good service
within its functional remit but provides no opportunity
to share information or practices across traditional boundaries*

Figure 2.5 *Vertical management systems structure*

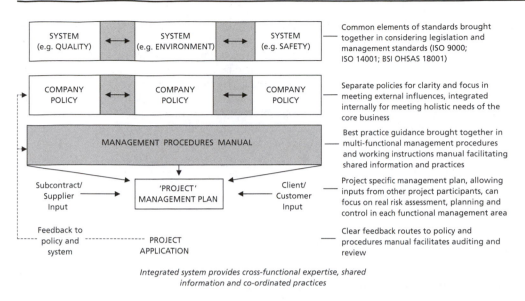

| SYSTEM (e.g. QUALITY) | ↔ | SYSTEM (e.g. ENVIRONMENT) | ↔ | SYSTEM (e.g. SAFETY) | Common elements of standards brought together in considering legislation and management standards (ISO 9000; ISO 14001; BSI OHSAS 18001) |

Common elements of standards brought together in considering legislation and management standards (ISO 9000; ISO 14001; BSI OHSAS 18001)

Separate policies for clarity and focus in meeting external influences, integrated internally for meeting holistic needs of the core business

Best practice guidance brought together in multi-functional management procedures and working instructions manual facilitating shared information and practices

Project specific management plan, allowing inputs from other project participants, can focus on real risk assessment, planning and control in each functional management area

Clear feedback routes to policy and procedures manual facilitates auditing and review

Integrated system provides cross-functional expertise, shared information and co-ordinated practices

Figure 2.6   *Horizontal systems structure*

Systems Elements

The key elements of any system, and in particular the systems developed and implemented within construction contracting organisations, are those which develop, drive, implement and incrementally improve organisational systems over time. They are a prerequisite to maintaining the successful business of the organisation over the long term. The key elements are:

- policy
- organisation
- risk assessment
- planning
- implementation
- auditing.

These key elements are commensurate with the key elements specified within international standards-based systems, for example BS EN ISO 14001: Environmental Management Systems:

- environmental management system
- environmental policy
- organisation and personnel
- environmental effects assessment
- environmental objectives and targets
- environmental management programme
- environmental management manual and documentation
- operational control
- environmental management records
- environmental management audits
- environmental management reviews.

Systems Documentation

A company system is configured to provide the broad corporate and operational structure for managing the whole organisation and its business in a holistic way. It exists to provide support to the core business, the assurance of quality of product and service, safety and to meet its legal, contractual and policy obligations. For any system to be effective it must do more than merely frame and structure the organisation's approach to its business, it must set standards and goals which are met through delivering its services or products and these must be formalised, or laid down, in company documentation against which organisational performance can be measured, reviewed and improved.

Generically, a management system might employ four levels of documentation which translate the company structure into operational management procedures. For a contracting organisation this means translation of the system into mechanisms which guide and control the work on a construction site. These are:

1. A management system manual
2. Sets of management procedures
3. Sets of working instructions
4. Project management plan.

Management system manual – this document, or series of documents, defines and explains the company policy, management responsibilities and organisation structure which is used by the company to ensure that it pursues and meets the specified strategic aims and objectives of its business. The manual identifies and describes, in outline, the overall framework of organisation and procedures through which the company will operate within the context of legislation, regulation and standards which impinge upon its business. In more simple terms, it explains the organisation of the organisation.

Sets of management procedures – these describe the principal management activities in each key area, or system, of organisational activity. They will translate the corporate perspective of the system into elements which form the basis for its project-based operations. For example, it might describe the management requirements for quality assurance, health and safety, contract planning or procurement of materials.

Sets of working instructions – these define the management requirements for specific functional project activities, for example quality control on site. Working instructions are generic in that they apply to any situation where a particular management function is to be carried out. They are also quite specific in that they are influenced by particular situations. In this latter situation further detail will be provided in the project management plans.

Project management plan – this is a plan which defines and describes the characteristics of a particular project and relates those characteristics to the specific management function to be performed. For example, management procedures and working instructions may provide generic descriptions of company health and safety management and the specific project management plan relates these to the particular project site. A clear example of this is the requirement for the principal contractor to produce a construction health and safety plan as part of meeting the requirements of the Construction (Design and Management) Regulations 1994.

As each principal contracting organisation will have its own way of doing things based upon those circumstances and situations which influence its business operations it is not appropriate to be overly prescriptive in this book. However it may be useful to outline a set of headings as an example of the key elements which might be considered when developing a manual, management procedures, working instructions and project management plans (see Figure 2.7). These are taken from Griffith and Howarth (2001). To see how these relate to specific management function applications the reader is directed to Chapters 10 and 11 where applications to health and safety management and environmental management are comprehensively detailed and explained. Figure 2.8 outlines a range of organisational and standards-based systems applicable to site management.

Organisational Management Systems Effectiveness

To ensure that a management system has an optimum chance of being effective there needs to be specific undertakings at corporate level and project level (Griffith and Howarth, 2001). At corporate

LEVEL 1	MANAGEMENT SYSTEM MANUAL (CORPORATE LEVEL)
	• Policy • Organisation • Responsibilities • Management Procedure References
LEVEL 2	MANAGEMENT PROCEDURES (CORPORATE & PROJECT LEVEL)

	• Company Service Management
	  (finance, legal, marketing, public relations)
	• Company Support Services
	  (quality, environmental, health & safety, personnel, training,
	  materials/plant procurement)
	• Project Management
	  (professional liaison, contract establishment, project planning and
	    resourcing, site control and supervision)
	• System Documentation
	  (document control, auditing, reviews, improvement)

LEVEL 3	WORKING INSTRUCTIONS (PROJECT LEVEL)
	• Induction • Risk Assessment • Control Mechanisms and Supervision • Site Meetings • Inspection Routines • Working Procedures • Record Keeping
LEVEL 4	PROJECT MANAGEMENT PLANS (PROJECT LEVEL)
	• Policy • Organisation • Responsibilities • Statutory Obligations • Method Statements • Risk Assessment • Site Rules • Record Keeping • Auditing and Review

Figure 2.7 *Levels of documentation for management systems
(adapted from Griffith and Howarth, 2001)*

level there must be:

- demonstrable commitment to the organisational policy and strategy
- a clear statement of the organisation's objectives and policy, circulated throughout the organisation
- employee ownership of the system management functions through involvement in their formulation, development and implementation
- the setting of goals and targets at both corporate and project level against which group performance can be measured
- the availability of adequate resources to facilitate system framework and operation
- appropriate and continuing education and training for management and the workforce in operational procedures
- on-going review and improvement to applications to enhance employee experience and expertise.

MANAGEMENT FUNCTION	SYSTEM TYPE	APPLICABLE STANDARDS	KEY DRIVER	PERFORMANCE INDICATORS
Project planning	company	none applicable	planning software used	business excellence
Cost planning	company	none applicable	cost modelling software used	business excellence
Human resources	company	Investors in People (IiP)	IiP model	IiP certification
Procurement (plant/materials)	company	none applicable	procurement software/ databases used	business excellence
Quality	standards-based	BS EN ISO 9000 EFQM*	management system specification	certification
Environment	standards-based	BS EN ISO 14000 EMAS**	management system specification	certification
Health and Safety	standards-based	BSI OHSAS 18001	management system specification	certification

* European Foundation for Quality Management.
** Environmental Management Audit Scheme.

Figure 2.8 *Range of organisational and standards-based management systems applicable to site management*

At project level, a management system will be supported by:

- the identification of key issues that need to be addressed on the project site
- risk assessment by and across the boundaries of the specific management functions
- development of action plans in response to identified needs
- distribution of good practice guidelines to functional staff
- determination of audit procedures between the corporate level and the project level
- briefing all project team members on system operation
- training in the use of procedures used to plan, monitor and control the management functions
- availability of practice notes on those good management procedures to be adopted
- preparation of item checklists for the monitoring of specific procedures
- creating self-audit and review documents for the activities of system supervisors
- providing guidance notes on potential actions when problems occur to encourage proactiveness
- making clear reference to corporate management and other support services where assistance may be needed.

Project, or Site, Structure

There are a great many contracting organisations within the construction industry, each serving their own particular markets. Such organisations will all differ in terms of their corporate structure, management procedures and working practices. Although the principal aims and objectives of a particular

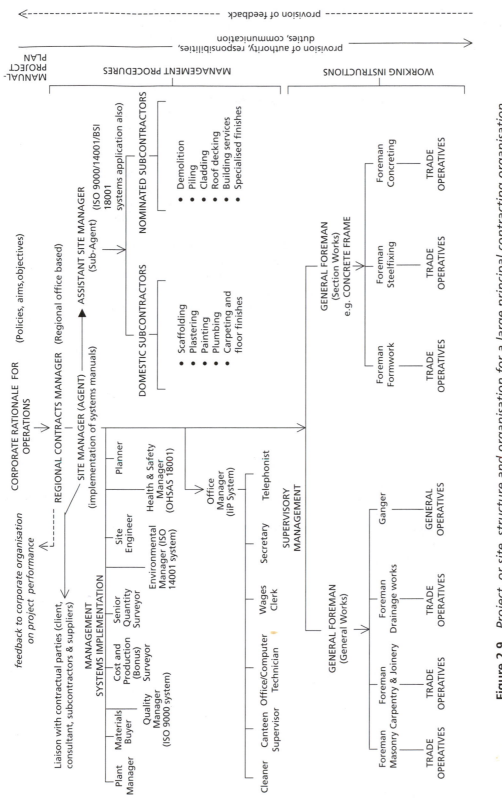

Figure 2.9 *Project, or site, structure and organisation for a large principal contracting organisation implementing management systems to functional specialisms and subcontracted operations*

company will determine its precise configuration and constitution and therefore, site organisational arrangement, all generally evolve and develop as the size of their business turnover increases. The larger the contracting organisation, the greater the tendency to decentralise operations to strategically determined geographic locations as contracts are won. The emphasis shifts from managing the works from one centralised office to delegation to sub-organisations or regional or area administrative offices. While a small contractor will operate within a controlled radius from its central, or head office, with little delegation, the large principal contracting organisation will give virtually autonomous power to each of its large project sites to effectively operate independently of, although always in liaison with, the central, or head, office.

Although the site organisational structure for a large construction project will invariably differ from project to project, the functions of the site team members are generally the same on each project and reflect the functional specialist departments operated at corporate level, as seen earlier in Figure 2.3. It is the corporate organisation's task to ensure that its structure is configured in such a way as to facilitate the translation of business aims and objectives into procedures and practices documented in its systems manuals and lay the foundation for a sound site-based structure such that effective management can operate.

Prescribing a set site structure in any situation and for any construction project is virtually impossible since there is such diversity in possible organisational structure. However, there are commonly accepted principles of organisation which should be considered when formulating any structure. Principally, there should be a clear line of authority identified from the site manager to all functional managers and supervisors. Everyone based on site must be clear as to their position, duties and responsibilities, for whom they are responsible and to whom they are accountable. These routes of authority and responsibility also form the key channels of formal communication and upward, downward and lateral feedback throughout the organisation. It is absolutely essential that these channels are clearly established at project commencement if the holistic structure is to be effective. Site management also needs to establish good links with the supporting corporate organisation and also to other parties involved in the project, for example, the client, consultants, subcontractors and suppliers. This is a primary role of the site manager.

Figure 2.9 shows the project, or site, structure and organisation for a large principal contracting organisation undertaking a major construction project. It indicates the management systems applications to functional disciplines and also the link to subcontract operations. In the figure it will be seen that the project arrangement is extensive, requiring the appointment of an assistant site manager to manage subcontract operations. A further assistant might be utilised as an intermediary between supervisory management and the site manager if necessary. Functional disciplines are translated from the corporate organisation into site-based functional activities, for example quantity surveying and management systems translated into site-based applications by, for example the project quality manager, using ISO 9000 management procedures. These procedures will be documented in the management procedures manual described previously. Similarly, supervisory management will invoke working instructions and working priorities, again described within the organisation's manual allied to the project management plan. It is such a structure at site level that needs to be in place to provide effective project management and lays the foundation for the management activities which follow throughout the book.

Summary

A carefully conceived and robust organisational structure is a prerequisite to the provision of effective management at corporate, directive and operational levels through to supervision of the siteworks. The company mission, policies, aims and objectives must be successfully disseminated throughout the organisation both at corporate and project levels. Management systems, or sets of procedures and working instructions which guide and direct functional management activities throughout the organisation, need to be clearly specified and effectively implemented. Types of organisation structure and associated management systems will, obviously, vary from company to company and project to project. Nevertheless, there are key elements of both which are standard, in particular with management systems which are influenced greatly by international standard specifications for development such as BS EN 1SO 9000, 14000, BSI OHSAS 18001 for quality, environment, and health and safety management respectively. As with almost all systems approaches, a core element is company documentation. Within the principal contracting organisation this is provided by a set of documents which incorporate manuals, procedures, working instructions and project plans. This chapter has introduced the importance of organisational structure and management systems and these aspects will be revisited in subsequent chapters as particular management concepts, principles and practices are examined.

References

Checkland P. (1993) *Systems Thinking, Systems Practice*. Wiley, London.

Griffith A. (1999) *Developing an Integrated Quality, Safety and Environmental Management System*. Construction Paper 108. *Construction Information Quarterly*, 1(3), 6–18.

Griffith A. and Howarth T.A.P. (2001) *Construction Health and Safety Management*. Addison Wesley Longman, Harlow.

Griffith A., Stephenson P. and Watson P. (2000) *Management Systems for Construction*. Addison Wesley Longman, Harlow.

Hamilton A. (1997) *Management by Projects: Achieving Success in a Changing World*. Thomas Telford, London.

Harris F. and McCaffer R. (1995) *Modern Construction Management*. Blackwell, Oxford.

Newcombe R., Langford D.A. and Fellows R.F. (1990) *Construction Management: Organisational Systems*, Vol. 1, Mitchell.

Pilcher R. (1992) *Principles of Construction Management*. McGraw-Hill, London.

Schermerhorn J.R. and Chappell D.S. (2000) *Introducing Management*. Wiley, London.

3 Effective and Efficient Site Management

Introduction

A construction organisation's performance can be measured in many different ways. Two of the most common ways are in terms of efficiency and effectiveness. There is a considerable difference between these two concepts.

Efficiency

The ability to '*get things done correctly*' is a mathematical concept – the ratio of output to input. An efficient manager is one who achieves higher outputs (results, productivity, performance) relative to the inputs (labour, materials, money, machines and time) needed to accomplish them. In other words, a manager who can minimise the cost of the resources used to attain a given output is considered efficient. Or, conversely, a manager who can maximise output for a given amount of input is considered efficient.

Effectiveness

Effectiveness is the ability to '*do the right things*', or to get things accomplished. This includes choosing the most appropriate objectives and the most efficient methods of achieving the stated objectives. That is, effective managers select the '*right*' things to do and the '*right*' methods for getting them done. For managers, '*the pertinent question is not how to do things right, but how to find the right things to do, and to concentrate resources and efforts on them*'.

This chapter investigates some of the key components for Construction Managers in the attainment of both efficient and effective methods of working towards predetermined objectives.

Within the following text, various basic principles of Construction Management are established, commencing with some background information on types of organisational structure and site managerial roles. However, many previous texts have details upon these topics; therefore within this chapter new concepts have been interwoven in order to broaden the reader's knowledge horizon. To this end, topic areas such as anthropocentric production systems and the attainment of 'best practice' have been included. This mix is designed to build upon established knowledge and then further develop it based upon lessons learned from the UK Manufacturing Industry and international research. These new broad concepts have value for the UK Construction Industry.

A case study is incorporated for the reader to study and answer the related questions with the main points for consideration being provided.

Workforce Organisation and Supervision

Post-World War II Historical Influences on the Construction Industry: 1945–Present Day

The 'post-war era' ushered in a catalyst for change, incorporating:

- society dissatisfaction with previous low standards of homes and workplaces
- start of overseas travel, experience and exchange of ideas from abroad

- greater surge to better and wider education for all – looking for higher expectancies
- growing workload for companies rebuilding UK after war damage
- better, more efficient building designs
- awareness of scarce water resources conservation
- growing awareness of 'Green and Conservation' issues in design etc.
- growing influence of professional bodies, i.e. RICS, CIOB, RIBA, ABE etc. in drawing up standards
- growing awareness of issues of quality in terms of assurance.

'Current' Nature and Organisation of UK Construction Industry

The Construction industry has rethought its position and long-term survival through recessions within the broad pressures related to the following influences:

- more customer-focused – providing value for money
- increasing competition for work
- recession of the late 1980s forced willingness to become 'flexible' in business
- emphasis of quality assurance – seeking repeat business from satisfied clients
- new procurements for greater 'Risk on Contractors'
- widening professional skills and knowledge required
- performance monitoring and control
- closer control of budgets
- faster project completion times yet retaining quality and cost control
- 'joint' venture and 'partnering' with complementary expertise to 'share-risk'
- working within 'niche' markets with repeat business clients preferably.

UK Construction Industry Response to the above Pressures

Traditionally, in order to survive, companies were required to make higher profits by whatever means contractors had at their disposal from each project regardless of how their relationship with the client evolved.

Currently the Construction industry is slowly realising that to survive in a competitive market, customer relationships, supplying or providing what the customer wants and, most of all, valuing the potential long-term customer by retaining a good relationship throughout with the object of continuous work and reasonable profit, are paramount. This indirectly facilitates predictable long-term company workloads, cash flow and long-term planning and changes, enabling the company to survive in a highly competitive market. Word-of-mouth from one satisfied client to another is the best marketing strategy, with a slogan similar to '*each project a successful outcome for both parties*' or, as the Latham Report (1994) states '*win–win!*' for both client and contractor.

Company Organisation in a Changing Market

Management theory states that 'organisational form' must be the servant or consequence of the company 'functions', and not the other way round! This implies that organisational structures must 'facilitate' or 'improve' company output and performance and therefore they must be flexible and easily able to change as a company's functions, output, markets, client needs change.

Organisational Structures

Entrepreneurial Structures (Figure 3.1)

Many construction firms commence life as a sole trader (entrepreneurial) organisation. They seek to extend the capacity and capability of the individual, who may be a bricklayer, plumber or a professional,

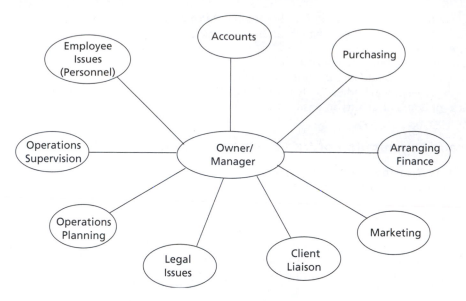

Figure 3.1 *Entrepreneurial structure*

such as an architect. Owing to this individual *"having had the initiative and taken the risk, gives the entrepreneur such dominance in the evolving organisation that everything depends upon him"* (Torrington *et al.*, 1989).

The main features of an entrepreneurial structure include:

- all activities are focused upon the owner/manager
- the entire enterprise is totally centralised with no clear demarcation of responsibility or autonomy
- this form empowers the owner/manager to control the firm during its early development until it reaches 'critical mass'
- the owner/manager is unlikely to possess a sufficient range of specialist knowledge to ensure the continued survival of the company if it starts to grow
- there must come a point when the owner/manager cannot manage efficiently and effectively, because of the many varied demands of the firm.

Thus an element of bureaucracy becomes inevitable.

Bureaucracies

These consist of levels of authority or hierarchies, where subordinates have a specifically defined role to deal with routine issues. When they are faced with these issues, they are 'programmed' to respond in a predetermined manner. Any issue facing them outside that role is referred to a superior. This ensures that the overall organisation can predict the behaviour of the individual. Only new or novel problems are referred to superiors who can concentrate their efforts upon them. The authority of the superior is described as legal or rational, as it is bestowed upon the holder of a particular office. This presumes that all participants will accept this validity.

That hierarchies are often synonymous with bureaucracies is not surprising in that a multi-layered system will evolve to ensure compliance with organisational requirements and to deal with new problems. Specialised 'departments' also evolve to deal with individual areas. The method of communication within hierarchies is controlled by the superior levels. Hierarchies are often perceived as pyramid shaped. This approach to organisational structures gives rise to the 'line and staff' type of structure, as depicted in Figure 3.2 (see also the work of Weber later in this chapter).

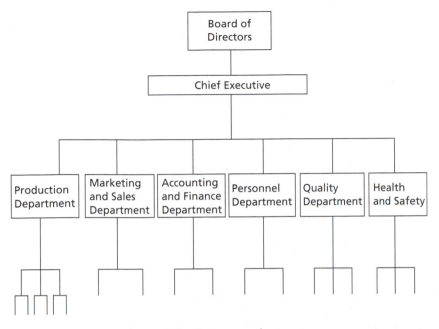

Figure 3.2 *Bureaucratic structure*

The bureaucratic or functional structure's main features are:

- the firm is organised around the required tasks to be performed to ensure that the distinctive competence of the company is maintained
- organisational control is exerted through the Chief Executive
- it utilises the concept of 'division of labour', i.e. specialists develop expertise
- clearly delineated relationships exist
- it is suitable for fairly static operational environments
- there could be a loss of the entrepreneurial spirit because of a concentration on specialist functions
- the holistic nature of the company could be lost in its drive for a sustainable competitive advantage
- there could be problems in trying to co-ordinate the disparate functions
- functional specialists may seek to further their own little empires at the expense of the company
- there is a danger that specialists will concentrate on short-term profit maximisation at the expense of long-term corporate survival.

Matrix Structures

In a matrix structure each member has a specialist function but each also has an overlapping competence with some other group members. However, when we place people to work in groups we must have an understanding of the effect of the group on behaviour.

A group will develop its own 'group norms'. The matrix structure has been developed to overcome the disadvantages of the entrepreneurial and bureaucratic structures. The system was first developed in the US aerospace industry because of government demands for a single project leader who would be responsible for the progress of each government contract.

A matrix structure is a combination of structures and, in construction, usually takes the form of function and project (or geographical division). Matrix structures do not just arise in large complex organisations but are also utilised in much smaller firms. A typical matrix structure is indicated in Figure 3.3. In the example provided in Figure 3.3, the site team for Project 1 are allocated from the specialist functions/departments. This group of staff then provides the team for project operations and management. This process would be repeated for any remaining projects.

Departments/ Functions / Projects	Planning	Finance	Plant	Health and Safety	Labour	Contracts Manager	Quality	Quantity Surveying
Project 1 Site Team								
Project 2 Site Team								
Project 3 Site Team								
Project 4 Site Team								
Project 5 Site Team								

Figure 3.3 *Matrix structure*

The main features of a matrix structure are:

- clear lines of authority and accountability when related to decision-making
- direct contact between specialists replaces bureaucracy
- development of managers via an increased involvement in the decision-making process
- optimum utilisation of skills and resources with maximisation of profit potential
- can be difficult to implement and sometimes the issue of dual responsibilities can cause some confusion.

However, this form does allow an organisation to react better to dynamic changing operational environments, as is the case with construction.

Leadership

The theories of both motivation and groups are closely linked to that of leadership. It is the leader who sets out to develop effective motivation for the workforce and the group, whether formal or informal, and this will either establish or undermine a leader.

Traditional Theory

The traditional theories of leadership are almost at the level of mythology, in fact they can be traced back to Plato. These suggest that leaders are born and not made. The characteristics of leadership are, in fact, natural personality traits within the human make-up, neither taught nor transferable.

While there may be an element of truth in this understanding, the flaw of the proposal is that the characteristics vary from leader to leader and have no measurable quality. Although intelligence, physical prowess, a wide breadth of interests, inner motivation etc. have all been shown as useful qualities, a lack of any of these has not disadvantaged certain acknowledged good leaders.

Styles of Leadership

Rather than accept traditional theories, styles of leadership can be identified:

- **Authoritarian** leaders strictly control the amount and nature of the work of subordinates. Information is strictly controlled which maintains the leader's authority (see Figure 3.4).
- **Democratic** leaders allow the group to determine its own work schedules and amounts, but direct the overall tasks (see Figure 3.5).

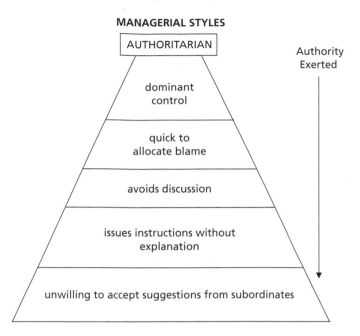

Figure 3.4 *Authoritarian management style*

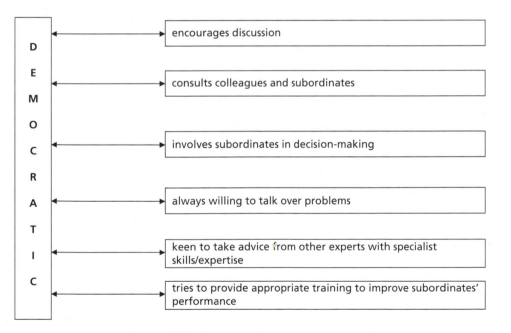

Figure 3.5 *Democratic management style*

- **Laissez-faire** leaders abdicate responsibility. Decisions have to be made by subordinates and their successes are not recognised (see Figure 3.6).
- In addition, the **charismatic** leader may be identified. This is an individual who attracts a personal following, not essentially based on work, but subordinates put great faith in the quality of their decisions.

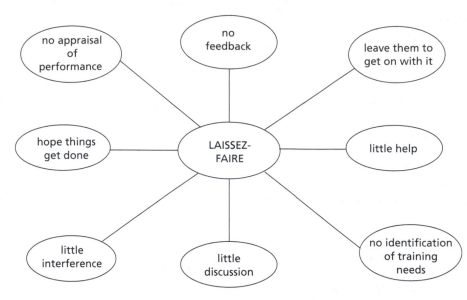

Figure 3.6 *Laissez-faire management style*

These characteristic types of leadership may be easily recognisable to the reader assessing their own leader. What is more difficult is to assess one's own style, or the most effective style to adopt.

Management Styles

Socio-economic factors affect leaders and the old idea was that managers are born and not made. Events of World War II placed all sorts of different types of people in leadership roles. Management techniques suggested you could train people to be managers.

The leader who is in effect a one-man orchestra is what we shall describe as an autocrat. The autocrat demonstrates the following characteristics:

- gives orders which they insist shall be obeyed
- determines policies for the group without consulting it
- gives no detailed information about future plans but simply tells the group what immediate steps it must take
- gives personal praise or criticism to each member on their own initiative
- remains aloof from the group for the greater part of the time.

Contrasted with this type of leader is the democrat who:

- gives orders only after consulting the group
- sees to it that policies are worked out with the acceptance of the group (this is critical for effective implementation)
- never asks people to do things without sketching out the long-term plans on which they are working
- makes it clear that praise or blame is a matter for the group
- participates in the group as a member.

The third type of leader is the laissez-faire. They do not lead but leave the group entirely to itself and do not participate.

These types may be subdivided and anyone who is acquainted with industry will be able to name examples of each of the following types of leader in the various companies they have worked in:

A Autocratic leaders
    (1) Strict autocrat
    (2) Benevolent autocrat
    (3) Incompetent autocrat
B Democratic leaders
    (1) Genuine democrat
    (2) Pseudo-democrat
C Laissez-faire leaders

These terms are largely self-explanatory. Strict autocrats are stern, strict but just, according to their principles. They do not delegate authority and each is a one-man show.

Benevolent autocrats resemble the foregoing in many respects but are afflicted by a conscience. No improvements in the material sense are too good or too expensive for their employees but they have to take what they get and like it.

Incompetent autocrats are completely unscrupulous and lie, bribe and bully, or takes any measures which they feel will help them to attain their goals.

Genuine democrats are the conductor of an orchestra rather than a one-man show and realise that their job is to co-ordinate the willing work of their employees. Authority is delegated and all levels of management feel sufficiently secure to consider the well-being of their subordinates.

The pseudo-democrat may aspire to be democratic but is too insecure to make a success of it and ends up by being not very different from the benevolent autocrat.

Defining Construction Site Management

The highly competitive market of the construction industry leads to great uncertainty and variability.

Illingworth (1993) concluded that "*A site manager, for example, conducts, controls, administers and is at the head of a construction team on site.*" The importance of this role was advocated by Mustapha and Naoum (1998): "*The site manager stands at the heart of the building process.*" Djerbarni (1996) demonstrated an understanding of the impact that the site manager has on the overall construction process: "*the role that a construction site manager plays is crucial to the success of their organisations.*" He further stated that "*there is a consensus on the important influences site managers exercise in ensuring the success of a project.*"

For site managers to be effective and efficient, they must have certain characteristics such as a sound knowledge of construction and the competencies to overcome daily obstacles. However, in order to become a proficient manager, there is a requirement for them to have some key basic managerial skills, as opined by Calvert *et al.* (1995): "*Building managers should understand and master the principles and methods involved in management.*"

Characteristics

Construction managers have to incorporate a number of key functions into their role: these are forecasting, planning, organising, controlling, motivating, co-ordinating and communicating. The seven functions have been recognised by various authors in this subject area, with Calvert *et al.* (1995) postulating that "*planning is perhaps the most important tool of management.*" The role of the site manager has been described as one of the most demanding jobs in the construction industry. Wakefield (1982) writes "*The site manager's job is the most arduous, demanding, responsible single function in the building process.*" This view is corroborated by Djerbarni (1996): "*site management is characterised by a high work overload, long working hours, and many conflicting parties to deal with, including the management of the sub-contractors and liaison with the clients.*"

Recent studies have highlighted that site managers have to be very dedicated individuals in order to be successful on site. They have to be an effective link between senior management and the on-site workforce as purported by Gunning (1998): "*the most important part of the site manager's job is the control of day-to-day operations using clearly defined targets to performance.*"

An extremely strong character is necessary and they must be proficient in all aspects of management, with qualities that ensure their survival and, ultimately, success on site.

Wakefield (1982) suggests that "*effective site management is concerned with leading and managing men. The site manager's job is primarily concerned with mobilising and developing human resources effectively and motivating them.*" Wakefield demonstrates his detailed knowledge of the site management function by making specific reference to personnel management. In order to achieve the desired outcome when dealing with personnel management, it is necessary to employ excellent communication skills. The site manager must be "*a skilled communicator, able to explain what is required*" (Trotter, 1982a).

The characteristics put forward are essential to good site management practice and, if any one of these qualities is lacking in a manager, they would have problems in performing their role. Harris and McCaffer (1995) define the personal traits that a site manager should possess: "*Managers, by implication, should be people with experience, understanding and vision, and should have enough confidence to delegate responsibility and stand by decisions.*"

There exists a high positive correlation between the performance of site managers and the productivity and ultimate success of construction projects.

It is undeniably true that the implementation of poor quality management practices results in the failure to deliver a quality product on time and within a specified budget.

Site managers perform a similar role and function to managers in other industries. However, the competitive nature of the construction industry adds to the pressures on the site manager. The influence that the site manager imparts is outlined by Trotter (1982a): "*Site managers operate at the point of productivity and therefore make or break the profit return.*"

Sir Michael Latham's (1994) publication was undertaken as a result of concern over the industry's competitiveness and the nature of its working practices. The final publication contained 30 main recommendations and many more minor ones, which were directed at improving the industry's productivity and competitiveness. The aim was to reduce conflict and litigation and to show the need for creating a long-term relationship between client and contractor, encouraging teamwork and co-operation with the objective of overcoming adversarial attitudes. Latham set a target of 30% greater efficiency by the end of the decade. These changes were to include quicker and more reasonable payment rights throughout the contractual chain. For the client, substantial efficiency could be achieved if buildings and design became more standardised. In return, prompt stage payments would help cash flow and profits which, in turn, would enable steady growth and lead to an investment in both skills and machinery. For the principal contractor, if trading arrangements changed and there was an increased rate of cash flow in relation to interim payments, then perceptually there could be a decrease in profit margins as one has had a detrimental effect on the other.

As a follow-up to the publication produced by Sir Michael Latham, four years later Sir John Egan (1998) produced a report with the Construction Task Force Team for the Deputy Prime Minister, John Prescott, on the scope for improving the quality and efficiency of UK construction (known in the industry as the Egan Report). The intention of this publication was to improve conditions and working principles throughout the construction field by the deployment of a set of improvement targets. There are five key drivers of change:

- committed leadership
- a focus on the customer
- integrated processes and teams
- a quality-driven agenda
- commitment to people.

In enforcing these key drivers of change, the report aims to accomplish annual reductions of 10% in construction costs and construction time, reduce defects and accidents in projects by 20%, increase predictability by 20% and to improve construction firms' turnover and profits by 10% per year.

While the entire report is of paramount importance to the construction industry, the review of its educational framework, qualifications and training programmes, namely 'more and better training', specifically relates to the supervisory function: "*The key grade on site is the supervisor. The UK has one of the highest levels of supervision on site internationally but one of the poorest records of training for supervisors. We invite the Construction Industry Training Board and other relevant National Training Organisations to consider the issue as a matter of urgency*" (Egan, 1998).

To try and predict the future role of site managers would be a difficult exercise but it is clear that education is going to play a primary functional role.

Research Conducted on the Role and Functions of Construction Site Managers

The views of fifty construction site managers were obtained and the rationale for conducting the study was to establish their opinions on the key functions performed and to assist in the identification of possible areas for improvement. All managers were asked to rank, from 1 to 7 in order of importance, the following skills that a site manager should possess in order to be proficient at their job as defined by Watson (2000) in "*Improving the performance of construction site managers*" (the seven functions of management):

- planning
- organising
- controlling
- forecasting
- motivating
- co-ordinating
- communicating.

The results are provided in Table 3.1.

They were also asked if the seven functions ranked in Table 3.1 were in fact the critical areas of activity for site managers. The results of the responses are provided in Figure 3.7.

Table 3.1 *Ranking of the seven functions of management*

Ranking by managers		
Most important	1	Communicating
	2	Planning
	3	Organising
	4	Co-ordinating
	5	Controlling
	6	Motivating
Least important	7	Forecasting

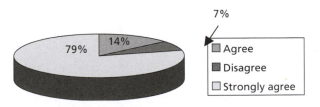

Figure 3.7 *Response to the seven critical functions. As can be seen, 79% strongly agree that these are the critical functions*

The site manager is responsible for directing and controlling all on-site activities within the imposed limits of the organisational hierarchy. Therefore, how they spend their time is of critical importance. Data has been collected from the fifty practising site managers upon how they spend their time on site. The results have been tabulated in Figure 3.8.

The results are interesting, with 25% of the sampled site managers' time being spent on administrative duties. The sampled site managers believed that there should be administrative staff on all sites to deal with these day-to-day functions, leaving them free to concentrate on managing the site more efficiently and effectively. They advocated the use of information technology (IT) as a means of reducing the time spent by site managers on the administrative function. From the field research on the utilisation of IT on construction projects, it is evident that full advantage is not being taken of IT in managing on-site activities. This is illustrated in Figure 3.9.

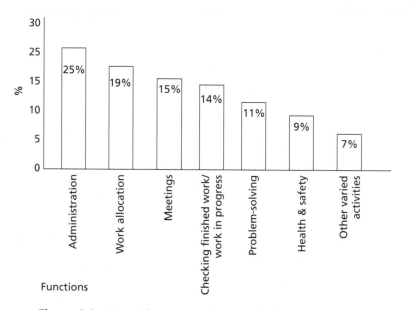

Figure 3.8 *How site managers spend their time on site (indicative only, because of the sample size)*

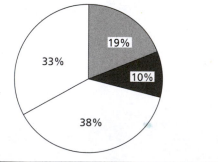

Percentage	Activity
10%	Administration (paperless system)
19%	Report writing
33%	Valuations
38%	Planning

Figure 3.9 *Utilisation of IT on construction projects*

The use of a paperless system through IT implementation on site can lead to a reduction in inefficient procedures and practices.

Table 3.2 provides a summary of the key issues advocated by site managers for attaining an efficient and effective construction site.

A question was posed asking them to identify specific areas of working practice which could be improved, and the results are tabulated in Table 3.3.

Managers were encouraged to comment on whether a more formalised systematic approach to site management would provide benefits in both efficiency and effectiveness on construction projects. The results are identified in Figure 3.10.

Analysis of the responses provided some interesting results. The three prime functions of a site manager were noted as:

- communicating
- planning
- organising.

Table 3.2 *Key activities of site management established from the field research*

Activity	Site managers' comments
Managing people	Treat all site personnel as you would like to be treated, with respect and courtesy
Managerial approach to rules	Managers must enforce the rules fairly and without any favour; they must be consistent
Innovation	Create an environment where ideas for improvement are accepted and the penalties for failure do not outweigh the rewards of success
Lead by example	The site manager must have the ability to work well with other people as part of a productive team
Plan all work activities	Plan all work activities with sufficient time to obtain the best solutions. All activities must be co-ordinated, with safe working practices being employed at all times
Communication	Make sure all site staff know what is required of them and that they have all the required information and resources to complete set tasks
Information Technology	Become familiar with the latest computer technology and make the best use of IT
Training	It is essential that a manager's performance and role are reviewed by their organisation: the effectiveness and efficiency of the manager can be improved by ensuring adequate training is provided to address any shortfalls
Delegation	Site managers should not be overwhelmed with tasks which, as a consequence, would lead to some vital aspects of their role suffering. They should be prepared to delegate but this must be linked to the training issue

Table 3.3 *Key areas for improvement*

Areas	Response (%)
Subcontractor co-ordination	22
Operative co-ordination	19
Communication	15
Improved availability of information	15
Material requisition	13
Plant organisation	8
Management organisation	5
Overall administration	3

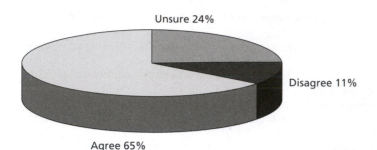

Figure 3.10 *Construction managers' responses relating to a more formalised approach*

The issue of motivation was ranked sixth out of the seven prime functions, although Wakefield (1982) identified this as a critical activity.

Six managers strongly agreed that the seven functions were in fact the critical activities, with forty managers agreeing.

The three most important factors considered by site managers to be 'critical success factors' in gauging their performance were:

- completion on time
- completion within budget
- attaining the specified quality level

with the two most important criteria by far being time and budget factors.

A very positive result was obtained regarding their views on having a formal systems approach when performing the construction site management function, with 65% in agreement.

When designing a formal system for site management, the ranking of the functions and the critical success factors should be addressed. This is necessary if both an efficient and effective system is to be attained.

Problematic Issues and Advocated Improvements

From the study, problem areas associated with site management have been identified. These have to be addressed in order to ensure that project objectives are achieved.

Problematic Issues

Health and Safety

One of the most problematic issues facing site managers is that of legislation relating to Health and Safety on site. All managers questioned underlined that safety on site was one of the most important aspects of their role. To comply with the standards expected, they suggested that a dedicated member of the organisation should deal with all aspects of Health and Safety, especially on medium to large-scale projects.

Obtaining Information

Every manager had problems with the communication flow between the site and the client/architect. It was mainly described as ineffective.

Further problems resulted from this lack of information flow because planning of a project is associated with the ability to forecast when problems may occur in the future. Without the relevant information being available, this aspect of planning is virtually impossible.

Technical Knowledge

With technology in the industry advancing at a rapid pace, site managers must enhance their knowledge of new techniques, improved materials and ways of overcoming problems. This would enable them to keep in touch with other professionals engaged in the construction process.

General Administration/Paperwork

This activity should not be the sole responsibility of the site manager. Within most firms contacted, managers had insufficient time to undertake what were sometimes very time-consuming tasks. Site managers in control of the entire site found that time spent in the office performing non-productive tasks would not add value to the project.

Advocated Improvements

Other areas were noted by respondents to the questionnaires but these were of an individual nature. Thus, the issues discussed are those aspects which were problematic to a large percentage of managers. The following outlines advocated solutions designed to address the previously discussed problems.

Training

Technical knowledge and the improvement of management skills have been identified as problems for which training could provide a viable solution. The provision could be in-house or specific courses designed to improve knowledge in aspects of their role and functions. The utilisation of IT within the construction industry is increasing at a rapid pace and managers require these basic IT skills. Most of the managers questioned during the study noted that they undertook training at every opportunity in order to improve their knowledge.

It is essential that a manager's performance and role are reviewed by their organisation; the effectiveness of that manager can be improved and any major problem areas pinpointed and addressed. Communication was the most important skill recognised by the managers. Addressing this management skill should improve both relations on site and information flow.

Delegation

Site managers should not be overwhelmed with non-critical tasks which, as a consequence, would lead to some vital aspects of their role suffering. At the same time, managers should not delegate work to subordinates who are unable to perform the role to a satisfactory standard. Therefore, training in the art of effective delegation is a vital activity. The site manager has to be the person taking the decision to delegate work and, if successfully done, this should assist in the satisfactory completion of site operations.

The role of the site manager has been claimed by themselves and various authors as the function that exerts the most influence on the potential success of a project. Therefore this important contribution should be recognised.

In order to overcome or reduce the problems faced by site managers, a generic model has been produced. It concentrates on areas of self-evaluation and the constant refinement of the way a site manager works, enhancing performance and ultimately adding to the success of a construction project. The model (Figure 3.11) has been compiled using research gathered from experts in their respective fields/positions and provides a means for addressing the strategic development of construction site managers. It further ensures that specific project aims are monitored and attained.

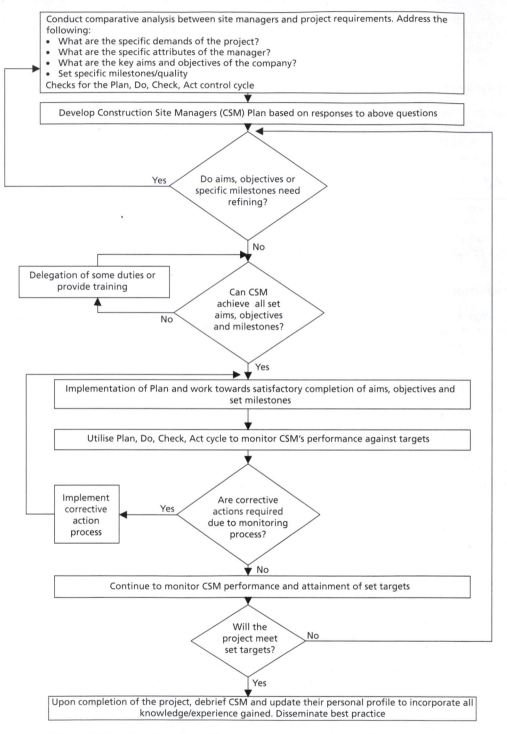

Figure 3.11 *Construction site managers' generic monitoring model*

Summary

Possible Developments

The future looks extremely prosperous for construction site managers and indeed the construction industry as a whole. There is a shortage of skilled workers throughout the entire industry and especially within the managerial hierarchy. Fewer academic managers are coming through the colleges and universities, which identifies an early warning sign that the existing managers will soon be in greater demand and, as such, will require payment to suit. While this sounds very attractive for the site managers, it does not help with the current problems they encounter on a daily basis. Contracts are normally under-staffed, the contract periods are usually too tight, the site-based paperwork is ever-increasing, the hours on-site are 'unofficially' above the working rule agreement, and the complexity of the design of the buildings and complex fixtures and fittings is constantly increasing. All of these external factors are causing the job role of the site manager to evolve. Over the past ten years, it has changed from one of a practical nature to a position of immense responsibility of a professional nature and, with ever new legislation and health and safety directives continually being introduced, the more time the site manager will have to spend on these issues.

It has been thoroughly highlighted in the earlier sections that continuous training is of paramount importance and the future looks no different, as the Egan Report (1998) states that *"More and better training"* is required, including the abolition of role ambiguity and the clarification of role expectations.

Organisation Development

Methods of Corporate Growth

The three main options which require consideration when choosing a strategy for growth are:

- internal development
- acquisition or merger
- joint development.

Like most strategic decisions, the choice between these methods is essentially a trade-off between a number of factors such as cost, speed and risk. How this trade-off is viewed in any one situation will depend not only on the circumstances of the company but also on the attitudes of those making the decision and the decision-making process. This should be apparent since companies within the same industry often have quite different and long-standing approaches to development, some always preferring to develop internally, others by acquisition.

Internal Development

This is a method of expanding by using reserves of profit within the business or by borrowing or issuing more shares and it has the following features:

- **Orderliness** – expansion is a smooth progression with no violent changes within the business.
- **Management** – the specialists required will already be available except for any new areas the company may be entering. It is possible, however, that the existing management may not be sufficiently expert to cope with an enlarged organisation. Also, if the old management was to continue there would be no encouragement to bring in new ideas, as there would be if 'new blood' was introduced.
- **Organisation's structure** – if the expansion is to be largely within the present scope of operation, little disturbance of the organisation's structure should be required.

For many organisations, internal development has always been the primary method by which strategy has developed and often there are some compelling reasons why this should be so. Very often, particularly with products which are highly technical in design or method of manufacture, companies will choose to develop new products themselves since the process of development is seen as the best way of acquiring the necessary skills and knowledge to exploit the product and compete successfully in the marketplace.

Although the final cost of developing new activities internally may be greater than by acquiring other companies, the spread of cost may be more favourable and realistic. This is obviously a strong argument in favour of internal development for small companies who simply do not have the resources available in the short term to develop in any other way. A related issue is that of minimising disruption to other activities. The slower rate of change which internal development brings usually makes it favourable in this respect.

Acquisition or Merger

One of the most significant reasons for developing by acquisition is the speed with which it allows the company to enter new product or market areas. In some cases, the product or market is changing so rapidly that this becomes the only way of successfully entering the market since the process of internal development is too slow in comparison. Another common reason for acquisition is the lack of knowledge or resources to develop certain strategies internally. For example, a company may be acquired for its knowledge of property speculation or a particular type of production system.

The overall cost of developing by acquisition may, in certain circumstances, be particularly advantageous. Companies going into liquidation may be a good buy.

The competitive situation may influence a company to choose acquisition. In markets which are static and market shares of companies reasonably steady, it is often a difficult proposition for a totally new company to enter the market since its presence would upset the equilibrium. If, however, the new company chooses to enter the market by acquisition, the risk of competitive reaction is reduced. The same arguments also apply when an established supplier in an industry acquires a competitor either for the latter's market share or to shut down its capacity to help restore a situation where supply and demand are more balanced and trading conditions are more favourable.

Sometimes there are reasons of cost efficiency which would make acquisition more favourable. This cost efficiency could arise from the fact that a company which is already established and running may already be a long way down the learning curve and have achieved efficiencies which would be difficult to match quickly by internal development.

The major problem with acquisition lies in the ability to integrate the new company into the activities of the old and this could provide a severe culture shock.

Methods of Merger/Acquisition

Mergers or acquisitions are the result of joining with or taking over other businesses. The methods of merger are as follows:

- The undertaking of one company is purchased by another, the former being dissolved.
- Two companies wind up and form a new company containing the combined undertakings. The shareholders of the two companies would receive shares in the new company in exchange for their old ones.
- One company buys the shares of another from the present holders. This is a take-over bid and results in the continuation of both companies (thereby preserving the goodwill of each) but one would be a subsidiary of the other.

Forms of merger or acquisition are:

- **Vertical integration** – this is the merging with other units operating at different points in the same chain of production. Thus a company can make certain of their supplies by purchasing or merging with suppliers or securing their market by acquiring retail outlets. This also has the effect of adding to the profit-making areas. This could be forward or backward in the supply chain, for example a speculative housing company moving forward into distribution/sales.

- **Horizontal integration** – this involves combining with those at the same level of production and similar activities.
- **Conglomerate** – this applies where a company enters an area completely unrelated to its present one. The intention would be the diversification of risk by not being entirely dependent on the fate of one industry. This form requires expertise in diverse companies and thus holds certain dangers.

Joint Development

The attractiveness of joint ventures could increase as environmental change accelerates. For example, given the rate of obsolescence of construction plant following technological innovation, it could well be advantageous for a company to consider becoming a marketing and distribution operation working on a joint venture basis with a manufacturer who is more specialised in the necessary field of technology (Johnson and Scholes, 1984).

Strategic Analysis and Corporate Planning

In order that the correct decision can be made regarding the type of growth, it is essential that the company understands the nature of its operational environment and the influences acting upon it. It is therefore necessary to carry out a strategic analysis. This consists of developing an understanding of the strategic situation of the organisation. What changes are going on in the environment and how will they affect the organisation and its activities? What is the resource strength of the organisation in the context of these changes? What is it that those people and groups associated with the organisation aspire to? How does this affect the present position and what could happen in the future?

The significant point is that differences in the nature of the environment call for different approaches to understanding and responding to the environment. In simple/static conditions, an organisation is faced with an environment which is not too difficult to understand and is not undergoing significant change. In these companies, their processes are relatively straightforward, their competition and markets are likely to be fixed over time and there may well be few of them. If change does occur, it is likely to be fairly predictable.

Organisations in complex dynamic situations are faced with environmental influences difficult in themselves to understand. In dynamic conditions, the environment is changing. Companies faced with technological advances, more sophisticated clients and an internationalisation of markets find that they can no longer make decisions based on an assumption that what has happened in the past will continue to happen in the future.

An analysis of the nature of the environment helps to provide guidelines on how to proceed with further analysis and where emphasis should be placed on managing the environment. Whichever approach is taken, the aim is to understand the environment in such a way as to generate a credible strategy in relation to the main forces in that environment.

The corporate planning process is a critical issue for building-related organisations. However, complicated bureaucratic corporate planning models are of little value to most construction organisations. Therefore a more simplistic, yet effective, methodology is required. Such an approach has been adopted within this chapter.

The construction industry is a fluctuating fragmented area of work. Barnard (1981a) describes the construction industry as a collection of industries with a mobile factory and variable workforce, constantly in varying conditions. Research by Barnard (1981b) and Bengtsson (1995) indicates that, in the construction industry, little use is made of corporate planning, especially in smaller businesses. This lack of use is mainly due to having a limited knowledge of the corporate planning process. The small building business manager does not have the time to engage in the process. In most cases, this can be attributed to 'crisis management' taking priority over long-term strategic needs.

Many question whether in the construction industry it is necessary to go to the extremes of corporate planning. The lack of management skills within the small, often craft-based, family businesses has prompted Lansley (1975) to argue that *"probably in no other industry is it more necessary for small firms to keep in touch with their market than in construction."*

Key Issues to be Considered

What factors must we consider? Management has often been described as getting things done through people, so obviously the human factor should be the first consideration. There is nothing more likely to produce a plan that will not succeed than failure to involve from the beginning the personnel whose jobs it will be to implement the plan. If the person charged with implementation has an input into the important decisions relating to plant selection, subcontractors and the methods to be employed, they will have a vested interest in the success of the plan. Corporate planning must therefore be a function at all levels with participation by everyone. Remember, consultation is an activity with two aspects – firstly it provides feedback and secondly it has a psychological element.

A further major consideration is timeliness (see Operations Planning in Chapter 5). Very few managers of building-related businesses have time to stand back and ponder their activities. If time is not available to anyone from within the company, they must look externally to a management consultancy. Barton (1983) suggests that "*Building managers are primarily concerned with maintaining the steady state – getting the job built on time and within budget.*" Therefore consultants should not be ruled out if they can provide the necessary staff in order to undertake the activity as quickly as possible. Vasser (1981) opines that "*Building is done with people.*"

Goals are indispensable to the effective management of construction companies. In a small business, the overall objectives of the firm are often implicit, e.g. survival, growth, being profitable or having innovative power. Corporate appraisal can help translate these long-term objectives into more specific and unambiguous goals at the level of individual projects.

Within the construction industry, employees do tend to work very closely as a team. In this kind of operational environment, shared objectives and group norms are very important. In Lansley's (1981) view, corporate planning is concerned with identifying the half a dozen factors which really matter to a business and of analysing these factors so that plans can evolve. Successful entrepreneurs do not make decisions based on hunches and intuition, rather they have a highly personalised, even idiosyncratic set of business principles. These principles provide insight into opportunities and enable them to form valued judgements more quickly than would other people.

There is no one correct way of formulating a corporate plan; much depends on the particular firm, its present state of development and the skills of its management. A common assumption among construction managers is that smaller firms are much the same as larger firms except for the volume and value of the business and for the number of employees: "*For this reason many smaller firms have looked towards the larger firms and towards text books and learned journals devoted to corporate planning*" (Lansley, 1981). There is, however, no place in the smaller construction firm for the highly bureaucratic rituals found in some large companies. The success of the smaller firm rests upon simplicity and flexibility: "*Most of us are born with a good deal of flexibility; it's a helpful trait*" (Passmore, 1994). This flexibility must be preserved at all costs. The smaller organisation will not have sophisticated techniques or specialist staff for corporate planning purposes. What is important is that the manager is committed to and personally involved in steering the planning process forward. They must make sure that the objectives and successful development of the firm are shared by all employees enthusiastically.

A Strategic Appraisal System for Construction-Related Businesses

The high number of construction ventures which have failed in the past leads Lansley (1981) to the conclusion that "*They were not organised for profit. Under normal trading conditions, they could never be successful. In times of surplus, out come the 'cowboys' who before they finally disappear drag down others who otherwise might have survived. So much for business judgement under pressure.*" Success often leads one to doing less of what one is best at. Trotter (1982b) postulates that "*The small successful builder suddenly finds himself in a larger league, having to negotiate finance, manage the cash flow, do detailed estimates and valuations, comply with a strict specification and probably have to contend with a quantity surveyor; little wonder small businesses go bust.*" There are dangers in over-trading, and under-trading too, for the small building business. This serves to prove that in both cases planning is a necessity. Planning must be based upon a sound strategic appraisal system. In this section we shall investigate such an appraisal system as advocated by numerous authors in the related fields of corporate planning.

Current Strategy

The first and perhaps most important area of corporate appraisal is the firm's existing objectives and strategy. Strategy is defined by Taylor (1977) as a proposed disposition of resources to achieve the place, time and conditions of business that meet the firm's objectives. Benes and Diepeveen (1985) argue that "*Corporate planning should be the construction firm's main tool of management.*" In the construction industry strategy should be analysed, according to Barnard (1981a), by considering the following:

"*(a) Geographic areas in which the firm will operate.*
 (b) Types and structures it will seek to construct, e.g. house or small factories.
 (c) Type of client it will favour, e.g. local authority or private bodies."

Harris (1977) urges us to recall "*the possible share and fluctuations of the market in future years.*"

Current Objectives

Barnard (1981a) suggests that current objectives should be considered under the following two main headings:

(a) economic objectives; and
(b) non-economic objectives.

Under the heading of economic objectives are placed growth of turnover, earnings, market share and the number of markets in which the firm operates, stability of annual gross turnover, gross profit, return on investments and the utilisation of scarce physical or human resources held by the firm. Non-economic objectives include internal political, e.g. retention of control by the existing owners, external political, e.g. to avoid intervention by central, local or other government bodies, to meet the reasonable aspirations of employees and to develop them to their full potential, to serve clients and the general community and to maintain a good reputation within the industry. The owner/manager's values influence their perception of these objectives. For example, Fryer (1985) notes that some managers have recognised the need to meet employees' needs and companies have made jobs more intrinsically satisfying by relating them to the work of Maslow *et al.* in the form of socio-technical systems.

Internal Appraisal

Following the initial investigation of the business and the determination of the current strategy and objectives, the next stage identified by Barnard is to implement an internal appraisal of the business. This is performed in order to determine the corporate identity or characteristics of the businesses. It will also assist in discovering organisational strengths and weaknesses. The analysis will identify whether certain areas of the host company require attention. Barton (1983) describes this process as ensuring that the organisation remains attuned to environmental constraints and opportunities.

The following list should be consulted, and given due consideration, when conducting a strategic analysis of a construction business:

- **Finance** – An assessment should be made of the reasons for variances from industry norms. Details of structure of ownership, sources of finance, liquidity, cash flow and growth record should be addressed. Steiner and Miner (1977) specify that "*the owner/manager must also identify whether she/he has sufficient capital or can obtain it, to see the strategy through to a successful implementation.*"

- **Sales** – Sales turnover, growth rate, profitability and market share obtained in the last five years should be investigated. Contracts should be analysed according to:

  (i) length
  (ii) value
  (iii) profitability.

Other aspects for consideration are competitiveness and success rates in the tendering process. Somerville (1981) points out that "*Once again, what is important to stress is the need to understand conditions of the future … the biggest factor effecting change is financial.*"

- **Marketing** – The effectiveness of the following should be analysed:

  (i) methods undertaken in locating work
  (ii) successful tenders and the number declined
  (iii) direct enquiries
  (iv) advertisements
  (v) image created by the firm.

The organisation should try to capitalise on its 'distinctive competence'. Birley (1979) identifies that the problem the business faces in respect of advertising is deciding which forms are most cost-effective. By this, Birley means which firms attract the most customers for the least cost to the business, least cost not necessarily referring to the cheapest total cost but to the lower cost per additional customer attracted.

- **Organisation, administration and personnel** – To analyse organisational structure, define the nature and system of management. Organisational structure should follow strategy, and also the mission of the organisation will influence corporate strategy. Three distinct types of mission are (Hambrick, 1983):

  (a) prospectors – this is an entrepreneurial organisation
  (b) defenders – this type of company is risk averse and does not welcome strategic change
  (c) analysers – this type of company will engage in change processes but only after a thorough analysis of all data.

The approach adopted will fundamentally affect the policies and strategies developed by the host organisation.

- **Production performance** – Analysis of this area can be difficult in a small business because of the scarcity of data with which to make comparisons. The first questions therefore, according to Jadrin (1976), are subjective and do not require another firm's data, but clients provide answers from the utilisation of questionnaires. The percentage of repeat customers and number of complaints are a good general comparative guide. Barnard (1981b) prefers more explicit comparisons, with ratios provided by companies conducting inter-firm comparisons. James (1979) advised the monitoring of productivity trends in sales related to capital employed, the added value per employee and the added value as a multiple of capital employed. An inspection of the physical resources should include age, condition and life expectancy (risk of obsolescence) and suitability of locations (proximity to customer base).

- **Purchasing** – In appraising this topic area, Barnard (1981a) identifies the following headings for consideration:

  (i) purchasing methodology
  (ii) cost of purchasing unit
  (iii) sources of supply
  (iv) stocks held
  (v) methods of controlling materials on the sites.

- **Competition** – The strengths and weaknesses of competitors should be analysed by:

  (i) company
  (ii) market share
  (iii) concentration and dominance
  (iv) resources and competitive strategies
  (v) geographical spread.

  "*An investigation should also be made of gaps open to entry, e.g. product or growth areas and customer dissatisfaction*" (Barton, 1983).

- **Pricing, estimating and surveying** – Blyth and Skoyles (1984) advocate that in competitive tendering a completed building rarely reflects the information provided at the tender stage. Therefore feedback information systems must be 'closed looped' and linked to a 'pricing and bidding strategy'. The establishment of the 'final account' should be closely monitored.

Barnard (1981a) advises that a complete strengths, weaknesses, opportunities and threats analysis be implemented in order to formulate the company's future strategies.

External Appraisal

The next stage is to conduct a strategic environmental appraisal for the host business. This empowers management to make informed decisions appertaining to diversification through a reasoned judgement of both the opportunities available and the possible threats to the business as it exists or may exist in the future. James (1979) proposed that "*The corporate appraisal highlights areas of the organisation and environment that are seldom considered in the day-to-day running of the business.*"

Barnard (1981a) advises identifying both the key factors for success (the critical success factors) and also those over which one has little control; this requires an investigation of the following:

(a) Geographical factors: forecast changes in the area of operations – opportunities for greater geographical coverage.
(b) Economic factors: economic and construction forecasts – anticipate government and private capital spending levels.
(c) Social and political factors: political stability and strength of government, government policies, finance, earnings, taxes, employment growth and regional aid. It is a fact of life that governments use the building industry as a 'barometer of the economy', as it is subject to wide variations during booms and recessions.
(d) Competition: ease of entry and exit, role of capital investment and trends in the competitive pattern.
(e) Opportunities and threats: the conclusion of the appraisal sets out the opportunities available to the company balanced by the possible threats from the operational environment and stakeholders.

An analysis such as the one suggested will identify where areas of insufficient information exist. It also highlights areas of the organisation and its environment which are seldom considered in the day-to-day running of the business. Perhaps, because of the very variable conditions that small construction business managers have to comply with, there are few comparisons which can be made accurately. Construction sites have certain production performance limitations, i.e. they are mobile factories, with different people, work, weather, clients, sites, geography, access and timing.

Taking all these variables into consideration, most authors strongly advise construction businesses to engage in the corporate planning process. The advantages are there to be seen and obtained.

Strategic Models for Construction-Related Businesses

According to Benes and Diepeveen (1985), to be realistic, corporate planning models utilised by construction-related businesses should have a planning horizon of one to two years. They claim that changes in society, in social climate and in business activity are now too frequent for long-term (five years plus) planning.

Any model employed for corporate planning purposes should be based upon flexibility. Koontz and O'Donnell (1974) note that *"flexibility in planning should reduce the risk of losses due to unexpected events."* The strategic planning process must enable a flexible approach within a construction operational environment.

Many complicated strategic planning models are in existence today. However, they are too complex to be of any real value for construction-related businesses. It is not the authors' intention to duplicate existing works but rather to develop a model that will be both applicable and advantageous to construction-related enterprises.

The model (Figure 3.12) is intended to aid strategic thought by focusing attention upon environmental issues.

Summary

The analysis generated by such an approach aids appraisal of changes that will take place and increase the time available for the introduction of new plans. It will, as pointed out by Barnard (1981a), *"help to create and structure an appropriate organisation."* Whatever the size of an organisation, it cannot escape the fact that it operates in a dynamic operational environment and must also negotiate credibility with its stakeholders.

The survival of construction organisations depends to a great extent upon flexibility and adaptability, and they must develop a competitive environmental perception. This may require changes in both organisational form and management, however, as pointed out by Hutchins (2001): *"Managing change has been subjected to a vast amount of research over the years. It is an area where much is known. Yet still major problems exist in the management of change within almost all organisations."* Construction firms are not excluded from this observation.

Networks

This section is concerned with what might be seen as an 'emergent' form of strategic development. It is a strategy of increasing relevance to construction managers who manage across boundaries – whether between or within organisations. Traditionally the competitive spirit has always dominated corporate thinking. Strategic management is seen as an entrepreneurial adventure in which firms must circumvent 'threats' and exploit 'opportunities'. Inter-organisational relationships are usually viewed as inherently competitive and antagonistic.

Here an alternative strategic approach is advocated in the form of 'coalitions'. These coalitions are a configuration of 'networks'. In summary, they offer an alternative methodology for construction managers, incorporating a strategic move from the old 'adversarial' model.

The old adversarial models based upon 'transactional contracts' are not suited to dynamic operational environments. They also have certain implications for the parties involved which are not always advantageous. Contrasted with the above is the new 'relational contract'. Within the construction industry, this type of relationship has been advocated within the Latham Report (1994). The report opines that *"Partnering is a means for construction companies to gain a competitive advantage."* Hellard (1995) advocates that *"the philosophy of teamwork and co-operation, not confrontation and conflict, is long overdue."*

The CBE talks of building lasting partnerships with customers and suppliers to ensure consistent quality and efficiency at all stages of the production process.

Networks – a Strategic Approach

The term 'network' has emerged from a variety of disciplines and there has consequently been no uniformity of its use. If one reviews the literature appertaining to 'networks', six definitions can be

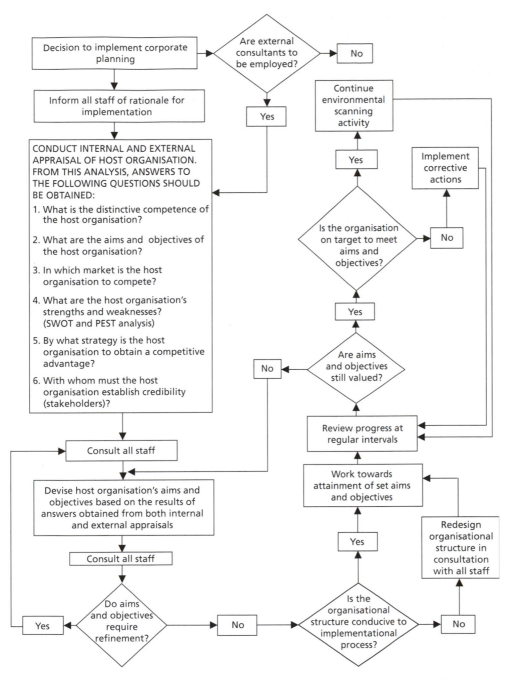

Figure 3.12 *A corporate planning model for construction-related business*

formed as follows:

- talking to people
- links between individuals or organisations
- lateral rather than vertical processes

- relational contracting rather than transactional contracting
- developing and maintaining a wide range of communication links
- a process of joint action between organisations.

The above may be summarised as operating under a 'collective strategic approach'. A fundamental requirement for entering into networking is the attainment of synergistic advantage(s). In order to obtain the noted synergy and avoid the possibility of failing in your strategic alliances, consideration must be given to the seven 'I's identified below:

- **Individual excellence** – Both partners should have something of value to contribute to the relationship.
- **Importance** – The relationship complements major strategic objectives of the partners, and therefore gets adequate resources, management attention and sponsorship.
- **Investment** – There is an agreement for long-term investment by all concerned, by which the partners demonstrate their respective stakes in the relationship and each other.
- **Interdependence** – The partners need each other, and this assists in maintaining the correct balance of power between all involved.
- **Integration** – The partners develop linkages and shared ways of operating so that they can work together smoothly. They build broad connections between many people at many organisational levels.
- **Institutionalisation** – The relationship is given a formal status, bolstered by a framework of supporting mechanisms, from legal requirements to social ties and shared values, all of which in fact empower a trusting environment.
- **Integrity** – The partners behave towards each other in honourable ways that enhance mutual trust. They do not abuse the information they gain, nor do they undermine each other (Kanter, 1989). "*With understanding comes trust and with trust comes the possibility for synergy*" (Hellard, 1995).

If, after due consideration by all concerned, it is agreed to form a network, the next stage in the implementational process is deciding upon what form of network to operate.

Network is a generic term and in fact four types of networks are available. Table 3.4 summarises the various types of networks. Table 3.5 provides a generic deployment guide.

Characteristics of Networks

Organisations must first establish the rationale for entering into a network. Secondly, the host company needs to establish which form of network best suits its strategic objective(s). Remember that entering into a formal network may require an organisational change in attitudes and possibly structure. Morgan (1993) claims: "*There is no one best way of organising. The appropriate form depends on the kind of task or environment with which one is dealing ... Different approaches to management may be necessary to perform different tasks within the same organisation and quite different types of 'species'*

Table 3.4 *Alternative types of network organisations*

Type of network	Number of organisations	Examples
Productive network	One	Lateral linkages, network between divisions
Bilateral network	Two	Joint ventures, strategic alliances, vertical disaggregation
Hub network	More than two	Turnkey projects, Keiretsu, broker-centred networks
Federation	More than two	Partnerships

Table 3.5 *Generic implementational guide*

Key issue	Advocated action
Personal commitment	Treat the collaborative venture as a personal commitment. It is people that make networks successful
Management time	Anticipate that the venture will consume management time. If this is not available, then networking is not a strategic option
Synergy	Remember that all partners must obtain something from the network. This will probably mean you have to give something up. Recognise this from the outset
Trust	Mutual respect and trust are essential. If you do not trust the people you are negotiating with, the networking concept with the parties involved is not possible
Legality	Make sure you tie up a tight legal contract. Don't put off resolving contentious issues
Flexibility	Recognise that during the course of collaboration, circumstances may change – be flexible
Mutual expectations	Make sure host company and partners have mutual expectations of the collaboration and timescale
Know opposite personnel	Get to know your opposite numbers at all levels of personnel
Culture	Appreciate that corporate cultures may be different. Don't expect partners to act/respond identically to you
Partners' interests	Recognise your partners' interests and independence
Corporate approval	Make sure you have corporate approval with corporate commitment to the network, then you can act with the necessary positive authority
Achievement	Celebrate achievement. It is a shared elation (motivational issue)

or organisation are needed in different types of environment." The application process may be problematic but the rewards are there to be obtained.

The advocation of networks as an evolutionary step in the corporate strategic thinking of construction companies is based upon the success of its implementation within the manufacturing and retail sectors. This success has been most profound in the area of 'partnership resourcing' and this would seem to be a specific area of network application that is very pertinent to a construction-related operational environment: "*The traditional approach has long been one of confrontation and negotiation, these practices are wasteful in the long-term*" (Lamming, 1993). It is time for a more radical approach to the way construction-related organisations formulate their strategic objectives. A new methodology should provide for careful consideration of all available options and hence must include networking.

Summary

Networking is applicable to construction-related organisations and could be utilised as a mechanism for obtaining/maintaining a competitive advantage in the most dynamic environment. However, the organisations need to give careful consideration to both the rationale for and means of obtaining a truly holistic view of the company and its interactions with stakeholder groups. The rewards of pursuing such a strategic initiative should prove to be most advantageous provided that the correct form of network is selected and the implementation process is fully resourced.

Best Practice Attainment

Turning construction organisations into world-class businesses presents one of the stiffest challenges for managers today. Yip (1992) notes that, because of its difficulties, being able to develop and implement an effective 'best practice' strategy is the acid test for any company. Hickman and Silva (1989) consider that "*Everyone recognises that an Olympic gold medal and a Nobel prize reward excellence. Such recognisable excellence extends to the business world.*" The excellence identified by the above authors is termed 'best practice'. Best practice essentially entails having the right production capability to make a profit by totally satisfying the customer, with high-quality services and products at the right price delivered at the right time. This criterion is not just applicable to a construction operational competitive environment. It means operating at standards equal to the best practice in the world, but is relevant not only to companies that export but also to those facing competition from companies in external countries.

The following provides a methodology for construction companies to obtain a sustainable competitive advantage in a dynamic operational environment: "*Companies all over the world face stiffer competition than ever before. National boundaries are becoming less and less of a constraint on competition and in many industries a single global market is now a reality*" (Anderson, 1994).

Adding to the difficulties of maintaining a competitive advantage in the above noted circumstances is the fact that methods used to control production have also undergone great changes. A retrospective look at the issues seen as important to managers over fifteen years ago would have encompassed:

(i) financial issues
(ii) overall business plan issue
(iii) marketing issues.

Very rarely would managers have been concerned with the production operations at the heart of the organisation. However, the world has changed and "*production excellence is being increasingly recognised as a central element in the overall success of a business*" (Anderson, 1994).

The production activity should provide a competitive weapon built around the construction organisation's 'distinctive competence'. Therefore the production activities of most businesses are now the focal point for management attention (see Chapter 5). Maskell (1992) identifies six specific areas for consideration around which a competitive operations strategy could be built:

(i) quality
(ii) cost
(iii) delivery reliability
(iv) lead time
(v) flexibility
(vi) employee relations.

Maskell suggests that there is room for optimism as surveys of Western companies have indicated considerable improvement in all six areas. In order to gain a competitive advantage, companies "*will have to strive for excellence in every area of their operations*" (Nunney, 1992). The challenge for construction companies is to develop a strategy that encompasses their operational activities, thus enabling the attainment of a sustainable competitive advantage (Maskell, 1992).

The thoughts of Maskell are corroborated by Hickman and Silva (1989): "*Running a successful business has always required smart strategy. Yet the business world has undergone changes… What we're moving towards is an integrated theory of production.*" The integrated theory advocated is the concept of a 'World Class Business' and this is only possible if a company operates at best practice level.

A Methodology for Adopting Best Practice

The first step in identifying how an organisation can adopt best practice is to have a complete understanding of where that company is now in best practice terms. This will require the company to implement a 'Strategic Analysis'. Johnson and Scholes (1993) consider that the strategic analysis encapsulates "*The Environment, Organisational Culture, Strategic Capability and Stakeholder Expectations.*" Once the organisation attains full comprehension of its true state, it must then establish where it is in relation to its competitors. This process will involve benchmarking various activities of the host organisation. Benchmarking should, however, be undertaken at the 'Strategic Business Unit Level'. Ross (1991) suggests that every function of a business can be benchmarked. The author further notes that the specific emphasis of benchmarking can vary, but in general companies utilise it to:

(i) obtain a better understanding of how they are performing in specific areas relative to other companies
(ii) characterise the strategies being followed by their competitors.

The following key questions must be addressed and answers provided by organisations seeking to be the 'best':
Is the company striving for:

(i) better products/services?
(ii) better organisation/management?
(iii) better information/communication systems?

The above are not mutually exclusive.
Obviously the ultimate value of the exercise will depend upon identifying the 'leading competitor' in the field – and in benchmarking against that company.

Generic Benchmarking

This compares the business processes of organisations regardless of the industry to which they belong. Some business processes are common to all industries, 'purchasing' and 'recruitment' being two examples. The advantages of generic benchmarking are that it breaks down the barriers to thinking and offers a great opportunity for innovation. It also broadens the knowledge base and offers creative and stimulating ideas. The disadvantages are that it can be difficult, time-consuming and expensive (McGeorge and Palmer, 1997).

Collaborative Benchmarking

The 'benchmarking' activity could be performed as a collaborative venture. There is no reason why 'best practice' cannot be shared between non-competing organisations. Figure 3.13 provides a methodology for conducting a collaborative 'benchmarking' activity.

Market Requirements Curve

A market requirements curve (see Figure 3.14) can provide the 'foci' for a construction company when developing its strategic objectives. It is also of value when conducting collaborative 'benchmarking'.
A construction company must first establish what the 'critical success factors' of its operations are. Secondly, it must gauge (benchmark) clients' requirements in these areas and check for correlation between actual performance levels in the critical success factors against required performance levels. The market requirements curve is useful in assessing where organisational effort (resources) need to be focused in order to maintain credibility with clients.
Harrison (1993) advocates the way forward as first, creating a vision for and of the company, secondly installing ownership of the issues facing the company, and thirdly planning and implementation of a continuous change process. However, above all the company must 'focus on the customer'.
Some of the techniques and issues to be addressed by an organisation in order to function at best practice level include the following. The first issue to be considered is that of Senior Management Commitment: "*There is no substitute for effective leadership by top management. The number of*

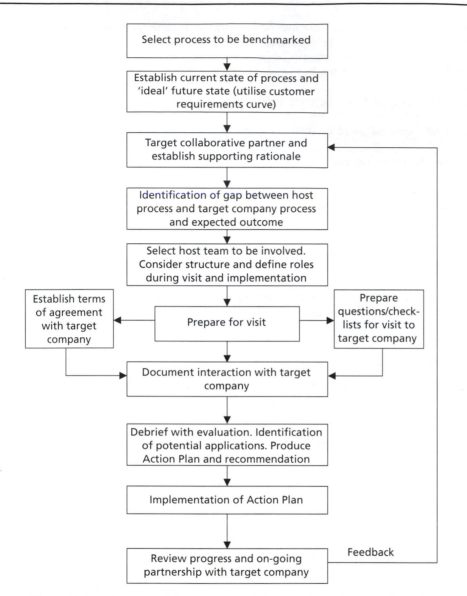

Figure 3.13 *A methodological model for collaborative benchmarking*

failures on the road to operating at best practice level blamed on the lack of management commitment suggests that not only is it needed but that it cannot be assumed and may be difficult to obtain in practice" (Nunney, 1992). Mundy (1992) recommends that *"People at all levels in the organisation must fully understand the organisational objectives and the timeliness of the objectives."* A consideration for organisations pursuing a strategy of best practice incorporates the four 'P's:

Purpose

People perform better when they understand the objectives of the company and team work is essential; therefore people should be trained to work in teams. An organisation must embrace the concept of 'empowerment' and people need feedback upon their performance as this affects intrinsic motivation.

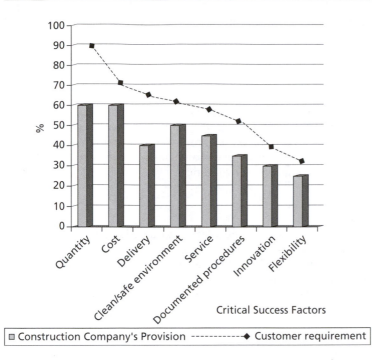

Figure 3.14 *Market requirements curve example*

Planning

Planning and monitoring require the setting up of a dynamic closed feedback loop. A further require-ment is that the Senior Management Team (SMT) must not think just in terms of reducing labour costs per service facility or product manufactured. Consideration should also be given to reducing the time taken for materials to undergo each production process. This, of course, requires planning; incorpo-rated within this should be a collaborative partnership approach towards suppliers with the attainment of long-term strategic benefits being the prime objective.

Process

The process(es) must be flexible and this may demand as much organisational effort in support facili-ties as it does in the operational activities. Design activities must be incorporated with the production function; this requires a holistic approach.

The processes will, in most instances, determine the service/product quality. It must be appreciated by all concerned that quality is everyone's business.

Performance Measurements

This requires simple dynamic systems. However, it is not the quantity of data that is important but its usefulness in the management decision-making process. In summary, it may be stated that the prob-lematic issues are not the technological-based ones but the 'people issues': "*Management acts to develop its people by caring for and training them*" (Hickman and Silva, 1989).

External and Internal Changes

The concept of continuous improvement is an important aspect of a best practice performer (Yip, 1992). Yip is stressing the dynamic nature of best practice operations. In order for an organisation to become and/or remain a world-class performer, it must consider both internal and external requirements.

A list of external issues for consideration are provided below from Mundie and Cottam (1993).

Competition

"*The trend towards economic liberalisation in general and privatisation in particular has had a major impact on business activity*" (Preston, 1993). As noted by Preston, competition has increased for most organisations. Therefore a company must fully understand the nature of its competitive environment. Organisations must operate a macro business strategy, not a micro one. Environmental scanning would provide the means for analysing a company's competitive environment.

Suppliers

The way an organisation deals with its suppliers must be as joint venture(s), i.e. partnership(s). The company and its suppliers must understand the synergistic advantages that are available to both parties. The function of purchasing is a vital activity if a company is to perform at best practice level.

The purchasing function must be elevated from its traditional clerical role in the organisation. Ruch (1992) states: "*Many companies have elevated the Chief Procurement Officer to top management status.*"

Environmental Factors

The environmental factors must be analysed to the extent that they provide either opportunities or threats to an organisation. The requirements of the environment may involve some physical alteration to the company.

Demography

Changes in the population may reflect in changes to market share, and this in turn may require the restructuring of the process of provision for the service or product, for example as promulgated by Waters (1991): "*It may be difficult to say which is the best process in given circumstances but factors which affect the choice include demand patterns.*"

Economic Factors

Similar to the demographic considerations are the economic considerations. The company must respond to varying changes in monetary value. This factor will also impinge upon the supply and demand for the service or product, subject to the rules of elasticity.

Technological Factors

Considerable interest has been shown in the implementation of new technologies within service sectors. Cumbers (1986) suggests this is because "*The effective management of technical change represents an opportunity for British management.*" These technologies are not just for internal consumption but enable organisations to communicate and interact on an international basis. This is certainly true for international procurement arrangements.

The internal changes that may be required can be to a certain extent viewed as changing the production systems from the traditional 'push system' to the modern 'pull system'.

Some of the issues raised under internal and external changes overlap each other. However, one must not forget that the main internal change required is people-orientated: "*Until recently, most top operations managers did not perceive the human organisation as a source of competitive advantage*" (Ross, 1991).

Employees, in the view of Kanter (1989), have an in-built resistance to change. Nevertheless, change may be necessary. For example, the type of organisational structure that the company utilises can affect its ability to function at best practice level: "*Lack of business success in many cases can be attributed to a persistence with an outdated organisation structure*" (Pilcher, 1986).

If change is necessary then, as promoted by Ross (1991): "*The key to making the transition work is in the employees' understanding of its necessity.*"

Usually groups of people in organisations recognise that work could be done more efficiently and/or effectively (Anderson, 1994). It is worth noting the thoughts of Ross (1991): "*To participate effectively in the global market place, the implementation of production technologies must be combined with a programme aimed at aligning organisational structure and culture, the role and flow of information and people resources, if enterprises are to exploit opportunities.*"

The above is a move towards 'Anthropocentric Production Systems'.

Requirements of Anthropocentric Production Systems (APS)

The Concept

Anthropocentric Production Systems are computer-aided production systems based on skilled work and human decision-making. The aims of APS are to:

- create a flexible, innovative and efficient organisation
- provide high quality, high productivity, high innovation capacity, increased flexibility
- create a maintainable, humane working environment.

A methodology for attaining the above is based upon an increasing knowledge of organisational behaviour, management and technology. The system adapts and adopts certain principles utilised by Japanese industry. APS is dependent on the following:

- technological innovations
- economic conditions
- social conditions
- other environmental conditions (e.g. the political will to implement APS technology).

The following components are of crucial importance to a successful implementational process:

- a flat hierarchy and empowerment of employees
- minimised division of labour (job redesign if necessary)
- appropriate training and education of the whole workforce
- integration of the various organisational functions, i.e. research and development, marketing, procurement and operations.

The Required Pre-Conditions for APS Implementation

There are two main obstacles to the current implementation of APS technology within most UK construction organisations. They are:

- uncertainty and risk aversion concerning the design and implementation of APS technology
- rigidities in existing technical, social and economic structures within organisations and their operational environments.

Implementation of APS technology may require some sweeping changes by organisations relating to existing technical, economic and social structures and their interaction with operational competitive environments.

Production Regimes

Although the development of computer-integrated manufacturing in European industry is still dominated by Tayloristic production concepts and models of full automation, there are a number of cases in which applications of computer-based production technology is associated with the redesign of organisational structure and process delivery mechanisms. These are predominantly in industries where competitiveness is based on high product quality and rapid product innovation. There is a strong positive correlation between quality of production and the use of APS technology. The whole organisational regime must be considered and factors for consideration should include the following:

- the decentralisation of planning
- the decentralisation of programming and control
- the integration of jobs and equipment.

The objective of applying APS technology within organisations is to produce semi-autonomous work groups. Companies with existing Tayloristic structures are introducing forms of participative management to reduce conflicts. For the long-term facilitation of the APS, there must be trust between workers and management, in fact to make this distinction could be argued against from an APS point of view.

Europe is far from implementing full APS technology. However, once the social constraints (people and organisations) have been overcome, the development should take place.

Workforce participation in the development of APS is a basic requirement. If APS technology is to become the strategic goal of an organisation and therefore senior management, they must understand that the implementation of any strategic or tactical goal is ultimately dependent on people.

Johnson and Scholes (1984) consider that *"Corporate objectives are usually formulated by senior members of the Board or even the Chief Executive. They are more likely to be handed down to lower levels."*

However, APS implementation requires a different approach, as noted by Massie (1981): *"Organisational objectives give direction to the group and serve as a media by which multiple interests are channelled into joint effort."* The point to note from the writings of Massie is that of joint efforts. By this, Massie means people at all levels within the host company working together. Management must not forget the need for consultation. People are the mechanism of implementation, therefore good industrial relations are a vital issue. This point is corroborated by Miles and Snow (1983) who postulated that *"Organisations do not have objectives but people have values."*

Organisation

Organisational structures are there, in a sense, to justify the authority of people within the firm. As noted by Glassman (1978): *"If appropriate authority does not accompany managerial responsibilities and duties, the manager's effectiveness within the organisation is impaired."*

Formal authority of a manager is when the authority is viewed as originating at the top of an organisation's hierarchy and flowing downward through delegation. However, the above noted that the real source of authority possessed by an individual lies in the acceptance of its exercise by those who are subject to it. Formal authority is therefore nominal authority (Herbst, 1976). APS technology requires that people be given responsibilities to perform. Management under the APS philosophy does not have to lose control because *"Management control can focus on the activities of the responsibility centres"* (Anthony, 1988).

People must be allocated adequate authority to be able to complete their tasks. The organisational structure must be conducive to the strategy. Figure 3.15 provides an implementational model for APS technology.

Information Required for the Implementational Process

As previously noted, the host organisation must conduct a strategic analysis to obtain a true picture of its present circumstances, opportunities and threats. After this has been accomplished the company must:

(a) Identify industry best practice, thus allowing benchmarking to position the company in relation to its competitors. The area of focus for the benchmarking activity should be the 'critical success factors' (CSFs) for the company.

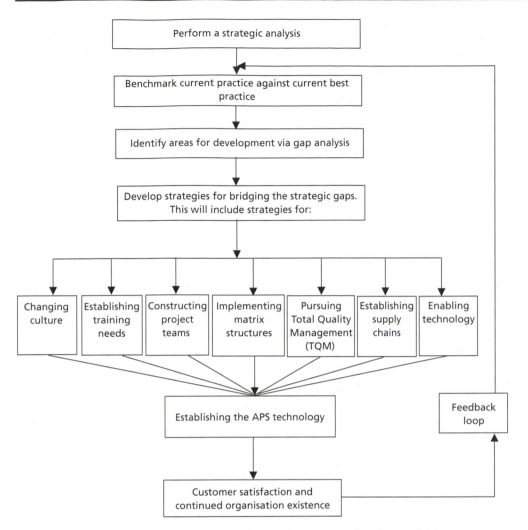

Figure 3.15 *A model for implementing APS technology within a construction company*

(b) Upon the identification of the CSFs and their analysis, Senior Management must then formulate a strategy to provide a platform for instituting improvements. This would require information in the form of the
  (i) development of objectives (shared objectives between both organisational and functional levels)
  (ii) development of performance measures (performance indicators) – these must be consistent with CSFs and market requirements
  (iii) development of a means of communicating and evaluating at all levels of the organisation.

The Benefits of Becoming a Best Practice Construction Company

The tangible benefits of being or becoming a best practice company are:
  (i) The provision of the product/service that meets customer specifications. This is possible because clear specifications would have been set during the contract review process. Many advantages would flow from this process.
  (ii) The tangible process(es) by which the product or service is manufactured/delivered would be the optimum process(es) for the service package or product.
  (iii) Time spent on corrective actions would be greatly reduced (or completely eliminated).

The intangible outcomes would include:

(i) Greater customer satisfaction from their personal contact with the organisation. Mundie and Cottam (1993) state that this would encapsulate *"peace of mind and enhanced self-esteem."*
(ii) Employees would have an operational environment and culture which benefited both them and the client(s) (motivational issues, e.g. extrinsic and intrinsic).

As a final point, it is worth noting the thoughts of Hickman and Silva (1989). The authors state that when an organisation becomes truly the 'best', it should be able to:

- *"satisfy customer needs, recognising that different customers have different needs*
- *gain a sustainable competitive advantage, keeping product/service differentiation in mind*
- *capitalise on company strengths, remembering that it takes time to develop them."*

Summary

Operating at best practice level provides competitive advantages for organisations. Construction organisations could adopt the manufacturing best practice model as a vehicle for empowering them to conduct a thorough analysis. This will ensure that the operational aspects of the company do not become a corporate millstone but remain a competitive weapon based on its distinctive competence. There is no conflict between the advantages of becoming a best practice performer and the strategic objectives of construction firms.

Management Theory

We must appreciate, right at the beginning, that there is no single, simple, quick way of explaining organisation theories nor, indeed, can we pick upon any particular theory, or group of related theories as representing a unified picture. This means that one cannot explain the behaviour of organisations, or elicit principles of management in one concept. As yet a General Theory of Organisations or of Management, like Einstein's General Theory of Relativity which attempts to explain a wide range of phenomena, does not exist.

In the following sections we shall try to unravel the tangle, select the more important strands and construct an overall picture of current management ideas.

Unravelling the Tangle – H. Fayol (1841–1925)

The first principles of management were set out in 1916 by Henri Fayol, a French industrialist. Unfortunately, a complete translation into English of his major work was not available until 1949. Therefore, his contribution was not widely known until comparatively recently. It is true to say, however, that many of his ideas are now widely accepted, almost without criticism, in business and management circles.

Henry Fayol was born in 1841 and at fifteen entered the Lycée at Lyon where he spent two years. From there, he passed to the National School of Mines at St. Etienne. At seventeen he was the youngest student and at nineteen graduated as a mining engineer.

He obtained a job as engineer with the French mining and metal producing firm of Commentary–Fourchamboult–Decazeville, spending most of his working life with this combine. He worked his way through general management to become managing director from 1888 to 1918. The success he brought to the business is one of the romances of French industrial history – the company was in a poor situation in 1885; when he retired in 1918 the financial position was excellent and the quality of staff exceptional.

So committed to his job was he that though he had the intellectual ability to think, write and lecture on his ideas of management, he did very little of this until his retirement. During retirement he wrote his book (*General and Industrial Management*), founded a Centre of Administrative Studies which influenced the French army and navy, and undertook a commission from the French Post Office to investigate the organisation. In the course of his work, he overhauled the Post Office from top to bottom and at the time of his death he was engaged in a similar task at the request of the French tobacco industry.

Fayol's Approach

Both F.W. Taylor, an American we shall consider later, and Fayol realised that the problem of personnel and its management at all levels was the key to industrial success. Both tried to apply 'scientific method' to this problem. But, while Taylor concentrated primarily on the operator or worker level, from the bottom upwards, Fayol concentrated, not unnaturally, on the managing director downwards.

Fayol's Doctrine

Fayol examined three aspects of management and attempted to define:

(1) the activities of the enterprise – what the enterprise does
(2) the elements of management – what management does
(3) the principles of management – a series of practical suggestions.

Activities of the Enterprise

Fayol stated that "*All activities to which industrial groups give rise can be divided into the following six categories*":

(1) technical activities (production, manufacture and adaptation)
(2) commercial activities (buying, selling and exchanging)
(3) financial activities (search for, and best use of capital)
(4) security services (protection of property and persons)
(5) accounting services (stocktaking, balance sheets, costs and statistics)
(6) managerial activities.

These six elements, he stated, would be found regardless of whether the undertaking was simple or complex, large or small. Most jobs would encompass the activities in varying measure; the largest managerial element would be present in senior jobs and the least (or even a complete absence) in direct production or lower clerical tasks.

Fayol even went to the lengths of producing charts to show the percentages of each activity to be found in particular jobs. For example, the job activities of a manager of a firm might be broken down into 40% managerial, 15% technical, 15% commercial, 10% financial, 10% security and 10% accounting.

M. Weber (1864–1920)

Weber's principal contribution to the study of organisations was his theory of the structure of authority. This developed from his enquiries into why people did what they were told. His major works and ideas were published in translation after his death, from 1947 onwards.

As far as organisations were concerned, he was primarily interested in the notions of 'power', 'authority', 'leadership' and 'bureaucracy'. It should be noted, however, that Weber was a sociologist, and the greater part of his work was concerned with much wider sociological enquiries into the many forms of social organisation in history.

Weber qualified as a lawyer and then became a member of staff at the University of Berlin. He remained an academic all his life, studying social organisation in history. He examined world religions such as Judaism, Christianity and Buddhism, and also the development of capitalism.

Weber's Ideas

To appreciate the points Weber makes, one must examine briefly his ideas about leadership, authority and obedience. When considering the three kinds of organisation – charismatic, traditional and bureaucratic – one must remember that Weber is presenting three models, and models can be an aid to understanding. A particular model may not fit precisely and an organisation may be partly charismatic, partly traditional or partly bureaucratic.

Weber used the term 'organisation' to mean the ordering of social relationships, the maintenance of which certain individuals took upon themselves as a special task. So, the presence of a leader and an administrative staff was a characteristic of an organisation. In fact, it was they who preserved the organisation. Basic to Weber's ideas was the notion that human behaviour was regulated by rules. The existence of a distinct set of rules was implicit in the concept of organisation. A leader and administrative staff had dual relationship to rules. Not only was their own behaviour regulated by them but they had the task of seeing that other members of the organisation adhered to the rules.

Weber distinguished between power and authority. Authority was a limited kind of power; authority only covered certain aspects. People accepted the authority of others if, and only if, they believed that:

(1) the orders were justified
(2) it was right to obey.

There were, however, three different kinds of belief that people had about orders and the providers of orders:

(1) obedience was justified because of the nature of the persons giving the orders – holy, sacred or charismatic
(2) obedience was justified because of a reverence for the past – we have always done things this way before
(3) obedience was justified because the person giving the order was acting in accordance with a set of rules already in existence and agreed upon.

We can conveniently continue with Weber's ideas of leadership. He identified three categories of leaders and the organisation types which are to be found with such leaders.

Charismatic Organisation

The word 'charismatic' was derived by Weber from the Greek 'charisma' – the supernatural quality found in some people which not only sets them apart from others but makes people in general, without knowing why, treat such leaders as possessing superhuman powers, e.g. knowledge. The basis of the authority of the leader is their special powers, and if these powers fail they may lose their leadership powers too.

The organisation structure is usually simplistic; decision-making is concentrated in the leader, delegation is limited to a small select band of intensely loyal staff (disciples) and personal obedience and devotion are the best ways of 'getting on'. Few rules and regulations exist; decisions are arbitrary and irrational. Everyone who considers the decisions they themselves can make in such a system will always measure the alternatives against what the leader would wish, approve of or need.

While the charisma lasts, the organisation preserves its original identity; but once it has gone or the leader leaves, unless a new leader with charisma succeeds, the organisation may change. Even a successful charismatic leader eventually has to lay down some rules and have deputies, and these are the beginnings of routinisation of charisma.

Traditional Organisation

Here the accent is on what has gone before. Precedent and usage are the basis of authority. The leader usually inherits the position and has authority from its status, being fixed by custom. When charisma is traditionalised it becomes part of the role, not personality.

The organisational form is feudal – the feudal system being the most developed – and can be found in the family firm where managerial positions are handed down from father to son. Ways of doing things are often justified in terms of precedent as a reason in itself. (Many supervisors will undoubtedly have had experience of situations of this kind.)

Bureaucratic Organisation

Weber wrote less about the bureaucratic leader than about the bureaucratic system itself, in fact he goes into this in great detail. We must be clear that Weber is not using the term in its common, quite critical sense to mean red-tape, top-heavy administration – bumbling inefficiency. For Weber, it is a type or organisation which is rational because such organisations have aims or goals that they try to achieve – a 10% increase in profit next year or a greater share of the market.

It is legal because authority is exercised by means of a system of rules and procedures laid down by various officials who occupy a particular office at a particular time. The whole system revolves around the fact that the entire structure, including management techniques, is largely determined without reference to named leaders or power holders. It is the office, not the person, which is important. The job of managing director is what we concentrate upon, not upon the person who holds the post.

Analysis of Bureaucracy

Weber never precisely defined bureaucracy but wrote a great deal about the role of the official in modern society. What made the official distinctive was:

(1) they had duties to perform
(2) they had facilities and resources provided by someone else with which to carry out the duties (as indeed the site worker is so provided for, but the site manager has a major distinctive advantage – authority).

As all officials had authority, all were involved in administration. Thus a wide variety of people could be 'officials' in this sense. Perhaps the closest we can come to a definition of bureaucracy in Weber's terms is 'an administrative body of appointed officials'.

Weber saw bureaucracy as the dominant form of the institutions of modern life. What intrigued him was the continuity of institutions of this kind.

The Basis of Legal Authority

According to Weber, legal authority depended on:

(1) the establishment of a legal code claiming the obedience of organisational members
(2) the code forming a system of general rules applied to particular cases
(3) the official obeying the law
(4) the obedience being due not to the person who holds the authority but to the law/rule/regulation which granted the official their position.

Structure of Bureaucratic Systems

(1) In the 'ideal' organisation, tasks would be organised on a continuous, regulated basis.
(2) Tasks would be divided into functionally distinct spheres – specialisation and division of work.
(3) Offices would be arranged in a hierarchy with job descriptions.
(4) Rules, technical or legal, would exist in a sufficiently complex form to create a need to train personnel to fill official posts.
(5) Great use would be made of written documents, making the office the hub of the organisation.

While authority systems might take many forms, the best, present and most efficient is the bureaucratic system. Not only was it important but it would continue to become more so as time went on. It had:

- precision
- continuity
- discipline
- reliability

which made it technically the most satisfactory organisational form.

Supplementary Points

Other points made by Weber are, briefly:

(1) Need for training – people must be adequately trained to do their jobs properly.
(2) Keeping of files and records – this ensures continuity and is essential for stability. If information about the organisation's business is transmitted orally, it is lost when the person with that knowledge leaves.
(3) Separation of personal and business life. A job is something apart from one's social life. As such, the organisation must provide space for work to be done, equipment and files etc. The worker should not be expected to provide anything of his own, such as tools, paper etc.

F.W. Taylor (1856–1915)

Another person who had an 'ideal' was F.W. Taylor. However, his ideal was no theoretical model, like bureaucracy: it was the ruthless, restless search for 'the best'. Taylor was obsessed with the idea of maximising efficiency in an organisation in order to maximise profits. He assumed that:

(1) People could be related to their work, rather like machines, and made as efficient as it was theoretically possible to make them.
(2) Properly used (money) incentives would get people to work harder to earn more.
(3) People would see the need to co-operate with management – the financial rewards from doing so would benefit the firm (more profit) and themselves (increased wages).

Many writers on organisations have tended to assume that the strength of the attachment of members is determined by their psychological feelings. Man, as an economic animal, responding directly to money (and other similar carrots), is much in the thoughts of people like F.W. Taylor.

Taylor is interesting, not only because he was one of the first in the field and also a practising manager, but because he is one of the founders of the now well-established management techniques we know as 'O and M' (Organisation and Method Study), Work Study, Time Study and the like. He could be described as the 'Father' of Scientific Management.

Scientific Management

Taylor's experience in the steel industry led him to believe that all was not well in modern industry. Managers approached their jobs in arbitrary rule-of-thumb ways. Workers were casual and lackadaisical in meeting their work commitment.

He considered that the main obstacle to efficiency was a failure by managers to find ways to co-ordinate and control workers' output and a failure to work out fair and satisfactory ways of paying the workers to ensure full co-operation and the desired output. In particular, he claimed that managers had not studied workers' methods of working to find better ways of doing jobs but had left the workers to do their jobs as best they could – with disastrous results.

Taylor recommended making management a science, resting on fixed principles instead of more or less hazy ideas. In particular, he set himself the task of devising methods of job study, control of work flow and incentives, and succeeded brilliantly. That is, he did what he set out to do. In hindsight, many of the troubles of the modern mass production industrial scene have origins, or so it seems, in his methods and their application.

In modern jargon, we would say that Taylor preached the doctrine of 'cost effectiveness'. Cost effectiveness implied control, and control is really the central pivot of Taylor's message.

The Four Principles

Taylor felt that 'maximum prosperity' was what every firm and every worker wanted. The necessity for management and workers to work together towards this common aim was self-evident but there were conflicts, strife and strikes. Why?

Taylor suggests three reasons: the workers feared that more output meant fewer jobs; poor-quality management resulted in workers 'going slow' to protect themselves; and, worst of all, inefficient ways of performing their allocated tasks.

Scientific management would overcome these obstacles and the following four principles were vital:

(1) We do not know what a fair day's work is.
(2) If the boss does not know, how can we expect the worker to?
(3) This is to be remedied by establishing each man's daily task, i.e. the output expected.
(4) Workers must be well paid if they meet the target and fined if they do not.

So that the workforce can earn this high rate of pay we must ensure those hired are up to the job physically and mentally. Proper training is to be given so that they become a 'first class' person. Promotion opportunities are to be made available.

From Taylor's work sprang Frank Gilbreth's method study, i.e. the study of the best method of doing a job. All this activity involved observation, recording and analysis of the results, and the techniques of observations and timing were the beginnings of time study.

Time study led to work study, the study of the work itself. In fact, in the USA 'work study' is known as 'time and motion' study, which is perhaps a more precise definition of the actual activity. By analysing work methods and materials used, the aims of work study are to:

(1) establish the most economical way in which the job can be done
(2) standardise this particular method, type of labour used, and materials/equipment needed
(3) establish the time needed by a properly trained and qualified worker to do the job, working normally at a defined level of performance
(4) instruct that the chosen method be followed as standard practice.

Taylor would have totally approved of this modern definition.

If there were conflicts, Taylor condemned them and said the only reason they occurred was because of unscientific management. His major failings were:

(1) Not to realise what motivates individuals to work – he thought that motivation was purely economic.
(2) He did not understand that people in groups behaved differently from individuals. He, in fact, did not like groups. Purely social relationships were superfluous and tended to reduce efficiency. Work groups were broken up and operatives separated so they could not distract each other with idle talk.
(3) He did appreciate fully the evils inherent in piecework systems, such as the sacrifice of quality or the taking of dangerous risks.
(4) The assumed existence of a world of perfect competition where maximum output and efficiency were always required.
(5) Conviction that he was right every time, when in fact he was not infallible. Mistakes, and costly mistakes at that, were caused by adopting his ideas too inflexibly.

Taylor's investigations only covered part of the operation of the firm's production. There was much he omitted. However, the main significance of Taylor's work was that it demonstrated the possibility and importance of a systematic analysis of business operations and of the scope for using 'scientific' methods in a new field.

As we have seen, he was instrumental in creating techniques for increasing efficiency which, despite the controversy at the time and since, have been progressively developed and are in widespread use.

Summary of Taylor's Work

Implicit Assumptions

People are essentially lazy, disinterested in the organisation, are basically irrational and incapable of self-control, and must be directed and controlled.

Implications for Labour Management

Management plans, organises, directs and controls. Workers contribute the functions of doing the job. This is no point/need for participation or involvement. The above leads to a 'Utilitarian Contract', i.e. a fair day's pay for a fair day's work.

A. Maslow

Maslow's premise was that there are two main sources of motivation to work: one is job satisfaction and the other is the financial reward gained. Maslow used a pyramid to illustrate motivation as a series of ascending needs (see Figure 3.16). A need acts as a motivator for as long as it remains unsatisfied. However, once a need has been satisfied, it ceases to be a motivator and motivation takes place at the next stage up in the 'pyramid of needs'. It is also important to recognise that levels of needs may change according to a person's present and future circumstances.

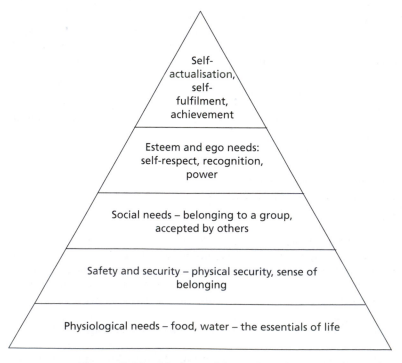

Figure 3.16 *Maslow's hierarchy of needs*

Main Points of Maslow's Theory

Psychological Factors

Workers need meaning and a sense of fulfilment.

Implicit Assumptions

In the right circumstances, workers are capable of self-motivation and control, and can achieve a sense of satisfaction and fulfilment from their own achievements in pursuing organisational goals.

Implications

Management's role is that of catalyst and facilitator concerned with providing opportunities for worker development, challenge and interest. Authority stems from work itself, not formal position. Some redistribution of power among management and employees is necessary.

The above leads to a psychological contract and this provides a high element of intrinsic (as well as extrinsic) reward for freely contributed involvement and commitment.

Table 3.6 shows how a construction company can implement the concepts of Maslow within an operational framework.

F. Herzberg

Herzberg's theory noted that there are demotivators as well as motivators. The effectiveness of the latter is dependent on the absence of the former. However, an increase in the acceptance of the former (i.e. a negative demotivator) is normally ineffective as a motivator.

In common with Maslow's theory, although it offers several interesting and useful insights, empirical findings do not support its widespread validity.

The findings of these studies, along with corroboration from many other investigations using different procedures, suggest that the factors involved in producing job satisfaction (and motivation) are separate and distinct from the factors that lead to job dissatisfaction. Since separate factors need to be considered, depending on whether job satisfaction or job dissatisfaction is being examined, it follows that these two feelings are not opposites of each other. The opposite of job satisfaction is not job dissatisfaction but rather no job satisfaction and, similarly, the opposite of job dissatisfaction is not job satisfaction but no job dissatisfaction.

Stating the concept presents a problem in semantics, for we normally think of satisfaction and dissatisfaction as opposites – i.e. what is not satisfying must be dissatisfying, and vice versa. But when it comes to understanding the behaviour of people in their jobs, more than a play on words is involved.

Two different needs of people are involved here. One set of needs can be thought of as stemming from their animal nature – the built-in drive to avoid pain from the environment plus all the learned drives which become conditioned to the basic biological needs. For example, hunger, a basic biological drive, makes it necessary to earn money and then money becomes a specific drive. The other set of needs relates to that unique human characteristic, the ability to achieve and through achievement to

Table 3.6 *Maslow's hierarchy of needs*

Hierarchy of needs	Construction company policy (example)
Self-actualisation	Jobs designed to provide variation, job rotation, staff training and development
Ego and esteem needs	Regular performance reviews, bonuses and fringe benefits
Social and belonging needs	Team work and a recognition of informal groups. Encourage involvement
Safety and security needs	Company pension and sick pay schemes. Good company safety record
Physiological needs	Good rate of pay, staff canteen etc. Bonus scheme

experience psychological growth. The stimuli for the growth needs are tasks that induce growth in the industrial setting; they are the job content. Contrariwise, the stimuli inducing pain-avoidance behaviour are found in the job environment.

The growth or motivator factors that are intrinsic to the job are: achievement, recognition for achievement, the work itself, responsibility and growth or advancement. The dissatisfaction avoidance or hygiene factors that are extrinsic to the job include: company policy and administration, supervision, interpersonal relationships, working conditions, salary, status and security.

A composite of the factors that are involved in causing job satisfaction and job dissatisfaction are shown in Figure 3.17. The results indicate that motivators were the primary cause of satisfaction and hygiene factors the primary cause of unhappiness.

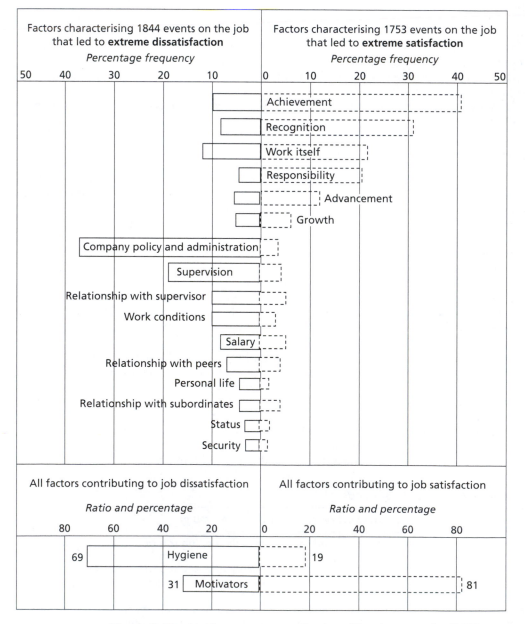

Figure 3.17 *Hygiene versus motivators (Herzberg et al., 1959)*

Table 3.7 *Herzberg's two factor theory (hygiene factors)*

Hygiene factors	Definition/example
Company policy and administration	Availability of clearly defined policies, degree of 'red tape', adequacy of communication, efficiency of organisation
Supervision	Accessibility, competence and personality of the line manager
Interpersonal relations	The relations with supervisors, subordinates and colleagues, the quality of social life at work
Salary	The total rewards package, such as salary, pension, company car and other rewards
Status	A person's position or rank in relation to others, symbolised by title, parking space, car, size of office, furnishings etc.
Job security	Freedom from insecurity, such as loss of position or loss of employment altogether
Personal life	The effects of a person's work on family life, e.g. stress, unsocial hours or moving house etc.
Working conditions	The physical environment in which work is done, the degree of discomfort it causes

Table 3.8 *Herzberg's two factor theory (motivator factors)*

Motivator factors	Definition/example
Achievement	Sense of bringing something to a successful conclusion, completing a job, solving a problem, making a successful sale. The sense of achievement is in proportion to the size of the challenge
Recognition	Acknowledgement of a person's contribution, appreciation of work by company or colleagues, rewards for merit
Job interest	Intrinsic appeal of job, variety rather than repetition, holds interest and is not monotonous or boring
Responsibility	Being allowed discretion at work, shown trust by company, having authority to make decisions, accountable for the work of others
Advancement	Promotion in status or job, or the prospect of it

Definitions of hygiene factors and motivators are provided in Tables 3.7 and 3.8 respectively.

Motivation as a Complex Process

Motivation factors vary and interact with both the individual and the situation.

Assumptions

(1) A person's motives are variable and complex.
(2) A person is influenced by the organisation and learns new motives.
(3) An individual's motives may be different in different roles.
(4) Individual motivation to perform depends not only on their motives but also on
   (a) the nature of the job
   (b) relationships with others
   (c) own personality, capability and experience.
(5) A person responds to different management styles at different times and in different circumstances.

People differ widely in what they seek from work.

Influences on Attitudes to Motivation

(1) External influences – community, values and norms, alienation and social class and family.
(2) Social comparison – perception of fairness and equity between colleagues in relation to the reward/effort balance.
(3) Expectations about valued rewards/outcomes, perceived match between expectations and reality.
(4) Individual differences – individual needs, skills intelligence, personality etc.

Implications

(1) Whether the outcome is intrinsic or extrinsic – if it is valued by the individual as an anticipated outcome of an action, it becomes an incentive for the action.
(2) Outcomes are valued to the extent that they satisfy physiological or psychological needs. If an outcome is not linked to satisfaction, it will no longer be valued and will not form an incentive.
(3) Employees will work hard to achieve organisational goals if they can be related to their own goals.
(4) Most lower-order needs (Maslow) are satisfied in modern society. Continuing satisfaction of higher-order needs does not necessarily diminish the desire to satisfy those needs. Hence, continuing motivation seems more of a reality if higher-order needs are engaged.
(5) Individuals capable of higher-order needs experience something they feel is worthwhile and meaningful. They need to work on meaningful jobs which provide feedback on the adequacy of their performance.

Social Factors

Workers are motivated by social needs.

Implicit Assumptions

People's attitudes, behaviour and performance are influenced by social structure. Their need to belong, to establish meaningful relationships with others and to gain group approval can have as much, if not more, effect than directives and incentives from management.

Implications

These include participative management style, emphasis on the group as the unit for analysis and design, good communication and involvement by the workforce in the management process. (This is very much in line with APS.)

Naoum (2001)

This describes the four phases of management theory development as:

- classical
- human relations
- systems
- contingency.

These are related within the time continuum indicated in Figure 3.18.

Classical theory encompassed the works of Taylor and focused upon formal structures and production measurements within a framework of extrinsic motivation. The human relations approach to management theory (organisational behaviour) incorporates the works of Maslow, Herzberg and Mayo (Hawthorne experiments, 1924–1932), who developed his theory on group dynamics.

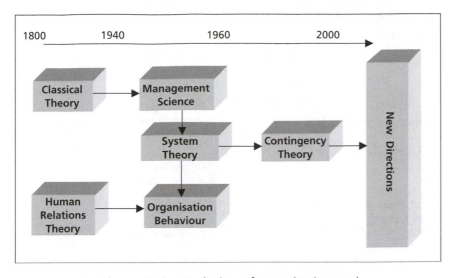

Figure 3.18 *Evolution of organisation and management theory (Naoum, 2001)*

Systems Theory

A systems approach, by its very nature, is made up of interdependent elements, as such actions which affect one element must affect others also, and actions of one element cause reactions on the other parts (King (1983) cited in Naoum, 2001). Figure 3.19 indicates the relationship of various organisational sub-systems to the overarching 'Organisational System'. The core business can be impacted upon by both internal and external influences. Each sub-system is a dedicated management function. Within construction firms, two distinct tiers of management will be developed within each sub-system. One will focus on specific management functions throughout the firm, while the second will focus at the operational site levels. Each aspect must support the other. Senior Management must fully appreciate the link between corporate objective setting and the impact this has at an operational level. Hence, construction organisations must have a fully functional 'operations strategy'.

Systems theory, like management theories, offers a way of understanding organisations. It is a model to aid explanation and analysis. The basis of systems theory is that analogies may be made with reference to biological organisms and even molecular structures. That these models have been considered appropriate for organisational analysis can be demonstrated by the widespread use of the terminology (if a rather less widespread knowledge of its true meaning).

Systems theory can be adopted at a number of levels. It offers systems as a way of thinking about any situation – a philosophy or paradigm. Examples of this may be seen in the economic system, the legal system and the political system.

Construction was fairly slow in adopting this approach. However, it will be seen that the characteristics of the industry (sites interdependently autonomous from head office) are perhaps particularly suited to this analysis.

At the root of systems theory is the understanding that any system exists in its environment. The relationship between a system and its environment is vital for its survival. This is demonstrated by the model depicted in Figure 3.20.

The terms input and output refer to the relationship to the environment, process to the 'work' done by the system. The definition of the boundary which delineates the organisation system from its environment is a crucial, if deceptively difficult issue.

The environment itself may be made up of any number of other systems and the individual system may comprise many sub-systems. The important point is that each of these systems depends for its survival on the relationship with its environment.

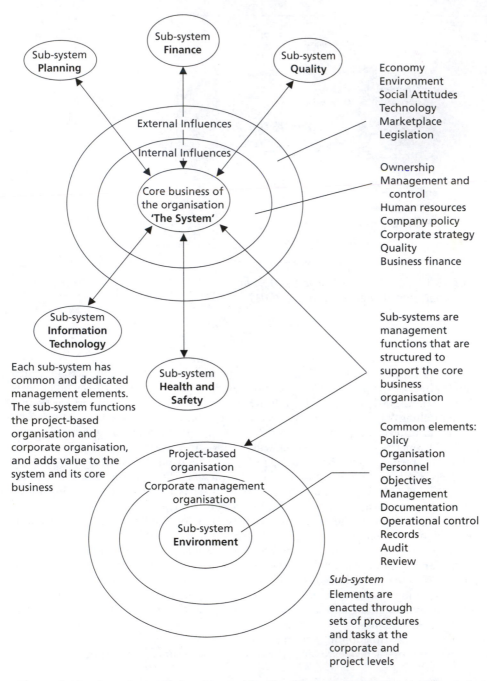

Economy
Environment
Social Attitudes
Technology
Marketplace
Legislation

Ownership
Management and
  control
Human resources
Company policy
Corporate strategy
Quality
Business finance

Sub-systems are
management
functions that are
structured to
support the core
business
organisation

Each sub-system has
common and dedicated
management elements.
The sub-system functions
the project-based
organisation and
corporate organisation,
and adds value to the
system and its core
business

Common elements:
Policy
Organisation
Personnel
Objectives
Management
Documentation
Operational control
Records
Audit
Review

Sub-system
Elements are
enacted through
sets of procedures
and tasks at the
corporate and
project levels

Figure 3.19 *Systems model – an organisation viewed as sub-system elements supporting the system and its core business (Griffith, et al., 2000)*

Figure 3.20 *Process model*

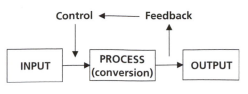

Figure 3.21 *Systems diagram*

The other important characteristic of a system is a method of feedback for control (Figure 3.21). This is a way of ensuring that the outputs of the system conform with the requirements proposed.

The noted boundary problem is further complicated in that a system can be said to be open or closed, and more recently 'soft'. This is of particular importance in organisational analysis as it may influence the success or otherwise of a concern.

Construction Example

As noted, systems analysis may be applied at a number of levels. In construction, the simplest starting point is that of the site. A construction project is a relatively easy subject to analyse as it is a temporary organisation specifically formed to complete a definable purpose. This fulfils the requirements as stated:

(1) Define the system
    (a) by a description of what the system is and what it does – the Primary Task
    (b) by the establishment of the boundary of the system – what is inside and what is outside the system.
(2) Identify the component parts of the system
    (a) the inputs to the system
    (b) the conversion processes which the system uses to transform the inputs into outputs
    (c) the outputs of the system, tangible or intangible
    (d) the feedback loops which complete the input–conversion–output cycle.
(3) Define the environment of the system
    This will include all those elements which will have an impact on the system and those impacted by it. This will consist of many other organisational systems and further 'higher-level systems'.

The primary task of a construction site can be easily identified as the completion of the specific project. It is a discrete aim. The boundary is problematic in that the involvement of consultants, subcontractors and various sub-systems creates a blurred edge. However, the site hoardings provide a useful physical distinction between the on-site and off-site; this may be used to define the boundary of the system.

Inputs to the system can be categorised as manpower, materials, plant, finance and energy. Information to head office and clients can also be considered as an input.

The conversion process is identifiable as the productive part of construction building assembly.

Tangible outputs are again straightforward: buildings and profits. Intangible outputs include pollution, waste, employment and wages.

The feedback loop will include the methods of checking the completed work against drawings, specifications and other design information. Consultants may be involved in this process.

The environment of the project will include head office, consultants, subcontractors, financiers, suppliers, legislative control bodies, the labour market, competitors, local people and environmental pressure groups.

If this analysis is used to examine the relationships between all the relevant systems, then a useful model can be developed. It should be noted that a common error in the use of systems theory is the overlooking of important elements in the environment. No single factor remains unchanged, and many are interrelated, such that change in one will result in a change to others.

Developments from Systems Theory

The development of systems theory, contingency theory, is based on the premise that for each environmental situation there will be an 'ideal' or at least most useful organisational form.

Contingency Theory and Summary

The original ideas on leadership characteristics have been seen to be flawed in that different situations require different leadership styles. Conditions of extreme emergency favour autocratic leadership, as on occasion do 'stagnant' situations. Democratic and participative management are undoubtedly more effective in the long-term as they encourage group coherence and a sense of belonging. Conditions external to the organisation may also influence the management style: a despotic political system would undoubtedly favour autocratic managers as participative organisations would not be able to function under such conditions.

Many different leadership styles can be identified within a particular organisation. Their effectiveness may be compared and the most appropriate chosen. Therefore contingency theory is based upon the premise that there is no one best way; one must give full consideration to the situation or problem in hand (Naoum, 2001).

Improving Morale

"*Indicators of morale, the degree of attachment to, or satisfaction with one's organisation, its goals, its traditions, one's work and the people one works with.*"

One motivational aspect with which all managers will be concerned is that of morale. This is something about which most individuals have an instinct but which tends to be difficult to define with any degree of precision. Many would define it as being equivalent to job satisfaction and indeed it has much to do with it, but others stress the group dimension which morale has in that it is often associated with phrases such as 'the morale of the troops was high'.

One possible definition for consideration in the context of the work situation might be 'Morale is the attitude of individuals and groups towards their work and working environment which serves to condition how well or how badly they perform'. This suggests that workers who like their work and the company they work for, and enjoy the companionship of their fellow workers, will be highly motivated and will wish to work co-operatively in the pursuit of a common purpose – such as achieving the goals of the organisation.

Things are rarely as ideal as this and part of the manager's role is to develop an awareness of the myriad of employee attitudes, feelings and sentiments which can colour their views and affect morale at any time.

Factors of Morale which can be Identified and Measured

There are within an organisation certain factors which may provide a valuable insight as to the state of morale. These include:

- labour turnover
- the level of absenteeism
- timekeeping
- the number of accidents
- the quality of work
- the quantity of work
- the number of grievances
- the number of disciplinary proceedings
- the number of disputes
- the level of participation in company schemes
- the utilisation rate of company facilities.

All these factors can be measured and the results tabulated to provide statistics which may be indicative of the level of morale. However, it must be remembered that such statistics can only be indicators, as other factors may be more significant.

For example, a high labour turnover may be an indication of low morale within a company but it may equally represent a reaction to some external factor such as housing shortage, poor education provision or changed personal circumstances. What labour turnover does provide is a trend, but the trend should be assessed in the context of the industry, both locally and nationally, and comparisons should be made with previous years. Turnover may remain low, however, in times of high unemployment, irrespective of the fact that morale may be at an all-time low.

The Main Influences upon Morale

Adequacy of Supervision

This is perhaps the most important factor affecting employee morale since it represents the point of contact between the organisation and the employee. Dissatisfaction with supervision can arise for a variety of reasons:

- a lack of leadership
- poor delegation
- poor knowledge of the job
- poor interpersonal skills
- a lack of caring and support
- inadequate communication skills
- expression of favouritism
- victimisation
- personality conflict
- excessive disciplinary measures.

The Job Itself

Most people find intrinsic satisfaction in a job which they particularly like and know they do well. Dissatisfaction arises when:

- there is too much work
- there is insufficient work
- the work is of the wrong calibre
- there is inadequate training
- the nature of the job is fragmented
- the job becomes deskilled
- there is little aesthetic pleasure.

Organisational Climate

Organisational climate – the 'feel of the workplace' – will be reflected in:

- the organisational structure
- the communication network
- the extent of participation and consultation
- the way in which change is introduced
- the team spirit engendered throughout the organisation
- the number of disputes.

How all the factors considered above could be represented on a 'Morale Factor Scale' is difficult to determine in that their degree of significance will change according to different circumstances and variable operating processes at any given time, both at the level of the individual and in the context of the organisation.

The individual will then act instrumentally in order to pursue those actions which are most likely to achieve the expectancies desired. This whole theory is based on the perception of the individual of both the expectancy and the valency. Indeed, the action itself may be perceived differently by different individuals. Therefore, this theory cannot be generalised in the organisation sense but must be considered for each individual case.

Extrinsic and Intrinsic Motivation

To draw the two themes together, it is possible to denote extrinsic and intrinsic motivators. The extrinsic ones are complex but may be influenced by management to achieve a specific aim. The intrinsic ones are bound up in the nature of the work and the perception of the individual to that work.

A final complication is that of the individual's perception. It is widely accepted that exactly the same circumstances will be perceived differently by different individuals. What may be a strong motivator to one is weak or non-existent to another. However, a further problem is that even one individual may perceive the same set of external circumstances in a different manner at different times.

Organisational Perspective

The intrinsic problems of addressing the psychological issues of individual employees has meant that managerial views of motivation have focused on that which they can influence. Maslow and Herzberg, together with Taylor's fundamental attitudes towards financial reward, provide a way of looking at externally developed motivation and are therefore widely accepted.

Psychological Views of Motivation

Far greater studies of both physiological and psychological motivators have been undertaken, with widely varying results. The basis is that of the individual's perception of the motivator. This is depicted in Figure 3.22 (from Lawler and Porter).

Lawler and Porter's model (Figure 3.22) recognised that, whatever the individual, values will act as a motivator. This motivating factor can vary with changing circumstances.

Group Dynamics

The construction industry, in common with others, relies to a great extent on the combined working of groups to function effectively. The design of a construction project of any complexity requires a combination of a number of consultants whose specialised expertise is often outside the abilities of one individual. In practice, even if one person could fulfil all the necessary roles, the time needed to carry out the task would become unreasonable.

Likewise, the management of construction projects, requiring skills in buying, subcontracts, planning, resource scheduling, contract and project management etc., necessitates group or team work.

On site, the notion of group work is embedded in the traditional craft skills methods of operating. From the ancient traditions of master craftsmen, journeymen and apprentices and labourers, the team or group working is still practised. Terms such as bricklaying and plastering gangs are still commonly used when people consider the industry.

Group Analysis

For the purposes of organisational analysis, groups can be categorised in two ways: formal and informal. The formal group is that which is described in terms of its organisational purpose. A design team would fall into this category, as would a committee. Informal groups are set up to satisfy their own ends: a football team, social club or a group of students (learning set).

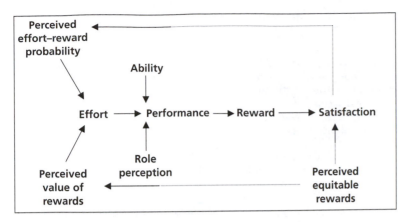

Figure 3.22 *The expectancy model of motivation by
Lawler and Porter (complex or process man)
(from Lawler and Porter, cited in Mullins, 1996)*

Common characteristics are that each has a primary aim or goal. In the formal case, that goal is introduced externally by the organisation or environment or by a member of the group. In informal groups, the goal is formed internally. This is one way of distinguishing the nature of the group. Whichever category the group belongs to, the efforts of the group are co-ordinated towards achieving the goal.

Individuals and Groups

It is important to recognise the motivation for an individual to join a group. People are social animals in that the greater part of life is spent in social interaction. Opportunities within groups include personal growth, the reinforcement of personal belief, safety – in psychological terms as well as physical – and other intrinsic rewards: 'belonging' for example. These motivations hold true for formal as well as informal groups.

Group Processes

One well-established method of group analysis (see Table 3.9) has identified the following group processes:

- forming
- storming
- norming
- performing.

- **Forming** relates to the process by which the characteristics and primary aims of the group are established. At this stage, many potential members may withdraw.
- **Storming** is the development of the above into a more recognised state, in which leaders and decision-makers are identified.
- **Norming** is one of the most important features of group process. It refers to the way in which 'hidden rules' are developed for behaviour. Acts become acceptable or unacceptable. Strangers may be easily identified by their lack of knowledge of a particular behavioural norm. The stronger the group, the more developed the patterns of behaviour.
- **Performing** is the positive actions by the group to achieve its particular aims. Again this may be externally set in an organisation or internally agreed upon.

Table 3.9 *Stages in the growth of group cohesion and performance*

Stage of development	Process	Outcome
1. Forming	There is anxiety, and dependence on leader; testing to find out the nature of the situation and what behaviour is acceptable	Members find out what the task is, what the rules are and what methods are appropriate
2. Storming	Conflict between sub-groups, rebellion against leader, opinions are polarised, resistance to control by group	Emotional resistance to demands of task
3. Norming	Development of group cohesion, norms emerge, resistance is overcome and conflicts are patched up; mutual support and sense of group identity	Open exchange of views and feelings; co-operation develops
4. Performing	Interpersonal problems are resolved, interpersonal structure becomes the means of getting things done, roles are flexible and functional	Solutions to problems emerge; there are constructive attempts to complete tasks and energy is now available for effective work

Group Pressure

Evidence of the pressure of groups is noticed in the need to conform. All groups will develop norms of behaviour. Individuals motivated to join a group will be strong enough to modify their own behaviour to accept those norms. Cohesion can be very powerful in even very small informal groups. Experiments have shown that individuals are prepared to accept group decisions known to be wrong, rather than rebel against the group. The fear of exclusion is the corollary of the strong motivation to join.

Organisation Analysis

For the purposes of examining an organisation, the importance of groups is to recognise that informal groups exist within organisations. These can be of benefit to the organisation by generating team spirit and informal co-operation, or they may be harmful (in organisational terms) as in a pressure group. Attempting to pressurise such a group from outside can be very damaging. If the organisational pressure succeeds, the group may disperse, but the effect on motivation and morale will be damaging. If the group resists, the authority of the organisation is called into question.

Networks

From the premise that all organisations are simply methods of co-ordinating the potential of individuals to undertake a certain task, networks can be a useful addition to the formal descriptions normally used. As they are set up informally, their shared aims are agreed upon and internalised. This has a great advantage over externally defined or imposed motivators. The development of the network is a secondary result of such action.

Networks in Construction

The working of informal groups within construction firms can be identified. On construction sites, a site 'identity' is quickly established. Workers involved on site will adhere to that identity for the length of the contract. In addition, inter-firm networks can be established between consultants' offices to share information etc. Partnering as advocated by Latham (1994) in construction may be considered as an extension of this trend. It is not just that such groups should be recognised and acknowledged but that they can be advantageous and therefore should not be discouraged.

Culture

Organisational culture is a vital aspect of change management and may be defined as:

> The pattern of basic assumptions that a given group has invented, discovered or developed in learning to cope with its problems of external adaptation and internal integration, and that have worked well enough to be considered valid and therefore to be taught to new members as the correct way to perceive, think and feel in relation to those problems.

The key theme of the noted definition is a coherent approach by group/organisation in the way people react within the corporate context.

Corporate culture can be formed/influenced by:

- societal or national culture
- influence of a dominant leader
- company history and tradition
- technology, products and services
- industry and its competition
- customers
- information and control system
- legislation and company environment
- procedures and policies
- reward and measurement systems
- organisation and resources
- goals, values and beliefs.

Implementing change processes, for example Total Quality Management (TQM), will more than likely require a cultural sea change for most construction organisations. The following points advocated by Bate (1994) provide a succinct summary of the key issues.

To achieve an integrated strategy of cultural change, the following need to be taken into account:

- The culture to be changed
  - Diagnosis to assess the current culture and identify how it can be changed
  - Not just collecting information – learning to think culturally and see the organisation from a cultural perspective
- Its origins and trajectory through time
  - This provides a safeguard against corporate amnesia and repeating mistakes
- The life-cycle of culture and the stage in the cycle that the culture has reached
  - Cultures evolve over time (although this is frequently interrupted by fundamental second-order change) and it is important to get a picture of the current stage in the evolutionary cycle
- The environmental context
  - Importance of relevance
  - Cultural lag
- The aims and ambitions of the parties involved
  - Recognition of political issues
  - Consideration of what can be achieved.

Attempts to borrow Japanese management methods and the expansion of Japanese companies manufacturing in other countries found that some techniques were inapplicable or inappropriate. From this experience, an examination of the context in which Japanese methods thrive was undertaken and the cultural differences highlighted.

It is worth noting, however, that the Japanese were quick to recognise this important difference and have successfully coped with the 'cultural shift' required by both management and employees.

Culture as a Context

An analysis has been made on the contextual discrepancies between Japanese and European (or US) companies. The context is the environment in which the organisation exists. As many of the topics presented have pointed out, an organisation cannot be considered in a vacuum, it is its relationship with its environment which governs its effectiveness.

The understanding of culture in this sense is the ideology, belief systems, norms of behaviour and social order which compose society. There are subtle differences between European countries but the Japanese culture is radically different. The deference to authority and compliance is far stronger in Japan, as is the sense of belonging which engenders co-operation. Both of these are based upon developments of ancient traditions: the Samurai warrior and the system of rice-growing. The sources of European culture are often more difficult to ascertain and may (for us) be considered more complex. However, the idea of traditional values and codes of behaviour developed over many centuries is the same. This is part of what has come to be known as culture.

It is only by identifying what would be considered as a 'norm' in a Japanese organisation that this contrast is highlighted.

Company Philosophy (taken from a large Japanese company)

Basic Business Principles

To recognise our responsibilities as industrialists, to foster progress, to promote the general welfare of society and to devote ourselves to the further development of world culture.

Employees' Creed

Progress and development can be realised only through the combined efforts and co-operation of each member of our company. Each of us, therefore, shall keep this idea constantly in mind as we devote ourselves to the continuous improvement of our company.

The Seven 'Spiritual' Values

(1) National service through industry
(2) Fairness
(3) Harmony and co-operation
(4) Struggle for betterment
(5) Courtesy and humility
(6) Adjustment and assimilation
(7) Gratitude.

These basic values, taken to heart, provide a spiritual fabric of great resilience. They foster consistent expectations among employees in work that reaches from continent to continent. They permit a highly complex and decentralised firm to evoke an enormous continuity that sustains it even when more operational guidance breaks down: "*It seems silly to Westerners,*" says one executive, "*but every morning at 8.00 am, all across Japan, there are 87,000 people reciting the code of values and singing together. It is like we are all a community.*"

Other anecdotal experiences are the MD or CEO sweeping the factory floor, no distinctions in dining areas or car-parking, and all workers (including directors) wearing the same uniform.

When these were first experienced by Westerners, they came as a culture shock. The experiences were different from the norms and standards which had been internalised in their own culture. The discrepancies between European and US cultures, although evidenced, were not seen as that important. Contrasts in European cultures have, until the recent attempts at the formation of a European

Union, been ignored, at least in the field of organisational analysis. That this affects the effectiveness of organisational and management styles is self-evident. The context of an organisation, including the cultural environment, is now considered important.

Organisations as Cultures

The development of a culture is important: it may be considered as an extension of the notion of the internalisation of organisational aims. The employee, when joining an organisation, implicitly accepts some of the norms and values of that organisation. This can be developed by making that acceptance a specific requirement and the repeated insinuation of those ideals. This can be supported by the actions and behaviour of the colleagues and peer group of the individual. If they accept fully these values, the new individual is likely to conform (see the previous sections on Groups). The advantage for the organisation is that this can to some extent provide motivation for employees. They can identify with the organisation and recognise their own contribution. In addition they will receive the benefits of a sense of belonging to a group with similar values and expectations.

Culture and Identity

It is possible to suggest a strong relationship between the concept of culture and the concept of personal identity. The ability to identify one's self with a particular group and identify others as of the same group is a basic human social need. In organisational settings, this may be exemplified not only through standards of behaviour and ideals but through uniforms and standard phrases.

Conclusions: Environmental Culture – Organisational Culture

Finally, it is worth recognising that the concepts of environmental culture and organisational culture are no more than convenient labels applied to denote some social phenomena which are empirically observed. Organisational culture may become pervasive. The growth and development of organisations across the world have led to a situation where some organisations may be seen as being dominant influences in the society as a whole.

However, within a construction operational environment, a post-modernist culture needs to be evident. In a post-modern organisation, where normal hierarchy does not exist and staff act according to agreed areas of expertise, the term for this approach is 'hecterarchy' in which very high levels of fluidity are maintained. This high level of fluidity is a basic necessity for organisations because *"too much is changing for anyone to be complacent"* (Peters, 1988). As construction organisations move to areas of increased complexity of service, there is a requirement to implement increasing hecterarchic ways of operating.

Mechanistic versus Holographic

A post-modernistic approach should embrace high levels of group work, each with a correspondingly high level of autonomy. The overriding linking force binding these empowered groups together is their organisational culture. This form more readily suits the reality of today's business environment because organisations, and the markets they operate in, are unpredictable and not linear. Building a shared culture and conception of the world takes a great deal of time and effort. Traditionally, in most organisations the existing culture is based on mistrust and the use of frequent sanctions by senior managers. Consequently, they manage their organisations in this light. However, post-modernistic ones think in terms of a 'circle'. They are encouraged to look for complexity and the interconnection of cause and effect. This demands a high level of staff participation and makes good management sense. The rationale for participation has been stated by Sayles (1989) as follows: *"When subordinates are consulted about and contribute to the change process, many benefits accrue."*

Morphostatic versus Morphogenic

Morphostatic processes are defined as those that support or preserve the present mode of operation, and include formal and informal control systems with the emphasis being placed on procedures. A more enlightened approach is adopted by the post-modern organisation where a morphogenic culture exists. Morphogenic processes are those that tend to allow for change and development and the exciting nature of change is always highlighted. This type of organisation encourages staff to be proactive.

Customer Focus

Construction companies have to be open to ideas about organisational improvement so they should be learning organisations. They must consider new management tools and procurement systems that have proved to be most advantageous in manufacturing and elsewhere. These tools and methods of operation can assist in differentiating an organisation from its competitors.

The need for construction organisations to rethink and redesign their operational processes is primarily driven by the following pressures:

- globalisation of the economy
- greater performance expectations from clients
- greater competition among domestic organisations
- continued restructuring of work practices
- industrial relations
- enterprise bargaining.

Today's customers are concerned with quality issues and this influences their requirements in respect of quality and project duration; they tend to require products or services in the shortest possible time. Establishing customer focused targets requires the attainment (assessment) of their needs. Construction organisations' targets have to be sufficiently radical to cause a real change in the customers' perception of what they consider to be value. Post-modern organisations place the customer at the centre of all activities. Thus organisational culture is a critical issue in its attainment of a sustainable competitive advantage.

Case Study – Motivating Site Personnel

Young Andrew Hill had Harry Hall puzzled. How can you accuse one of your best bricklaying site foremen of loafing on the job? Yet that's what Harry had to do. Ever since being made a Site Manager and working on numerous projects with Andrew, Harry had been noticing him. In fact, long before Harry became a Site Manager he had noticed the way Andrew had worked and not worked.

An exceptionally friendly and talkative young man, Andrew bothered Harry by his frequent practice of stopping his whole gang to tell them a joke or a story. It didn't bother Andrew that it was during working hours or that anybody might be watching. He just seemed to enjoy telling stories and being the centre of his gang's attention. The trouble was that the rest of his gang enjoyed him too. That's why it was tough to clamp down – there was no chance of using group pressure on him, for Harry had tried with a marked lack of success.

Soon after he took over the site, Harry was determined to straighten the gang out but he waited a couple of months before moving in. As Andrew was such a friendly guy, Harry was sure that he'd have no problem motivating him to get on the ball. The gang was on a group incentive. It shouldn't be hard, Harry thought at the time, to get them to see how much they were losing by standing around and talking. He waited for an opportune time to launch his campaign. One afternoon, Harry heard the whole gang burst out laughing while he was standing nearby checking a job. As Harry walked over, Andrew was holding the floor, telling one of his escapades. Andrew held the complete attention of the gang – all were enjoying his performance. None seemed to mind that Harry had joined the group.

Andrew acknowledged Harry's arrival with a smile and a nod but continued with his story. Harry interrupted by asking if anything was wrong.

Andrew caught the inference of the question and, speaking for the gang, he assured Harry that nothing was wrong and explained that he was telling them about his last date. Sensing that Harry was annoyed, he also added that it could wait and that they would get busy. Andrew, without question, was the informal leader of the gang over Harry. He had not the official designation of Harry but the gang followed him just as surely as if he carried the Site Manager's title.

Andrew's gang produced extremely well. When they worked – and that certainly was most of the time – they couldn't be equalled in their output. Harry had no quarrel with their total production; they were comfortably ahead of the other gangs doing the same work. However, the frequent non-scheduled breaks for story-telling did bother Harry. He reasoned that their production could be even better if they would stick to business. Their non-productive time could be converted into badly needed production output. Further, if they kept busy, they wouldn't be setting a poor example to the other gangs and the rest of the company. Andrew was the pacesetter for the gang so Harry decided to start with him. The next morning, he asked him to stop by the site office during the break period.

Harry started out by explaining that he wanted to talk about idle time. No sooner were the words out of his mouth than Andrew asked if their production record was slipping. Harry had to acknowledge that it wasn't but he said *'think of the money you could be making if you'd cut out the story-telling and keep that gang busy.'* *'What good is money if you can't enjoy it?'* Andrew asked. *'I've seen too many people, my old man included, sweat their whole life to rake in money. When they finish, what have they got? A lot of miserable years and no way of knowing how to enjoy what's left. No sir. Harry, life is too short to spend every minute of it trying to make more money. I want to enjoy myself – now, while I can. Tomorrow will take care of itself. We'll keep up our end of production but don't ask me to go around with a long face trying to squeeze out that last drop of blood. We've got a good production gang. We'll see you get a fair day's output.'*

'I didn't mean for you to drive them,' apologised Harry. *'I just meant that the time you spend horsing around could be put to better use. You could be making more money without working any harder – just steadier. You can still enjoy working but you'll never get ahead if you don't take things more seriously. We're all here to produce. It hurts like hell to have your gang sitting around talking when they should be working.'*

'OK Harry,' Andrew said. *'We'll quieten down. If we aren't keeping up our end of production, you let us know.'*

The conversation had not gone exactly the way Harry had wanted it to. He did not anticipate that Andrew would react the way he did to the chance of making more money. Things did improve for a while but in a week or so the old pattern repeated. Harry decided that if he couldn't get to Andrew, he would work around him. One at a time he arranged to talk to Dave, Steven and Tony. The fifth member of the gang, Ken, was relatively new and Harry didn't think he should involve him in the discussions.

The reactions of Dave, Steven and Tony were the same. They all felt that it would be very nice to be making more money. On the other hand, the incentive earnings were pretty good – better than those of the other gangs. Besides, it was a lot of fun working with Andrew. Each said they would try to buckle down and cut out the talking. However, the pattern of response repeated itself – first some improvement would be made and then they gradually went back to their old habits. Harry was convinced that it wasn't malicious. They were doing a fine job on production. It was frustrating, though, to keep knocking your head against a wall without getting any results.

As much as anything, Harry was worried about the appearance of things. He often wondered how he would explain how the gang could sit around laughing and talking if the Company Director should happen to visit the site while Andrew was spinning one of his yarns. Even if they were producing well, it would be hard to justify a party on site.

Harry was puzzled by his failure to motivate any of the gang. Why didn't they respond to the carrot he held out for them? Isn't everybody interested in more money? He remembered how pleased his wife was with the pay increase he got when he was made Site Manager. Yet Andrew showed no interest at all and the rest of them were interested but apparently not enough to put any pressure on him to stop horsing around. If a Manager couldn't motive his personnel with money, what in heaven's name was left?

Since it was obvious to Harry that he couldn't solve this problem by holding out financial incentives to Andrew and that his chance of forcing the issue with disciplinary action was not good, he searched his brains for alternatives. Maybe he could use Andrew to spark higher productivity with other gangs.

There was no denying that Andrew was a leader and that when he worked he took everyone with him. The trouble with that approach was the risk of having him spread his practice of playing around. At least the way it was now, he infected only one gang. Perhaps containment was a wiser policy. However, if Andrew could stimulate higher production it would really help. *'With Andrew's seniority, he couldn't be forced into a lesser position,'* Harry thought. *'You'd have to motivate him and I'm right back where I started. It's a puzzlement.'*

Andrew had Harry baffled.

Questions

1. What might be some of the reasons to explain why Harry failed to motivate Andrew to buckle down?
2. If the rest of the bricklaying gang really wanted more money, why didn't they keep working when Andrew stopped to tell them one of his stories?
3. Harry felt that if a manager couldn't motivate personnel with money, there wasn't much left. How else could a manager motivate his personnel?

Main Points of Case Study Questions

Question 1

- Harry had a pure 'Tayloristic' managerial approach. He felt that the primary driving force for motivating employees was the 'carrot of extra financial remuneration'. However, this was not the main consideration for Andrew.
- Andrew had sufficient monetary rewards to satisfy Maslow's lower-order needs of 'Physiological, Safety and Security'. Therefore these failed to act as motivators when applied by Harry.
- Andrew was more interested in the higher-order need of 'Social and Belonging'. The social interaction and acceptance by the gang were of importance to Andrew. The above is corroborated by Herzberg's theory of the 'hygiene factors' of salary, working conditions and job security, which were all satisfied.
- The 'extrinsic' approach to motivating Andrew was doomed to fail because this wasn't at the centre of his value system.

Question 2

- Again, the gang had an acceptable level of extrinsic reward. Even though they all said they would like more money, the question is somewhat invalid – we would not expect employees to say 'No, I would not like more money'.
- An important point to note is that of Harry not taking action as soon as he became Site Manager (he left it for 'a couple of months'). This sent out a clear message to the gang: 'this is acceptable behaviour'. Therefore it became part of their 'group norms'. Harry should have acted right away. The gang were probably puzzled now that he was not accepting their behaviour, as this demonstrates inconsistency.
- Harry also failed because he did not include all of the gang's members in his approach when addressing the gang; he left out Ken as he was a new member. Harry had not appreciated the Forming, Storming, Norming and Performing aspects of 'group dynamics'. The group becomes a 'Sociotechnical System'. Further, Harry should have spoken to the group, including Andrew, altogether. Failure to do this just meant nothing had changed and that's why the group reverted back to previous practice.
- The social interactions of the group were highly valued by all members.
- Anthropocentric Production Systems advocate the 'empowerment' of employees in attaining set targets with the minimal division of labour. This gang were meeting targets in fact better than all other gangs. This they achieved by working as a cohesive group with bounded acceptable group norms of behaviour. Remember 'Organisations do not have objectives but people have values' (Miles and Snow, 1983). Therefore what is valued by personnel motivates them.

Question 3

- Harry failed to acknowledge the intrinsic approach to motivating staff. This would include the higher-order needs of Maslow and the motivators of Herzberg.
- Harry needs to change his thinking from that of employees belonging to the model of 'economic man' and consider a more appropriate 'social man' approach.

Final Comment

Do you consider that Harry really had a problem with Andrew's gang? After all, productivity was high. Was the problem really Harry's, because of his own 'insecurity'?

Should he have been prepared to defend his staff if Senior Management had commented on their standing around talking?

Summary

Within this chapter various management concepts and tools have been advocated for improving the efficient and effective utilisation of corporate resources. They have been presented to enable the reader to better understand and learn from the deployment of such conceptual tools. If this can be achieved the construction industry will stand to benefit and will be more likely to attain the goals advocated by both Latham (1994) and Egan (1998). It is also important to remember that the presented 'generic' models are there to be adapted to suit individual construction enterprises but at the same time remaining within the overall objectives of the model.

It is worthwhile for the reader to take the time to work again through the case study provided above – 'Motivating Site Personnel' – as it allows the reader to consider much of the theory covered within a practical context.

References

Anderson E.J. (1994) *The Management of Manufacturing Models and Analysis*. Addison Wesley, Wokingham, ISBN 9–201–41669–7.

Anthony R.N. (1988) *The Management Control Function*. Harvard Business School Press, Boston, MA, ISBN 0–87584–184–8.

Barnard R.H. (1981a) Survival or Success, *Chartered Institute of Building*, Occasional Paper No. 25.

Barnard R.H. (1981b) A Strategic Appraisal System for Small Firms, *Building Management and Technology Journal*, September, pp 21–4.

Barton P. (1983) *Building Services Integration*. Spon, London, ISBN 0–419–12030–0.

Bate P. (1994) *Strategies for Cultural Change. Butterworth-Heinemann*, ISBN 0–750–60519–7.

Benes J. and Diepeveen W.J. (1985) Flexible Planning in Construction Firms, *Construction Management and Economics*, 3, 25–31.

Bengtsson S. (1995) *Building Technology and Management Journal*, Chartered Institute of Building, November, pp 32–4.

Birley S. (1979) *The Small Business Casebook*. HarperCollins.

Blyth D. and Skoyles E.R. (1984) A Critique of Resource Control, *Building Technology and Management Journal*, February, pp 29–30.

Calvert R.E., Bailey G. and Coles D. (1995) *Introduction to Building Management*, 6th edn. Butterworth-Heinemann.

Cumbers M. (1986) Management of Technology, *Management and Development and Advisory Service*.

Djerbarni R. (1996) The Impact of Stress in Site Management Effectiveness, *Construction, Management and Economics*, **16**, 281–93.

Egan J. (1998) Rethinking Construction, *Report of the Construction Task Force to the Deputy Prime Minister, John Prescott, on the Scope for Improving the Quality and Efficiency of UK Construction*. Department of the Environment, Transport and the Regions, London, p 29, ISBN 1–85112–094–7.

Fryer B. (1985) *The Practice of Construction Management*. HarperCollins, ISBN 0003830306.

Glassman A.M. (1978) *The Challenge of Management*. Wiley, Toronto, ISBN 0–471–02767–7.

Griffith A., Stephenson P. and Watson P. (2000) *Management Systems for Construction*. Addison Wesley Longman, Harlow, ISBN 0–582–31927–7.

Gunning J.G. (1998) Site Management – Mainly a Process of Control, *Building Management and Construction*, **21**, 29–31.

Hambrick D.C. (1983) Some Test of the Effectiveness and Functional Attributes of Miles and Snow's Strategic Types, *Academy of Management Journal*, **26**(1), 5–26.

Harris F. (1977) *Modern Construction Management*. Beckman Publishing, ISBN 0–846–40636–5.

Harris R. and McCaffer R. (1995) *Modern Construction Management*, 4th edn. Blackwell Science, Oxford.

Harrison M. (1993) *Operations Management Strategy*. Pitman, ISBN 0–273–60119–9.

Hellard R.B. (1995) *Project Partnering: Principle and Practice*. Thomas Telford, London, ISBN 0–7277–2043–0.

Herbst P.H.G. (1976) *Alternatives to Hierarchies*. Kluwer Academic, ISBN 9020706322.

Herzberg F., Mausner B. and Snyderman B.B. (1959) *The Motivation to Work*. Wiley, New York.

Hickman C.R. and Silva M. (1989) *Creating Excellence*, Unwin Hyman, London, ISBN 0–04–658252–5.

Hutchins T. (2001) *Unconstrained Organisations*. Thomas Telford, London, ISBN 0–7277–3016–9.

Illingworth J.R. (1993) *Construction Methods and Planning*. Spon, London.

Jadrin J. (1976) *Business Practice for Construction Management*. Asbridge Management College.

James F. (1979) Six Tasks Necessary for Corporate Planning, *Chartered Institute of Building Journal*, April, pp 32–6.

Johnson G. and Scholes K. (1984) *Exploring Corporate Strategy*. Prentice-Hall International, London, ISBN 0–13–2959240.

Johnson G. and Scholes K. (1993) *Exploring Corporate Strategy*. Prentice-Hall, London, ISBN 0–13–297441–X.

Kanter R.M. (1989) *When Giants Learn to Dance: Mastering the Challenges of Strategic Management and Careers in the 90s*. Simon and Schuster, ISBN 0–04–440670–3.

Koontz H. and O'Donnell C. (1974) *Essentials of Management*. McGraw-Hill, London, ISBN 0070353719.

Lamming R. (1993) *Beyond Partnership Strategies for Innovation and Lean Supply*. Prentice-Hall, London, ISBN 0–13–143785–2.

Lansley P. (1975) *Flexibility and Efficiency in the Construction Industry*. Ashridge Management College.

Lansley P. (1981) Corporate Planning for the Small Builder, Trent Lecture, *Building Technology and Management Journal*, December, pp 7–9.

Latham M., Sir (1994) *Constructing the Team – Final Report of the Government/Industry Review of Procurement and Contractual Arrangements in the UK Construction Industry*. HMSO, London, ISBN 0–11–752994–X.

Maskell B.H. (1992) Performance Measurement for World Class Manufacturing, *BPICS Control*, October/November.

Massie J.L. (1981) *Managing – A Contemporary Introduction*. Prentice-Hall, ISBN 0–13–550327–2.

McGeorge D. and Palmer A. (1997) *Construction Management – New Directions*. Blackwell Science, Oxford, Chapter 4, ISBN 0–632–04258–3.

Miles R. and Snow C. (1983) Some Tests of Effectiveness and Functional Attributes of Miles and Snow's Strategic Types, *Academic of Management Journal*, **26**(1), 5–26.

Morgan G. (1993) *Images of Organisation*. Sage, London, ISBN 0–8039–2831–9.

Mullins L. (1996) *Management and Organisation Behaviour*, 4th edn. Pitman, London, ISBN 0–273–61598–X.

Mundie P. and Cottam A. (1993) *The Management and Marketing of Services*. Contemporary Business Services. Butterworth-Heinemann, ISBN 0–750–60789–0.

Mundy K. (1992) Making the Right Choice, *Logistics Today*, September–October.

Mustapha R.H. and Naoum S. (1998) Factors Influencing the Effectiveness of Construction Site Managers, *International Journal of Project Managers*, **16**, 1–8.

Naoum S. (2001) *People and Organisation Management in Construction*. Thomas Telford, London, ISBN 0–7277–25017.

Nunney D. (1992) *Integrated Manufacturing*. Department of Trade and Industry, London.

Passmore W.A. (1994) *Creating Strategic Change*. Wiley, New York.

Peters T. (1988) *Thriving in Chaos: Handbook for a Management Revolution*. Macmillan (now Palgrave)/Harper & Row, ISBN 0–333–45427–8.

Pilcher R. (1986) *Principles of Construction Management*. McGraw-Hill, Maidenhead, ISBN 0–07–084061–X.

Preston J. (1993) *International Business*. Pitman, London, ISBN 0–273–601482.

Ross D. (1991) Aligning the Organisation for World-Class Manufacturing, *Production and Inventory Management*, Second Quarter.

Ruch W.A. (1992) *Fundamentals of Production and Operations Management*. West Publishing, New York, ISBN 0–314–92852–9.

Sayles L.R. (1989) *Leadership Managing in Real Organisations*. McGraw-Hill, London, ISBN 0–07–055017–4.

Somerville D.H. (1981) Cash Flow and Financial Management Control, *Chartered Institute of Building, Surveying Information Services*, No. 4, ISSN 0140 649X.

Steiner G.A. and Miner J.B. (1977) *Management Policy and Strategy*. Collier Macmillan, London, ISBN 0024167401.

Taylor B. (1977) *Corporate Strategy and Planning*. Heinemann, ISBN 043491911X.

Torrington D., Weightman J. and Johns K. (1989) *Effective Management: People and Organisation*. Prentice-Hall, ISBN 0–132–44344–9.

Trotter P. (1982a) The Site Manager, *Building Technology and Management*, **26**, 7.

Trotter P. (1982b) *Building Technology and Management Journal*, December, p 17.

Vasser E. (1981) Building from the Boardroom, *Building Technology and Management*, March, pp 12–13.

Wakefield N. (1982) Give Site Managers a Chance, *Building Technology and Management*, **20**, 8–9.

Waters C.D.J. (1991) *An Introduction to Operations Management*. Addison-Wesley, Wokingham, ISBN 0–201–41678–6.

Watson P. (2000) CIOB Managing People, *Construction Manager Magazine*, January.

Yip G.S. (1992) *Total Global Strategy*. Prentice-Hall, Englewood Cliffs, NJ, ISBN 0–13–357658–2.

4 Planning and Programmes

Introduction

This chapter examines the management function of *construction planning*. It also looks at the relationships between planning and the management functions of *progress monitoring* and *control*, which are addressed in a subsequent chapter. There are essential requirements on all construction projects to plan, monitor and control activities against the project's scheduled duration. Planning is a functional process carried out at specific phases of the project: pre-tender, pre-contract and contract. Central to this process is the development and implementation of *construction programmes*, which show, in diagramatic form, the proposed methods and sequence for undertaking the siteworks. This chapter introduces the concepts and principles of planning and its application to programmes, essential to the effective time management of all construction projects.

Construction Planning

Purpose of Construction Planning and its Basic Concepts

Planning is a process of determining, analysing, devising and organising all the resources necessary to undertake a project. The core element of planning is the establishment of a programme which reflects the planning process in relation to real time. In practical terms, construction planning, and its ancillary elements *progress monitoring* and *control*, is the total process of determining the method, sequence, labour, plant and equipment required to undertake a building project, the provision of materials and services, the formulation of an operational programme to carry out the works and the development of progress monitoring mechanisms around the plan which facilitate control of those works.

The most important objective, from the principal contractor's standpoint, is to obtain and maintain the necessary volume and speed of output, and ensure quality, i.e. to give the client value for money, with the best use of resources and time, while giving greatest economy for the principal contractor.

Construction planning requires the detailed consideration of each item of work – *operation* or *groups of operations* – which combine to form the works, so that management can derive the following information:

- The total contract period, necessary for determining the organisational overhead costs which will be expended in undertaking the works
- The labour requirements including breakdown of skills, size of gangs and ratio of trade operatives to general operatives
- The requirements for materials and components, including delivery schedules, handling and storage
- The requirements for plant and equipment, including breakdown of type, capacity, utilisation, servicing, and attendance or stand-by labour.

It is clear from the identification of these four requirements that construction planning not only serves to provide information for the construction phase on site – *pre-contract* and *contract planning* – but also for the tendering phase – *pre-tender planning*. Before looking at these aspects in detail, it is important to outline the basic requirements for, and principles of construction planning.

Construction Planning and its Relationship to Progress Monitoring and Control

Construction planning, in broad terms, is concerned with two fundamental aspects:

(i) Programming the Works

This involves the planning of the works over the anticipated duration of the project (developing a programme), in relation to its requirements with full knowledge of resource needs and availability. The programme provides a quantifiable basis, or yardstick, against which progress can be measured.

(ii) Progress Monitoring and Controlling the Works

Progress monitoring follows on from the programming of the work and compares the work undertaken against the programme. Progress monitoring should facilitate analysis, from which the constant evaluation of resources is made to meet the on-going demands of the works, thus enabling control. Monitoring and control are covered in a later chapter.

The fundamental requirements for any planning and monitoring and control approach are therefore:

- a plan, or programme, to which management and operatives are working
- a set of effective arrangements implemented which guide management and operatives to work to the programme (management procedures and working instructions).

These two requirements must be integrated as both are fundamental to project success. It is no use having a detailed and accurate programme if procedures are not in place to use it as a monitoring, control and action-taking mechanism. Developing a set of effective control arrangements is a difficult task and can be much more problematic than developing the programme, so much so that progress control is often given little attention during the planning process, whether it be at the pre-tender, pre-contract or contract stage. In practical day-to-day terms, time spent developing a comprehensive programme will be time wasted if that programme does not form the core of the progress-control mechanism. Interestingly, inaccuracies in a programme will not prove disastrous provided there are effective procedures in place to cope with deviations from the plan, as some examples in this chapter will illustrate.

In any planning and progress monitoring and control approach there must be an early warning that problems are likely to occur. The early warning of problems which affect progress is absolutely essential. Comprehensive and frequent reassessment of the programme for the remaining works, in particular through *short-term* planning, will assist in identifying relevant early warnings. An effective control mechanism will maintain the momentum of working to programme and act as a prompt to management and supervisors alike. Management will therefore be proactive rather than reactive to events.

One of the most erroneous aspects of construction planning is the unfounded belief that once a programme is followed the only thing necessary for an activity to commence on time is for the preceding activity to have started and finished on time. In reality, of course, there are a great many things which must occur but which the programme does not show. For example, labour, plant and materials must be procured and deployed, drawings may still have to be issued or specifications finalised. In practice, the detailed consideration of 'what actually has to happen' before any operation can start is essential.

Construction programmes frequently incorporate brief descriptions of activities or tasks. They often show that activities start and finish neatly and that they are discrete with little if any interrelationship with others. This is overly simplistic and explains the main reasons why programmes are often poorly implemented. An effective construction programme is not simply dependent on incorporating a large number of highly detailed activities, but rather on a detailed understanding of what the practical work entails within each activity and who has to do what before the remaining activities can commence. Before construction planning can even commence, the method of construction and the outline sequence of activities that will be involved must have been determined. The development of a well thought out and logical construction sequence, represented by an accurate *method statement*, is therefore a crucial early step in the planning and control processes.

The Benefits of Planning and Progress Monitoring and Control

Construction management can benefit from effective planning and progress monitoring and control in many ways. The following list, although not exhaustive, presents many of the potential advantages of developing and implementing an effective mechanism for time management. It can:

- determine the potential sequence of works
- provide management and supervisors with an effective and realistic works programme
- enable the assessment of labour requirements
- enable the determination of material requirements
- facilitate the scheduling of plant and equipment
- show the commencement and completion of programmed activities
- suggest a flow of work to avoid resource imbalance
- rearrange project priorities in the light of changed circumstances
- prevent work overload or underload
- allow the estimation of activity duration and completion dates
- highlight potential underresource to meet programme needs
- determine the material stock available for use
- check materials on order
- show when further orders should be confirmed
- show when subcontractors are needed on site
- show deployment and labour strengths of subcontractors
- provide information to head office on the project's progress.

Construction Planning: Preliminary Considerations

Before embarking on the development of any planning mechanism for the production phase, three fundamental questions should be asked:

(i) What kind of planning mechanism is required?
This is principally determined by the phase of the project, i.e. pre-tender, pre-contract or contract.
(ii) Who will use the mechanism?
This is predominantly determined by management level, i.e. site management, first line supervisors or corporate management.
(iii) Why is the mechanism being used?
This is principally determined by the specific purpose, e.g. estimating or method sequencing.

Consideration of these three questions will enable a broad choice to be made on the type of planning mechanism to best meet the specific project phase, management level and particular application.

Outline Classification of Planning Requirements

All construction projects are in some ways different and therefore require specifically designed planning mechanisms. In terms of its concepts, procedures and practices of programming all projects come within one of the following classifications:

- Simple – those projects with straightforward planning requirements because of the undemanding characteristics of the works

- Complex – those projects with complicated planning requirements because of particularly demanding characteristics of the works; for example special techniques of construction, difficult site conditions, restrictions to operation or tight work schedules.

The basic category into which the project falls has a fundamental and profound effect on the type and nature of planning mechanism which is considered. While a simple project may rely on one master programme to control all the works, a complex project may require a much greater degree of detail, monitoring and analysis, which necessitates *short-term*, or *sub-programmes*, reflecting particular parts of the project, to be developed.

Programming

In simple terms, when a principal contractor compiles a tender for a contract the total work is divided into work sections and operations. Consideration is given as to how the operations can be carried out, what labour, materials and plant will be required, how long the operations will take and the cost of each operation and section of work.

It is important that all programming is carried out to the conceptual programme that was envisaged at the time of tender. In some cases there can be a considerable time lag between tendering for and the award of a contract, so the project information may not be fresh in the minds of planning and production management. In addition, planners, estimators and contracts managers are likely to be working on a number of projects simultaneously and they will need to reappraise themselves of a specific project.

To ensure that the work is actually carried out in accordance with the pre-tender plan, the most effective method is to compile a programme which covers all contract work and includes all operations, or groups of operations in sections, specifies their anticipated duration and shows how they interrelate with others. Development of this programme requires the detailed consideration and careful planning of each operation such that all operations can be carried out in a logical sequence and with the most cost-effective use of resources.

To develop any programme a contractor must determine the following:

(i) Method – the best method for carrying out the work in all its parts, or operations. This requires a sound understanding of the technology needed to construct the building or structure.
(ii) Sequence – the most practical sequence in which the operations should be carried out. This links technology with the management of the construction process.
(iii) Volume of work – the amount of work involved in undertaking each operation.
(iv) Resources – the labour, plant and materials needed to carry out each operation.
(v) Presentation – the method of presenting the above information, i.e. the type of programme.

To achieve the above the contractor will need to follow a satisfactory programme development sequence: formulation of a *method statement*; *build-up of operational durations*; development of a *programme calculation sheet*; and preparation of an appropriate *contract or master programme*. Details of these elements, together with an approach which can be adopted, are shown later in the chapter.

Once the programme is established, the contractor will be in a position to calculate, with accuracy, the rate of progress required for each operation and groups of operations, and to determine resources appropriate to the on-going requirements of the operations across the project. The contractor is also now in a position to agree the intended programme with the parent company or corporate organisation who must support the project, the client's professional consultants who will provide contract documents in the form of drawings, specifications and probably a bill of quantities, and production management on site who will order plant and materials to resource the works. In addition, management will have a yardstick or standard against which performance, or *progress*, can be measured and evaluated.

Flexibility in Planning

To conclude this introductory section on the basic concepts of construction planning it is important to highlight the balance which should be maintained between rigidity and flexibility in planning. Some measure of rigidity in planning is important because it gives the essential structure to a programme and maintains the principles that are applied to support its concepts. However, a measure of flexibility is also essential because, at some time, progress will deviate from that which has been planned and an effective programme must accommodate this. A highly rigid programme will quickly become disorientated and therefore useless if there is any major change or disruption. Any mechanism must be inherently adaptable so that any change in project design, method of construction or operational sequence can be introduced into the programme and carried out during progress monitoring and control without affecting the principal aims and objectives of the project.

Construction Programmes

Planning Phases

The techniques of programming applied to construction management can, as stated earlier, be divided into three parts to meet particular planning phases of a project:

(1) Pre-tender: the planning of the project such that the principal contractor's *tender price* is based on a realistic construction programme.
(2) Pre-contract: extension of pre-tender planning to provide the contractor with a *contract*, or *master, programme* for the siteworks.
(3) Contract, or short-term: subdivision of the contract, or master, programme to facilitate *short-term planning*, which may be undertaken by the contractor before, although more often during, the siteworks. This aspect will be covered in a later chapter, as short-term programmes form an effective progress monitoring and control mechanism in their own right.

(1) Pre-Tender Programming

A pre-tender programme, prepared by the construction planner in consultation with construction management, is used often by the estimator in preparing the tender price for the contract. This programme should determine:

 (i) the overall construction period
 (ii) the approximate labour requirements
(iii) the approximate materials and plant and equipment requirements.

To formulate the programme, specific project documentation and information will need to be assembled and analysed. These include the following:

- Conditions of Contract – likely to include clauses which affect the duration of the project
- Bill(s) of Quantities – will include the elements of work to which time must be apportioned in calculating the unit rates used in the tender
- Specification and Drawings – will provide the details of the work from which the construction methods, sequence of operations and operational duration will be determined, together with materials requirements
- Project Correspondence – may raise specific matters which could affect the programme
- Site Visit Report – may highlight conditions on the site or within its environs which could affect the programme.

Construction programmes at the pre-tender planning stage have to satisfy the requirement to present an overall picture of the project. The programme has to be accurate and with sufficient detail such that should the contract be awarded, the contractor is confident that the project can be completed within the envisaged overall timeframe. Unfortunately time spent planning at the pre-tender stage can be rendered abortive should the contractor's tender be unsuccessful, so the contractor will prepare a 'reasonably accurate' programme yet one which will require subsequent development at a later stage in the process. However, the estimation of *contract preliminaries*, or the organisational overheads incurred in maintaining the site such as site staff, accommodation, services and facilities, are directly affected by the contract's duration. The preliminaries cost, while varying from one contract to another, can amount to around 5%, or a great deal more on some projects, and therefore the contractor needs a reasonably reliable assessment of project duration.

Examples from two projects illustrate the potential implications for the contractor when difficulties with pre-tender aspects arise. On the first contract, a new £18 million office development with a contract period of 28 months, the project preliminaries averaged approximately £29,000 per month (4.5% of the contract value). Difficulties in maintaining progress resulted in the contract period overrunning by 6 weeks and an additional preliminaries cost of approximately £30,000 was incurred. In purely arithmetical terms the additional cost would be nearer £44,000 (£29,000 per month × 6 weeks), but the contractor managed to alleviate a proportion of this by removing from the site some preliminary items which had become unnecessary during the latter stages of the contract. It was likely that some of the additional preliminaries cost would be offset by contractual claims made by the contractor.

On the second contract, a £6 million industrialised building with a contract period of 16 months, the preliminaries cost averaged just over £11,000 per month (approximately 3% of the contract value). Better than expected progress was made in erecting the steel portal frame and external cladding, resulting in the project being completed 5 weeks ahead of programme and showing a preliminaries cost saving of approximately £14,000.

Although in the above two examples the outcome figures are small in terms of both absolute monetary values and as proportions of total project costs, in real terms the figures represent to the contractors involved a deficit or loss in the first example and additional profit in the second example. This is why contract progress is so fundamentally important to construction management. The estimator, in consultation with the contract planner and contracts manager, will therefore base the tender on the pre-tender programme so that the estimation of project duration is based on a tangible calculation and is not merely a best guess. Should the contract be subsequently awarded, the pre-tender programme must then be reassessed and becomes part of the process of *pre-contract planning*.

As a fundamental objective of the pre-tender programme is to ascertain the overall contract period, essentially it need only comprise the main construction elements or groups of operations. The contract is therefore divided into these elements for analysis and presentation on the programme. Division on a typical building project may present the following major work elements, which, of course, incorporate the many individual operations and tasks required to complete the project:

- Demolition
- Excavations and siteworks
- Foundations
- Structural frame
- Superstructure enclosure
- Floors
- Roof coverings
- Internal walls/1st fix joinery
- Services/2nd fix joinery/finishes
- Fittings
- External works.

The duration for the main elements and the operations that combine to form these elements will be based on the quantities of work given in the bill, together with items measured from the drawings. A subsequent section in this chapter shows how programmes are built up based on calculated operational durations. The experience of the estimator is essential in determining the pre-tender programme since the duration calculated will only be approximate and is usually 'rounded-off' to arrive at the practical working duration as shown on the programme. Experience suggests that it is a mistake to be over-precise in making these calculations. For example, an operation that is calculated to take 4.25 days to complete will, in practice, usually extend to a full week. It is therefore more realistic to show one full week on the programme. This rounding-off technique will be used in determining the duration of all the operations or elements shown in the pre-tender programme.

(2) Pre-Contract Programming

The principal objective of pre-contract planning is to provide a *contract*, or *master, programme* for the siteworks. To formulate this programme, all the information that was assembled for the development of the pre-tender programme will be required, but in greater detail and involving deeper analysis. At the tender stage, the estimator, planner and contracts manager will have been concerned with producing a reasonably accurate overall programme from approximate quantities whereas, at this stage, a more comprehensive and searching analysis must be made of the bill of quantities. Elements, or groups of operations, will again be determined but this will be done with an eye for greater detail to ensure that all work items, and in particular those which will be critical to the progress of the works, are fully considered when formulating the programme.

Unfortunately, the way in which work elements are measured in a bill of quantities bears little resemblance to the way that they are programmed. This not only applies to the way in which the quantities are expressed, that is, the units of measurement used, but also to where the items measured in the bill are actually to be found on the project site. This apparent ambiguity means that the planner must take-off, or measure from the drawings, those elements or operations which are thought to be critical from the programming perspective, so that both the volume of work and its location are known with certainty.

Method Statement

Following the detailed scrutiny of the bill of quantities to determine the major components of the works, consideration must be given to the method and sequence of construction. This is achieved by compiling a *method statement* – a summary list of major operations, outlining the construction methods to be used and the principal labour, materials and plant resources needed, presented in a logical sequence of construction. The method statement will be compiled by the project planners, in close consultation with construction management staff, in particular the contracts manager and the site manager if appointed at the pre-construction stage, as they have the greatest practical experience of construction methods and sequencing and will be able to focus this towards the specific project.

Outline, or Draft Target, Programme

Before the detailed contract, or master, programme is formulated, an outline programme should be drafted. This will list those major operations given in the method statement. A time-related scale is presented, representing calendar dates and/or project weeks against which any conditions imposed by the client, such as completion date, partial completion date, and other information such as holiday periods, can be shown. This basic outline programme can then be used as a framework or template for developing the contract, or master, programme.

Contract, or Master, Programme

The contract, or master, programme that is subsequently produced, following the *build-up of operational duration* and development of a *programme calculation sheet* (both stages subsequently described) represents the working programme. This programme will be used on site to progress the works and hence there is no room for major errors or anomalies – it must be an accurate and realistic representation of the anticipated construction activity. The most important task of site management is to ensure that the contract period does not exceed the time given in the tender and, ideally, a shorter contract duration is desirable. In addition to showing accurate operational durations, the programme should consider resourcing requirements and availability. For example, a build-up calculation may have included an excavation task based on the use of a particular type of excavator and using a specific size of bucket. It is therefore, essential that a comparable machine is actually available on site to meet the programme. To ensure that this happens, resource schedules plan the major resources to meet the contract programme.

The Consideration of Method Study and Work Measurement

In developing the contract, or master, programme, it is relevant to consider the concepts of method study and work measurement in relation to those operations which benefit from very detailed analysis before their durations are calculated. Method study was used successfully on two recent examples of construction projects, first on a deep basement excavation where the method of dig and the cycle of earth removal were critical to maintaining progress, and second, on a multi-storey project where the sequence of reusing formwork for concreting operations was essential to maintaining the programme. In practice, consideration of method study – *finding the best way of doing the job*, and work measurement – *finding how long the job should take*, will rely largely on the experience of the planner, contracts manager and site manager, and on the records of past projects involving similar operations. The concepts are particularly applicable where certain operations are critical to progress and where, perhaps, a new or novel method of construction is being considered about which the contractor has little experience. The underlying principle is one of understanding more fully the requirements of the work and appreciating their effects on the total project.

The Importance of Critical Operations and Sequencing

Critical operations are those within the programme that directly affect overall progress. In developing the contract programme, these operations should be identified and analysed in detail in terms of the construction methods used to carry them out and their sequencing interrelationships with other operations. A simple example might be on a contract where the structural frame is steel, and where the programme scheduling the erection operations is critically dependent on the fabrication and delivery of the steel members. Again, in practice, the determination of critical operations will be a function of the experience of planning and management staff, with reference to the sub-programmes of intended siteworks by subcontractors and material and component delivery schedules.

The sequencing of construction operations is largely the product of experience. Critical operations will be determined in the method statement and these set the framework within which all other operations must be programmed. While all operations may not be shown in the contract programme, they will be enveloped within sections of work which are shown on the programme. To obtain details of these operations one would refer to sub-programmes, for example a subcontractor's fabrication and delivery programme.

Contract planning does not finish once the contract, or master, programme is implemented at the commencement of the siteworks but, rather, the programme serves as a basis for progress monitoring control techniques used during the course of the project – *short-term planning*. Short-term planning is not merely a simple sub-set of contract planning, it is an important planning and progress control application in its own right and for this reason occupies a separate section in a later chapter.

To develop a contract, or master, programme it is important to appreciate how operation durations are determined and from where the information needed originates.

Build-up of Operational Duration for Contract Programmes

The build-up of programmes, or determining the duration of construction operations, is a process of establishing the quantity of work involved in each item, determining the time required to complete the item by considering the labour and plant needed, and depicting the calculated duration on the programme.

Establishing the quantity of work is easily ascertained from the bill of quantities or by measuring from the drawings, with due appreciation of those aspects highlighted in the preceding section of this chapter. Determining the time required to undertake an operation is a more complicated matter. There are, principally, two ways in which the labour and plant rates needed to make the calculation for durations can be acquired. These are:

- From *experience* – the contractor will, over the years and from many projects, build up a library of information based on experience and feedback of data, such that labour and plant work rates, or outputs, often referred to as 'constants', are known
- From *building price books* – these present labour and plant constants for many operations in all fields of construction, although it should be remembered that such constants may not be as accurate, in the context of a particular contractor's method of working, as those derived from experience. In practice, calculations rely upon selection by the estimator of the most appropriate constants having regard to the particular work and site conditions.

It is essential that estimators and planners build up a library of programming information because the accuracy and value of programmes is mainly dependent on the accuracy of the constants used in the build-up.

Calculating Durations for Labour Inputs to the Programme

The calculation of the duration of an operation is made by dividing the *volume of work or quantity* by the *constant*, and converting the rate into the *unit of measure* used in the programme. This can be illustrated in outline with a simple example:

Suppose a bricklayer had to build a one-brick thick (215 mm) wall, 50 metres long by 2 metres in height. How long would this take?

Volume of work: $50\,m \times 2\,m = 100\,m^2$
$100\,m^2$ of 215 mm brickwork = 12,000 bricks approx. (quantity)
A bricklayer can lay 50 bricks per hour (labour constant)

Therefore $\dfrac{\text{volume of work (quantity)}}{\text{work output}} = \dfrac{12{,}000}{50} = 240$ hours

Convert the rate into unit of measure used on the programme, i.e. week (with 6 days, or 47 hours).

Therefore, duration $= \dfrac{240}{47} = 5.11$ weeks for one operative

---

In simplified terms, this duration could be converted into 2.55 weeks for 2 operatives or 1.28 weeks for a gang of 4 operatives. This choice would be made through the labour deployment determined in the short-term programme within the overall parameters for that particular work item as shown in the contract, or master, programme. In addition, the constitution of the work gang, i.e. the ratio, in a situation such as that given, would be 4 to 1 or 4 to 2 depending on the requirements of the work, where one or two general operatives are deployed to assist the four trade operatives.

Calculating Durations for Plant Inputs to the Programme

In the same way that the build-up of duration involving labour requires knowledge of work constants, so too does the build-up of durations involving plant resources. Work rates or constants for construction plant may be obtained from the following sources:

- From *experience* – as described previously
- From *building price books* – as described previously
- From *plant manufacturers and suppliers* – manufacturers and supplying dealers of construction plant and equipment will be able to provide approximate guides to rates of performance. However, such rates are likely to be on the optimistic side, since manufacturers and dealers are in business to sell their machines and they may, therefore, present their product in the most favourable light. Nevertheless, such information is extremely useful to the estimator.

In addition to the above, it must also be borne in mind that the true output of any piece of construction plant will vary, possibly considerably, depending on such factors as:

- the skill and experience of the operator
- the mechanical efficiency of the particular machine in use
- the waiting time for associated operations, such as hauling
- the ground conditions on the site
- the weather conditions.

The principles involved in building up a duration for an operation where construction plant is involved can be illustrated in the following simple example:

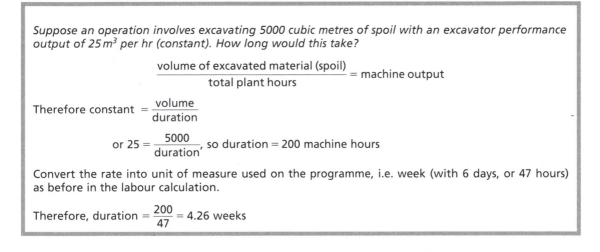

Suppose an operation involves excavating 5000 cubic metres of spoil with an excavator performance output of 25 m³ per hr (constant). How long would this take?

$$\frac{\text{volume of excavated material (spoil)}}{\text{total plant hours}} = \text{machine output}$$

Therefore constant $= \dfrac{\text{volume}}{\text{duration}}$

or $25 = \dfrac{5000}{\text{duration}}$, so duration = 200 machine hours

Convert the rate into unit of measure used on the programme, i.e. week (with 6 days, or 47 hours) as before in the labour calculation.

Therefore, duration $= \dfrac{200}{47} = 4.26$ weeks

A different calculation may have to be made where the plant item must be chosen to meet a particular operational timeframe given in the contract programme. For example:

Suppose the same excavating operation as above, i.e. 5000 cubic metres, is to be excavated in 20 working days. What is the required machine output to achieve this?

Total plant hours = 20 (days) × 8 (hours/day) = 160 hours

Therefore required output is $= \dfrac{\text{volume of excavated material}}{\text{total plant hours}}$

$= \dfrac{5000}{160} = 32$ m³ per hour

The required working output, 32 cubic metres per hour, can be checked against: the contractor's library database; a building price/plant handbook; or a manufacturer's performance specification list – and an excavator which meets the output rate requirement can be chosen.

Computing calculations for construction plant items is, perhaps, not as simple as those for labour intensive items. Where, for example, excavating operations are undertaken, spoil is usually removed from site by lorries. The determination of haulage might therefore be considered in conjunction with the selection of the excavating equipment. Again, this can be shown in a simple example:

Suppose an excavator with an output of 36 cubic metres per hour is being used and the spoil must be removed by lorries with 3 cubic metre capacity to a spoil tip with a 20 minute cycle time. How many lorries will be required?

The number of lorries required $= \dfrac{\text{cycle time}}{\text{loading time}}$

Cycle time $= 20$ mins

Loading time $= \dfrac{\text{machine output}}{\text{capacity}} = \dfrac{36 \text{ m}^3}{3 \text{ m}^3} = 12$ lorry loads/hour

or 5 mins loading time

Therefore number of lorries $= \dfrac{\text{cycle time}}{\text{loading time}}$

$= \dfrac{20 \text{ mins}}{5 \text{ mins}} = 4$ lorries

When calculations for haulage are made, it is often useful for site management to display the cycle of operations on a *multiple activity chart*. In this example it would involve: the time taken to load the lorry; and the time taken to travel to the tip, off-load and return – for the four lorries over a one-hour cycle.

Programme Notes

The calculations made during the build-up of operation durations should be systematically recorded and referenced, i.e. *programme notes* should be retained. These may be referred to during the contract should remedial action to any operation become necessary, and be available at the end of a contract for use in developing programme build-ups for future projects.

Programme Calculations where Unknown Quantities are Involved

When calculating the duration of operations, situations will occur where the amount of work cannot be quantified and so determining the anticipated duration by calculation is difficult. This may occur in building services work, for example. Usually on the basis of experience, the planner will, in consultation with the services consultant, make an estimation based on an alternative unit of measurement, e.g. per room, per floor, per house, etc. This, ostensibly, intuitive duration will be depicted in the programme as a block of time integrated with the calculated durations of the contractor's own works. The contractor is primarily interested, in practice, with the subcontractors completing their works on or before the last date shown on the overall programme. Works conducted within the scheduled duration do not usually interfere with the main contractor's programme, but any overrun may have a significant effect by, for example, hindering progress on a critical item of work.

Preparation of a Contract, or Master, Programme

This section traces the principal steps in the preparation of a complete contract or master programme, describing each task in the development sequence and including diagrams typical of those used widely within the building production management processes.

The Tasks of Contract Programme Development

Preparation of a full working contract, or master, programme for a building project involves detailed consideration of the following tasks:

 (i) abstract *quantities* from the bill of quantities
(ii) develop a *method statement*
(iii) compile a *programme calculation sheet*
(iv) draw up a *contract*, or *master, programme*
 (v) consider, if appropriate, the contribution of *short-term programming*
(vi) profile *labour and plant requirements*.

GROUNDWORKS

Foundations

Excavate topsoil for preservation, average depth 0.25 m	325 m²
Excavate to reduce levels not exceeding 0.25 m maximum depth	12 m³
Excavate foundation trench, over 0.30 m wide, maximum depth not exceeding 1.00 m	37 m³
Excavate foundation trench, over 0.30 m wide, maximum depth not exceeding 2.00 m	8 m³
Filling to excavations, average thickness over 0.25 m, arising from excavations	30 m³
Remove surplus spoil from site	15 m³
Compact bottom of excavations	70 m²

Earthwork support

Earthwork support, maximum depth not exceeding 1.00 m, width of excavation not exceeding 2.00 m	110 m²
Ditto, but maximum depth not exceeding 2.00 m	25 m²

INSITU/PRECAST CONCRETE

Insitu concrete

Concrete in foundations	22 m³
Concrete in reinforced ground floor slab, not exceeding 150 mm thick	42 m³
Concrete in base of service ducts, not exceeding 150 mm thick	5 m³
Ditto, but in walls	9 m³
Concrete in reinforced floor slab, not exceeding 150 mm thick	89 m³

Figure 4.1 *An example page of an abstract of quantities used in the preparation of a contract programme*

METHOD STATEMENT					
CONTRACT NO : **1312**		TITLE : DOMESTIC DEVELOPMENT PROJECT			
DURATION : 32 WEEKS		EDITION: 001		PLANNER : R J SMITH	
NO	OPERATION	METHOD	PLANT	LABOUR	REMARKS
02	Excavate topsoil for preservation	Machine reduce level and deposit on temporary topsoil heap	1 no multi-purpose tractor : excavator dozer	Machine operator 1 banksman (labourer)	Hired machine
03	Excavate to reduce levels	Machine reduce level and remove to holding spoil heap	- ditto -	- ditto -	- ditto -
04	Excavate foundation trench not exceeding 1.00 m depth	Machine excavate and deposit at side of trench, load and remove surplus spoil to tip	As above, with backhoe, 450 m bucket	Machine operator 1 banksman 1 labourer	Additional labourer to stand by for cable ducts crossing excavations
05	Excavate foundation trench not exceeding 2.00 m depth	Machine excavate (as before described)	- ditto -	- ditto -	- ditto -
06	Filling to excavation, arising from excavations	Machine backfill from holding spoil heap	- ditto -	- ditto -	- ditto -
07	Remove surplus spoil from site	Machine load to tipper lorry, remove to tip	Multi-purpose tractor 1 tipper lorry	Machine operator 1 banksman	Hired haulage

Figure 4.2   *An example page from a method statement used in the preparation of a contract programme*

Carrying Out These Tasks

(i) Abstracting the volume of work, or quantities – the volume of work for each operation is abstracted from the bill of quantities (see Figure 4.1). Items may be summarised under the sub-headings of *building element*, e.g. groundworks, *work section*, e.g. foundations, and *operation*, e.g. excavate oversite to reduce levels. It may also be necessary to measure quantities from the drawings to ensure that the full volume and location of the work in any item are accurately determined. The reasons for this were described earlier in this chapter.

(ii) Development of the method statement – an example page from a typical method statement, as defined and described previously, is shown in Figure 4.2. The method statement follows the format of a simple spreadsheet and presents information in columns, typically under the headings of: *operation number*; *operation (description)*; *method (of construction)*; *plant (requirements)*; *labour (requirements)*; and *remarks*.

(iii) Compiling programme calculation sheets – an example page from a typical set of programme calculation sheets, summarising the build-up of rates for each operation as described previously, is shown in Figure 4.3. A spreadsheet format is used, arranging information under the headings: *operation number*; *task reference*; *operation (description)*; *quantity*; *unit*; *constant (work output)*; *hours (computed)*; *weeks (converted)*; *labour gang (make up)*; *plant (type specified)*; *net duration*; and *remarks*.

When estimating the duration of operations it is essential to be practical as well as accurate. For example it would be practical where an operation had a calculated net duration of 3 hours to allow half a day on the programme, or where an operation had a net duration of 6 hours to allow a full day. In practice, the net hour duration of all operations would be adjusted for the purpose of practicality on the contract programme. In addition, such adjustments also serve as extra safeguards to flexibility within operation durations to meet any unforeseen delays to progress. Both operatives and management quite naturally think and plan in terms of days and half days, rather than in hourly durations.

PROGRAMME CALCULATION SHEET											
CONTRACT NO : 1312			TITLE :								
DURATION : 32 WEEKS			EDITION : 001				PLANNER :				
NO	OPERATION	QUANTITY	UNIT	RESOURCES	CONSTANT	HOURS	WEEKS	LABOUR	PLANT	NET WEEKS	REMARKS
01	PRELIMINARIES Temporary site set up	-	-	Labour	-	-	-	1 Jnr 2 Gen op	-	1	
02 03	OVERSITE EXCAVATIONS Excavate topsoil & preserve Excavate to reduce levels	325 12	m²} m³}	Multi-purpose tractor dozer	0.04 0.04	2.43	0.06	1 gen op	Multi-purpose tractor dozer	0.01	
04	Excavate foundation trenches maximum depth not exceeding 1.0 m	37	m³	- ditto-	0.59	22.0	0.55	2 gen op	- ditto -	0.60	
0.5	Excavate foundation trenches maximum depth not exceeding 2.0 m	8	m³	- ditto -	0.95	7.6	0.19	2 gen op	- ditto -	0.20	
06	Filling to excavation, arising from excavations	30	m³	- ditto -	7.00	4.29	0.11	1 gen op	- ditto -	0.20	
07	Remove surplus spoil from site	15	m³	Multi-purpose tractor dozer Tipper lorry	5.00	3.00	0.08	1 gen op	Tractor dozer 1 lorry	0.10	
LEGEND : Joiner (Jnr) General Operative (Gen op)											

Figure 4.3 *An example page from a programme calculation sheet used in the preparation of a contract programme*

(iv) Drawing up the contract, or master, programme – the most commonly used technique for presenting programmes is the *Gantt*, or *bar*, *chart* (see Figure 4.4). Alternative methods exist to present operational durations, for example line of balance diagrams, to sequence projects involving repetitive units of construction such as housebuilding, or networks for more complex construction projects, but for the vast majority of building projects the bar chart is the preferred method of programming. It is not the intention here to describe and discuss the various techniques of programming. Much literature exists in the subject area and the reader is directed to those sources. An example of a simple bar chart is shown subsequently. Adopting the standard convention for bar chart programmes, the *solid bars* that occupy the top half of each operation space represent the calculated and adjusted duration. The bottom half of the operation space is left blank so that progress can be recorded directly beneath *programme* as work proceeds on site. A bar chart may be far more sophisticated than the one depicted in this example, particularly where computer-derived programmes are developed. In such programmes, critical and non-critical operations may be systematically distinguished, duration *floats* (leeway in criticality of sequence) can be identified, and *resource levelling (balancing)* can be incorporated into the programme or shown separately to augment the programme, as shown in Figure 4.5.

(v) Consideration of short-term programming – although not essential at the stage of developing the contract, or master, programme on all projects, it is advantageous in the longer term for the contractor to look ahead at the likely requirements for short-term programming. Short-term programming is particularly applicable to complex construction projects where programmes are, for ease in use, subdivided into section programmes and where the duration of some operations are so long that an element of planning throughout their duration must take place. The concept, principles and techniques of short-term programming follow in a subsequent chapter.

(vi) Profiling the labour and plant requirements – it is standard practice to present a profile of the labour and plant requirements in the lower portion of the bar chart or on a separate programme formatted in a similar way to the contract programme. In the relatively simple example shown,

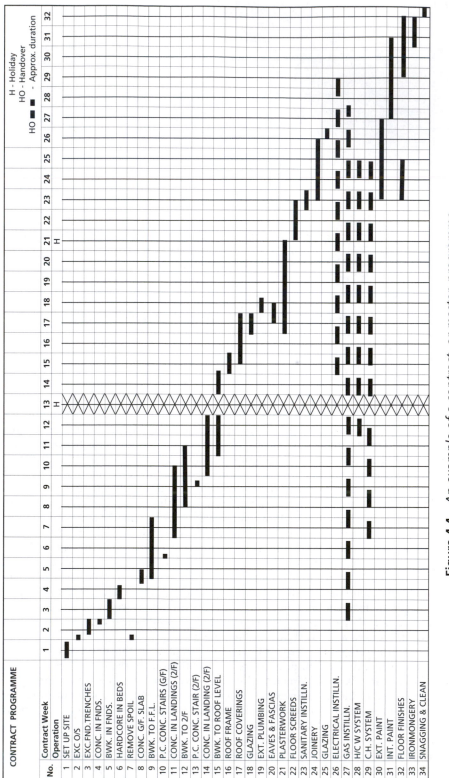

Figure 4.4 An example of a contract, or master, programme

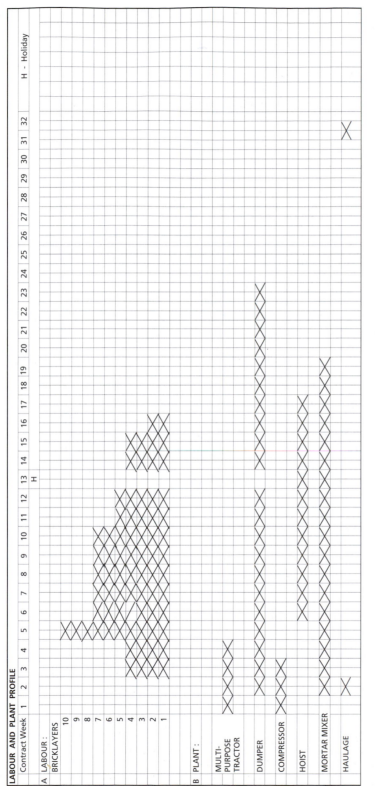

Figure 4.5 *An example of a resource programme*

a vertical block bar chart is used to depict the labour requirements for principal building elements, while the plant requirement over the contract is shown by means of a bar (horizontal) chart. Any preferred format of presentation may be adopted.

(3) Contract Programming

Using the contract, or master, programme compiled during pre-contract planning, contract planning develops sub-programmes which are used to guide stages of the project as the work proceeds. Therefore, the contract planning and programming stage focuses on interim and on-going planning mechanisms used to monitor and control the construction works. Planning of this type is often referred to as *short-term* planning. Short-term planning produces sub-programmes which can depict, for example, stages or sections of the works or particular elements of the works which require insight to a level of detail beyond that which can be shown on the contract, or master, programme. Greater details and explanations of short-term planning and programmes follow in the subsequent chapter which addresses progress and control.

Types of Construction Programmes

The practice of construction planning can differ widely among principal contracting organisations. While some utilise relatively uncomplicated planning methods to depict construction operations, others favour more detailed and complex methods to reflect the interrelationship of operations and their sequence. The many methods available to the contracting organisation to plan construction works can be confusing, and selecting a method appropriate to the project and its type of work can be far from clear. Construction projects differ in their characteristics – for example, type of works, complexity, scale, duration, location, specialisation and resourcing – and such aspects impart often conflicting demands for any particular planning approach.

In general, two construction planning methods are adopted in practice by most contracting organisations. These are *bar charts* and *networks*. Networks are usually adopted by the organisation for the purpose of providing highly detailed programming for a project, with bar charts used at the project site for clarity and ease of communication in conveying to site personnel the basis of the site-works programme.

Bar Charts

Bar charts are the most commonly used method of planning for all stages of the planning process and on projects both small and large. Construction operations, singularly or in groups forming elements, are listed vertically, with a bar representing those operations drawn horizontally on a time scale where the length of the bar is the planned duration of that operation. Operations drawn on the vertical axis are enumerated to a sequence used throughout the contract documentation for consistency. Bars reflecting activity durations are drawn on the horizontal axis and related to contract weeks and also to calendar dates to make the chart real time orientated. The completed bar chart is a graphic representation of the operations to be carried out within the total contract duration. While simple bar charts depict operations and their durations, more sophisticated charts will show breakdowns of key construction operations into sub-operations, non-productive periods during the contract such as holidays, simple links between overlapping, or parallel, operations, and may be developed to show labour and resource deployment (see Figure 4.6).

The major advantage of the bar chart is that it is simple to prepare relative to other types of programme, is easy to read and update and therefore becomes a useful communication tool for management on site. The simplicity and flexibility inherent to bar chart formation allow the concept to be applied to short-term planning and works progress monitoring in addition to contract planning. The principal disadvantage with the bar chart is its inability to reflect detail. On simple construction projects this may not be problematic but on larger and more complex projects the bar chart may be insufficient as a detailed planning tool. Complicated and involved projects are likely to have many and

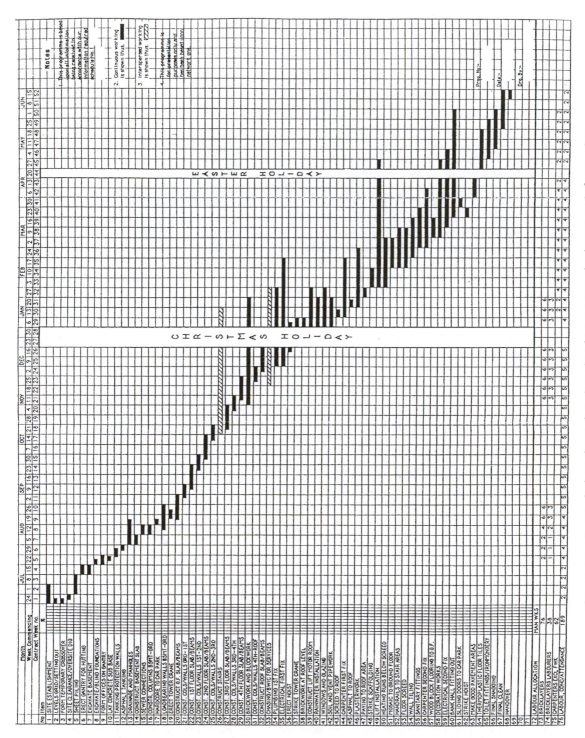

Figure 4.6 *Example bar chart showing simple resource deployment information*

varied construction operations which need to be reflected in a graphic representation where the inter-relationship of operations and their sequence are manifest. A common method of planning used to overcome the restrictions and limits of the bar chart, and also to augment use of the bar chart, is network analysis.

Networks

Network analysis is the generic name given to a group of planning techniques. Such techniques were originally conceived in the mid-1950s and applied to military developments. Network analysis is a useful construction planning method which allows interrelationships of operations to be clearly shown. This is not possible with the bar chart method as highlighted earlier. Networks are often presented as one of two types: as *'arrow diagrams'* or as *'precedence diagrams'*. The two methods essentially achieve the same goal, but with different means of depicting the information.

In simple terms, an arrow diagram consists of activities and events. An activity is a construction operation or process. All activities start and finish at an event, which is a point in time. It can be the junction of two or more activities. An arrowhead represents the end of the activity but the arrow length does not represent the duration of that activity. Activities are drawn in sequence representing the logical order of the construction operations, determined by those activities which must be completed before another can begin. Some activities can commence simultaneously, in which case a dummy activity is shown to indicate any dependency on other activities. In precedence diagrams a box, or node, represents the activity, with arrows linking the boxes and showing the logical relationship between the activities. The size of the boxes and length of the arrows bear no relationship to activity durations. Information concerning the activity duration, together with its start and finish dates and other relevant information, is shown in the box. Although similar, precedence diagrams have advantages over arrow diagrams, primarily in the amount and clarity of detail that can be conveyed within the diagram. The construction and analysis of both types of network are similar.

For a network diagram to work it is essential to study the sequence of operations very carefully. This should be undertaken by planning and construction personnel who have experience and a good understanding of the type and nature of the project. The whole of the project must be considered when preparing a network diagram. The first stage in the preparation of the network diagram is to determine all the activities that are required. The amount of detail that is looked at depends on which stage of the project the analysis is being carried out. A pre-tender analysis would not be as detailed, but could later be broken down into more activities as explained previously in this chapter.

The key steps in producing a network are:

(1) to determine the activities, i.e. the construction operations
(2) to determine the logical relationship between the activities
(3) to determine the duration of each activity, the start and finish times of each activity and the flexibility available between activities
(4) to determine the resources required to undertake the construction.

As each activity is considered the effect on other activities should be addressed, so that it is known which activity has to be completed before this activity can start and which other activities cannot start until this activity is completed. Also which activities have no logical relationship with this activity and can therefore take place at the same time. It is convention to show the duration of activity flowing from left to right. Dummy arrows, which are drawn in broken lines, are used to complete the logical links.

The next stage is to estimate the time required for each activity. This can be done from experience of the work, past records and work-study, as described earlier. This information can be added to the diagram. The duration can then be added to the diagram at the relevant activity arrow. The earliest start time of each event can be calculated and written in a box to the left of the event node by selecting the longest path. Once all these have been calculated the latest event times are calculated working from right to left, so that the latest time for the final event is the same as its earliest time, i.e. the project duration.

The critical path on the diagram is identified as the activities where earliest and latest event times are the same. The relationships of the other activities can also be seen and appreciated (activities with float).

On the non-critical activities, i.e. those which do not coincide with the critical path, a 'float' is highlighted. This is the spare time available within the various activities. Their effect on the arrow diagram is different. For instance, additional resources on this project could be obtained by an independent float which would not affect the rest of the network, whereas activities with interfering float would have a 'knock-on' effect on the float of other activities. Once this has been determined the resources can be checked to the different activities and duration can again be amended if resources are limited at any time along the project duration.

Network Analysis Diagrams draw attention to the activities on the programme which are on the critical path. This makes them very good for monitoring and controlling the progress of a project. It should be noted that although the critical activities are highlighted and warrant more attention, the non-critical activities should not be given less attention otherwise they will become the critical activity. For the sequence of steps involved in compiling a simple arrow network diagram see Figures 4.7 to 4.10.

For site use the network is usually converted into a bar chart which is used for short-term control and shows operations at scheduled times, as shown in Figure 4.11. The progress is plotted on the chart and then transferred back to the network to enable the main project to be updated. As an alternative the arrow diagram can be drawn to a time scale with the activities and scheduled times. Progress is then marked directly onto it and directly compared with the duration planned for each operation.

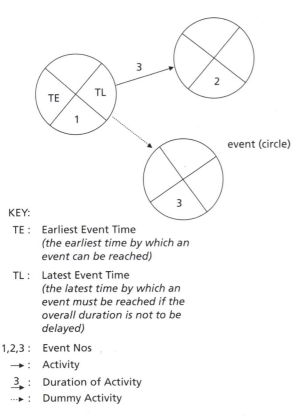

KEY:

TE : Earliest Event Time
(*the earliest time by which an event can be reached*)

TL : Latest Event Time
(*the latest time by which an event must be reached if the overall duration is not to be delayed*)

1,2,3 : Event Nos

→ : Activity

3 : Duration of Activity
→

⋯▶ : Dummy Activity

Figure 4.7 *Nomenclature for compiling an arrow network diagram*

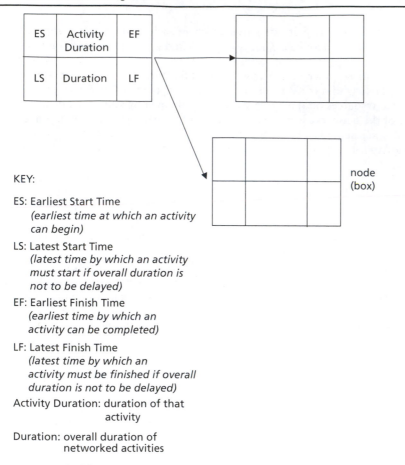

KEY:

ES: Earliest Start Time
   (earliest time at which an activity can begin)

LS: Latest Start Time
   (latest time by which an activity must start if overall duration is not to be delayed)

EF: Earliest Finish Time
   (earliest time by which an activity can be completed)

LF: Latest Finish Time
   (latest time by which an activity must be finished if overall duration is not to be delayed)

Activity Duration: duration of that activity

Duration: overall duration of networked activities

———▶ : Activity

Figure 4.8 *Nomenclature for compiling a precedence network diagram*

To Compile an Arrow Diagram:

(i)   identify the network arrangement
     (information from method statement and programme calculations)

(ii) number all 'event' circles and insert durations

(iii) insert TE and TL times
     (for TE, work from start to finish, insert largest figure)
     (for TL, work from finish to start, insert smallest figure)

(iv) identify 'critical path' and 'overall duration' for the project *(the critical path, or CP, is where TE = TL)*

(v)   convert the arrow diagram into a bar chart format

To Compile a Precedence Diagram:

(i)   identify the network arrangement *(as above)*

(ii) identify all activities as nodes *(boxes)*

(iii) insert ES, EF, LS, LF in the boxes
     (for ES, work from start to finish, insert largest figure)
     (for LF, work from finish to start, insert smallest figure)

(iv) identify the 'critical path' and 'overall duration' for the project

(v)   convert the precedence diagram into a bar chart format

Figure 4.9 *Steps in compiling arrow and precedence diagrams*

The re-analysis of networks should be carried out at regular intervals (weekly or fortnightly), so that delay on the critical activities can be seen and then action taken as necessary to correct this. Importantly, the frequency of re-analysis should be commensurate with the nature of the works and be focused towards the monthly project progress meeting where the contractual parties will review the progress of the works in detail.

There are several reasons why a programme (arrow diagram or precedence diagram) may need to be re-analysed to assess a delay. Some of these will be the fault of the contractor, others are beyond its control. Variations can have a greater effect on a project than simply increasing the quantities in the bill. They will probably have an effect on the contract duration, in which case the contractor would claim for an extension of time. From the network diagram the effects of the delay can be clearly seen, as activities that were non-critical may now have become critical as a result of a variation or delay. This will clearly show the knock-on effect for the whole programme including the new finish time.

Network analysis for major and complex construction projects is a detailed and exacting planning activity involving both manual and computer manipulation of operational project data. For greater depth of insight and applications the reader is directed to *Management Techniques Applied to the Construction Industry* (Oxley and Poskitt, 1996). In addition, a wide range of books provides comprehensive information on planning and programming (Pilcher, 1992; Harris and McCaffer, 1995; and Illingworth, 2000).

ACTIVITY	DURATION	PRECEDING EVENTS
A	10	–
B	4	–
C	7	A
D	9	B, G
E	6	C, D
F	2	D
G	12	A
H	8	E
J	20	–
K	5	J
Finish		H, F, K

Example network for small work project with 10 activities, given specified durations and preceding events

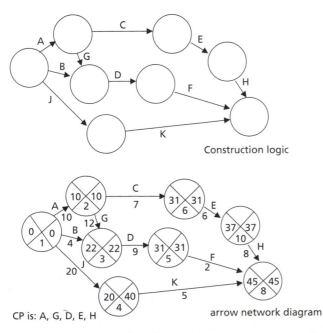

Construction logic

CP is: A, G, D, E, H

arrow network diagram

Figure 4.10 *Example of a simple arrow network diagram*

ACTIVITY	DURATION	WEEKS
A (1 to 2)	10	
B (1 to 3)	4	
J (1 to 4)	20	
G (2 and 3)	12	
C (2 to 6)	7	
D (3 to 5)	9	
K (4 to 8)	5	
F (5 to 8)	2	
E (6 and 7)	6	
H (7 and 8)	8	End 45 weeks
		0 10 20 30 40 50

Figure 4.11   *Conversion of an arrow diagram to a simple bar chart*

Computer Applications for Planning and Programmes

Computing applications to planning and programming are extensive nowadays, and many software packages are available to assist the construction planner. Software is user-friendly and can provide a variety of forms for presenting information. Bar charts and networks can be drawn on screen, linking many and complex activities and depicting the information in various formats to suit the nature of the application and user.

Software packages allow bar charts, arrow and precedence networks, resource scheduling, time and cost spreadsheets and tables to be configured relatively simply. In addition, information can be transferred from diagrams into project data for use in diaries and reports. Information can also be accessed easily and communicated rapidly through the use of e-mail systems. The advantages with such facilities are the speed at which complex plans can be developed and the ease with which they can be updated and modified to accommodate the dynamic nature of the construction processes.

It is not intended in this book to develop in-depth descriptions of specific computer applications or software packages. Other textbooks provide that information. However, a useful introduction is presented by Stephenson in *Management Systems for Construction* (Griffith *et al.*, 2000). In worked examples using Primavera Suretrax software, the task of generating a project network is explained together with the computer on-screen frames to highlight the plan's development.

The case study programme examples shown in Figures 4.12 to 4.15 were compiled using contemporary software and illustrate the type and clarity of detailing which can be typically achieved. Principal contractors will have their own methods of going about compiling their plans and networks, and use off-the-shelf or tailor-made software to suit their requirements. It would be inappropriate therefore to prescribe presentation software package applications and recommendations.

Case Study Planning Examples

Figure 4.12 shows a draft target programme in bar chart form for a major city centre office development project. It can be seen that the works comprise 100 key activities, broadly grouped into eleven major sections of work: preconstruction; design approval; preparation/enabling works; substructure; superstructure; envelope; services and finishes – core; services and finishes – core stairs;

Draft Target Programme

#	Title	Quantities	Resource
	PRE-CONSTRUCTION		
1	Award of Contract - 8th May 2000		
2	Contractor's Mobilisation		
3	Start on Site - 19th June 2000		
	DESIGN/APPROVALS/PROCUREMENT		
4	Planning Approval Granted		
5	Novate Design Team		
6	Design Development		
7	Building Regulations Submission		
8	Work Package Procurement		
	CONSTRUCTION		
	Preparation/Enabling Works		
9	Dilapidation survey		
10	Erect hoarding to Arundel Gate		
11	Erect demountable fencing to Union Street		
12	Establish site offices & welfare facilities		
13	Locate & mark existing services		
14	Clear site / remove trees / pay & display machines		
15	Initial setting out		
16	Grout mine workings		
	Substructure		
17	Bulk excavation to formation level inc. batters	380 m3	
18	Import stone fill / piling mat	963 m3	
19	CFA piling GL A-E / 10 towards GL A-E / 1	103 Nr	5 Nr / day
20	Test & trim piles		
21	FRC lower crane base		
22	FRC strip founds for ret. wall GL A	85m x 1m x 0.45m dp	50 m / wk
23	FRC ret. wall GL A inc. Access ramp	77 m2	25 m2 / wk
24	Tanking to ret. wall	77 m2	100 m2 / wk
25	Backfill to ret. wall		
26	FRC pile caps	49 Nr	2 Nr / day
27	FRC ground beams	333m x 1m x 0.65m dp	70 m / wk
28	Underslab drainage/service entries		
29	Prepare & FRC ground slab	1155 m2	300 m2 / wk
30	FRC column encasement		
31	Car park level blockwork		
	Superstructure		
32	Erect lower crane & commission		
33	Structural steelwork		
34	Metal decking	8407 m2	250 m2 / day
35	FRC suspended floors	8407 m2	
36	PCC stair units		
37	Dismantle lower crane		
	Envelope		
38	Independent scaffold		
39	Lift roofing materials		
40	Profiled metal roof coverings	868 m2	150 m2 / wk
41	Plantroom louvres	48 m2	
42	Flat roof coverings	953 m2	100 m2 / wk
43	Level 6 roof handrail	69 m	100 m / wk
44	Parapet coping	466 m	150 m / wk
45	Blockwork inner leaf	3710 m2	50 m2 / wk / b'layer
46	Facing brickwork	3710 m2	35 m2 / wk / b'layer
47	Wall cladding - Panels	990 m2	100 m2 / wk
48	Low level soffit	185 m2	200 m2 / wk
49	Car park entrance roller shutter	1 Nr	1 wk
50	Windows	616 m2	100 m2 / wk
51	Reflective glazing / curtain walling	602 m2	100 m2 / wk
52	Main entrance canopy		
	Services & Finishes - Core		
53	Blockwork partitions - Core / stairwells	6047 m2	50 m2 / wk / b'layer
54	Raised access floor / plywood lining	541 m2	100 m2 / wk
55	Services 1st fix installation		
56	Fire protection - board		
57	Joinery 1st fix inc. IPS framing		
58	Plumbing 1st fix installation		
59	Plaster / Render	250 m2	100 m2 / wk
60	Services 2nd fix installation		
61	Lift installation		
62	Joinery 2nd fix inc. vanity units / cubicles		
63	IPS panelling & sanitary ware	152 m2	
64	Ceramic tiling inc. entrance lobby floor	300 m2	50 m2 / wk
65	Suspended ceiling	323 m2	
66	Mist coat walls	250 m2	
67	Final fix all trades inc. reception desk		
68	Decoration	250 m2	
69	Vinyl floor coverings / Entrance mat	323 m2	
70	Test & commissioning		
71	Final clean		
	Services & Finishes - Core Stairs		
72	Services installation		
73	Joinery installation		
74	Plaster	600 m2	
75	Suspended ceiling	516 m2	
76	Handrail / Balustrades	86 m	
77	Decoration	600 m2	
78	Vinyl floor coverings	516 m2	
79	Final clean		
	Services & Finishes - Escape Stairs		
80	Block & screed built up floor	684 m2	
81	Services installation		
82	Joinery installation		
83	Handrail / Balustrades	264 m	
84	Make good stairs / Final clean		
	Services & Finishes - Open Plan Offices		
85	High level 1st fix services installation		
86	Plaster / Dry lining & skim	8137 m2	200 m2/wk/gang
87	High level 2nd fix services installation		
88	Under floor 1st & 2nd fix services installation		
89	Clients Access - Data Cabling 1st & 2nd fix		
90	Suspended ceiling	6303 m2	200 m2/wk/gang
91	Raised access floor	6376 m2	200m2/wk/gang
92	Joinery / Carpentry installation		
93	Decoration	8137 m2	
94	Final fix all trades		
95	Floor finishes	6349 m2	600m2/wk/br
96	Test & commission		
97	Final clean		
98	Access for Tenant Fit Out		
	External Works		
99	Drainage		
100	Hard/soft landscaping		

Seq Library 1 — Key data — Preconstruction — Design/Procure — Enabling Works — In Progress — Client Access

Notes:	Drawn by:	Date Drawn: 08-Jun-00	Programme No:	Sheet:

Figure 4.12 *Draft target bar chart programme*

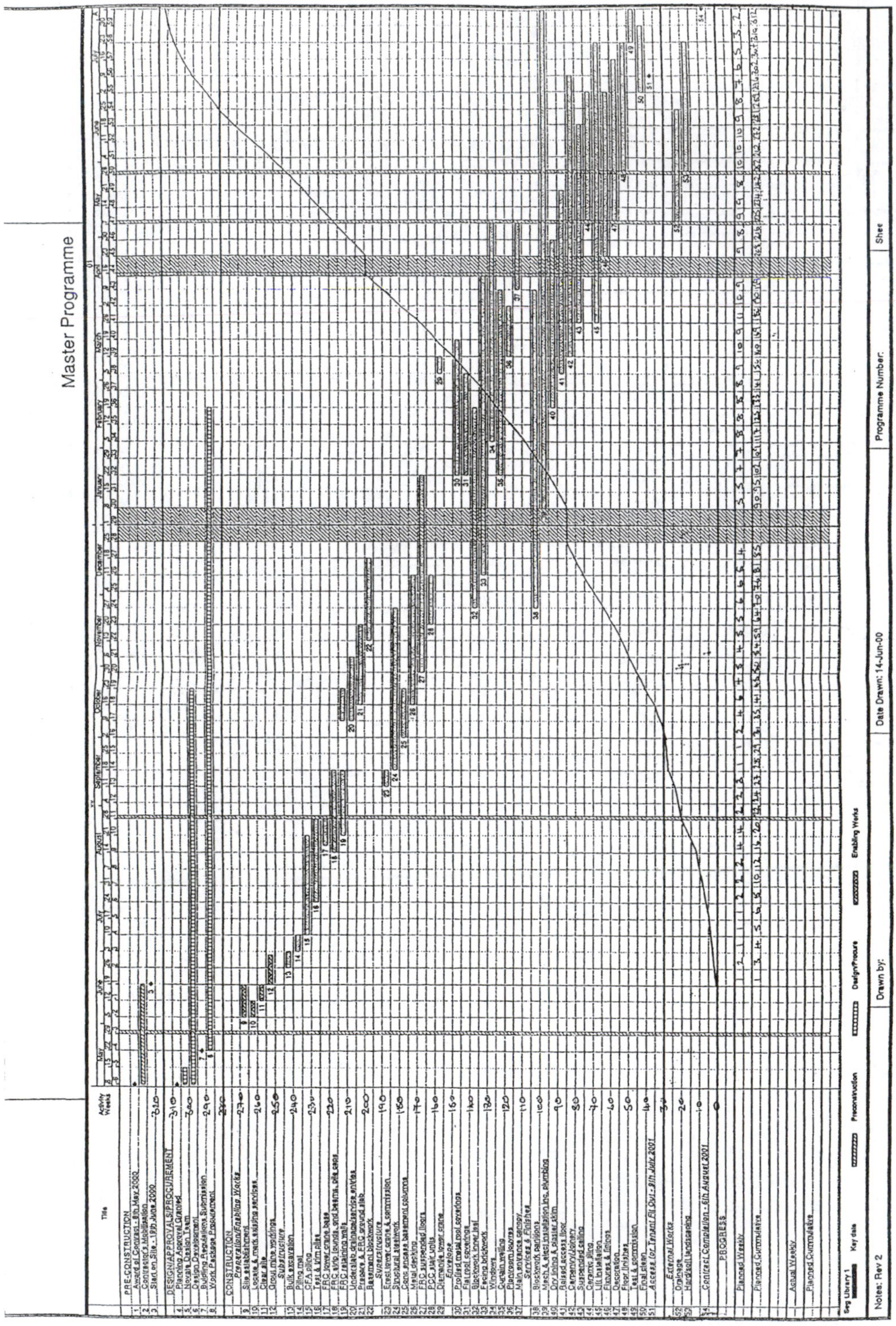

Master Programme

Figure 4.13 Example master programme

Prog Ref	Activity	I	j	Days	Type of Activity	Nett Completion Date	Actual Completion Date	Nett Completion Date	Actual Completion Date
1	Set up Site/Erect Hoardings	1	2	7	Normal	24/06/2000	24/06/2000	30/06/2000	30/06/2000
2	Setting Out	1	3	1	Normal	24/06/2000	24/06/2000	24/06/2000	24/06/2000
3	Form Access to Site	3	4	2	Normal	25/06/2000	25/06/2000	26/06/2000	26/06/2000
4	Prelim. Trench for Piling	4	5	4	Normal	27/06/2000	27/06/2000	30/06/2000	30/06/2000
5	Sheet Piling	5	6	16	Normal	01/07/2000	01/07/2000	16/07/2000	16/07/2000
6	Form Gantry for Offices	6	7	5	Normal	17/07/2000	17/07/2000	21/07/2000	21/07/2000
7	Excavation Works	6	8	7	Normal	17/07/2000	17/07/2000	23/07/2000	23/07/2000
8	Establish Offices on Gantry	7	8	7	Normal	22/07/2000	22/07/2000	28/07/2000	28/07/2000
9	Construct Sub Base/Founds	8	10	3	Normal	29/07/2000	29/07/2000	31/07/2000	31/07/2000
10	Brick Protection for Tanking	10	11	7	Normal	01/08/2000	01/08/2000	07/08/2000	07/08/2000
11	Drainage/Sewer Connections	10	12	21	Normal	01/08/2000	01/08/2000	21/08/2000	21/08/2000
		12	18	9	Lag	22/08/2000	22/08/2000	30/08/2000	30/08/2000
12	Work to Asphalt Tanking	11	13	7	Normal	08/08/2000	08/08/2000	14/08/2000	14/08/2000
13	Work to Basement Slab	13	14	7	Normal	15/08/2000	15/08/2000	21/08/2000	21/08/2000
14	Work to Columns in Basement	14	15	4	Normal	22/08/2000	22/08/2000	25/08/2000	25/08/2000
15	Loadbearing Bwk up to G.F.	14	18	14	Normal	22/08/2000	22/08/2000	04/09/2000	04/09/2000
		18	19	4	Lag	05/09/2000	05/09/2000	08/09/2000	08/09/2000
16	Construct High Level Founds	15	17	7	Normal	26/08/2000	26/08/2000	01/09/2000	01/09/2000
		17	18	0	Dummy	01/09/2000	01/09/2000	31/08/2000	31/08/2000
17	Hardcore/Conc. Car Park Slab	18	19	7	Normal	02/09/2000	02/09/2000	08/09/2000	08/09/2000
		19	20	7	Lag	09/09/2000	09/09/2000	15/09/2000	15/09/2000
18	Work to G.F. Slab and Beams	18	20	14	Normal	02/09/2000	02/09/2000	15/09/2000	15/09/2000
19	Cols and Walls Ground-First	20	21	7	Normal	16/09/2000	16/09/2000	22/09/2000	22/09/2000
20	Construct Upst and R.C.Walls	20	30	98	Normal	16/09/2000	16/09/2000	22/12/2000	22/12/2000
21	First Floor Slab and Beams	21	22	14	Normal	23/09/2000	23/09/2000	06/10/2000	06/10/2000
22	Cols and Walls First-Second	22	23	7	Normal	07/10/2000	07/10/2000	13/10/2000	13/10/2000
23	Construct Stairs	22	30	77	Normal	07/10/2000	07/10/2000	22/12/2000	22/12/2000
24	Second Floor Slab and Beams	23	24	14	Normal	14/10/2000	14/10/2000	27/10/2000	27/10/2000
25	Cols and Walls Second-Third	24	25	7	Normal	28/10/2000	28/10/2000	03/11/2000	03/11/2000
26	Third Floor Slab and Beams	25	26	14	Normal	04/11/2000	04/11/2000	17/11/2000	17/11/2000
27	Cols and Walls Third-Fourth	26	27	7	Normal	18/11/2000	18/11/2000	24/11/2000	24/11/2000
	Cure and Strike 2nd Floor	24	32	21	Normal	28/10/2000	28/10/2000	17/11/2000	17/11/2000
28	Superstructure Brickwork	32	33	63	Normal	18/11/2000	18/11/2000	19/01/2001	02/02/2001
		32	34	21	Lead	18/11/2000	18/11/2000	08/12/2000	08/12/2000
		33	35	14	Lag	03/02/2001	03/02/2001	16/02/2001	16/02/2001
29	Fourth Floor Slab and Beams	27	28	14	Normal	25/11/2000	25/11/2000	08/12/2000	08/12/2000
30	Cols and Walls Fourth-Roof	28	29	7	Normal	09/12/2000	09/12/2000	15/12/2000	15/12/2000
		29	30	7	Lag	16/12/2000	16/12/2000	22/12/2000	22/12/2000
31	First Fix Services	34	35	56	Normal	09/12/2000	09/12/2000	02/02/2001	16/02/2001
		34	41	28	Lead	09/12/2000	09/12/2000	05/01/2001	19/01/2001
		35	42	28	Lag	17/02/2001	17/02/2001	15/03/2001	15/03/2001
32	Roof Slab and Beams	29	31	14	Normal	16/12/2000	16/12/2000	29/12/2000	12/01/2001
		31	33	21	Lag	13/01/2001	13/01/2001	02/02/2001	02/02/2001
		29	36	7	Lead	16/12/2000	16/12/2000	22/12/2000	22/12/2000
		31	37	14	Lag	13/01/2001	13/01/2001	26/01/2001	26/01/2001
33	Rainwater Pipework	36	37	21	Normal	23/12/2000	06/01/2001	26/01/2001	26/01/2001
		37	39	0	Dummy	26/01/2001	26/01/2001	25/01/2001	25/01/2001
34	Work to Lift Room	31	38	28	Normal	13/01/2001	13/01/2001	09/02/2001	09/02/2001
		31	39	14	Lead	13/01/2001	13/01/2001	26/01/2001	26/01/2001
		38	40	7	Lag	10/02/2001	10/02/2001	16/02/2001	16/02/2001
35	Glazing/Curtain Walling	31	59	28	Normal	13/01/2001	13/01/2001	09/02/2001	09/02/2001
	Period for Scaffold Strike	59	60	21	Normal	10/02/2001	10/02/2001	01/03/2001	01/03/2001
36	Carpentry First Fix	41	42	56	Normal	20/01/2001	20/01/2001	15/03/2001	15/03/2001

Figure 4.14 *Data spreadsheet for compiling network programme*

(continued overleaf)

Prog Ref	Activity	I	j	Days	Type of Activity	Nett Completion Date	Actual Completion Date	Nett Completion Date	Actual Completion Date
		41	43	14	Lead	20/01/2001	20/01/2001	02/02/2001	02/02/2001
		42	44	14	Lag	16/03/2001	16/03/2001	29/03/2001	29/03/2001
37	Screed/Asphalt to Roof	39	40	21	Normal	27/01/2001	27/01/2001	16/02/2001	16/02/2001
38	Plastering	43	44	56	Normal	03/02/2001	03/02/2001	29/03/2001	29/03/2001
		43	45	14	Lead	03/02/2001	03/02/2001	16/02/2001	16/02/2001
		44	46	14	Lag	30/03/2001	30/03/2001	12/04/2001	12/04/2001
39	Lift Installation	38	63	84	Normal	10/02/2001	10/02/2001	03/05/2001	10/05/2001
		63	58	42	Lag	11/05/2001	11/05/2001	21/06/2001	21/06/2001
40	Heating Element under Screed	45	46	56	Normal	17/02/2001	17/02/2001	12/04/2001	12/04/2001
		45	47	21	Lead	17/02/2001	17/02/2001	08/03/2001	08/03/2001
		46	48	7	Lag	13/04/2001	13/04/2001	19/04/2001	19/04/2001
41	Mosaic at Ground Floor Level	60	61	14	Normal	02/03/2001	02/03/2001	15/03/2001	15/03/2001
		43	64	28	Lead	03/02/2001	03/02/2001	01/03/2001	01/03/2001
42	Stair Balustrading	64	65	28	Normal	02/03/2001	02/03/2001	29/03/2001	29/03/2001
		65	48	21	Lag	30/03/2001	30/03/2001	19/04/2001	19/04/2001
43	Screed to Floors and Stairs	47	48	42	Normal	09/03/2001	09/03/2001	19/04/2001	19/04/2001
		47	49	14	Lead	09/03/2001	09/03/2001	22/03/2001	22/03/2001
		48	50	28	Lag	20/04/2001	20/04/2001	17/05/2001	24/05/2001
44	Sliding Doors to Car Park	61	62	14	Normal	16/03/2001	16/03/2001	29/03/2001	29/03/2001
		62	58	77	Lag	30/03/2001	30/03/2001	14/06/2001	21/06/2001
45	Carpentry Second Fix	49	50	56	Normal	23/03/2001	23/03/2001	17/05/2001	24/05/2001
		49	51	7	Lead	23/03/2001	23/03/2001	29/03/2001	29/03/2001
		50	52	7	Lag	25/05/2001	25/05/2001	31/05/2001	31/05/2001
		49	66	7	Lead	23/03/2001	23/03/2001	29/03/2001	29/03/2001
46	Wall Tiling to Toilets	66	67	28	Normal	30/03/2001	30/03/2001	26/04/2001	03/05/2001
		66	68	7	Lead	30/03/2001	30/03/2001	05/04/2001	05/04/2001
		67	69	7	Lag	04/05/2001	04/05/2001	10/05/2001	10/05/2001
		69	52	21	Lag	11/05/2001	11/05/2001	31/05/2001	31/05/2001
47	Decoration Works	51	52	56	Normal	30/03/2001	30/03/2001	24/05/2001	31/05/2001
		51	53	14	Lead	30/03/2001	30/03/2001	12/04/2001	12/04/2001
		52	54	7	Lag	01/06/2001	01/06/2001	07/06/2001	07/06/2001
48	Plumbing Fittings	68	69	28	Normal	06/04/2001	06/04/2001	03/05/2001	10/05/2001
49	Electrical Fittings/2nd Fix	53	54	49	Normal	13/04/2001	13/04/2001	31/05/2001	07/06/2001
		53	55	21	Lead	13/04/2001	13/04/2001	03/05/2001	10/05/2001
		54	56	7	Lag	08/06/2001	08/06/2001	14/06/2001	14/06/2001
50	Bank Hall Fitting out	70	58	63	Normal	13/04/2001	13/04/2001	14/06/2001	21/06/2001
51	Thermoplastic Flooring	55	56	35	Normal	11/05/2001	11/05/2001	14/06/2001	14/06/2001
		55	57	21	Lead	11/05/2001	11/05/2001	31/05/2001	31/05/2001
		56	58	7	Lag	15/06/2001	15/06/2001	21/06/2001	21/06/2001
52	Final Snag and Clean	57	58	21	Normal	01/06/2001	01/06/2001	21/06/2001	21/06/2001
	Contractors Contingency	58	69	56	Normal	22/06/2001	22/06/2001	16/08/2001	16/08/2001

Figure 4.14 *continued*

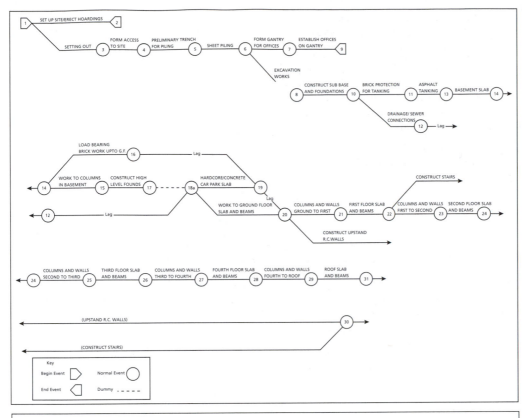

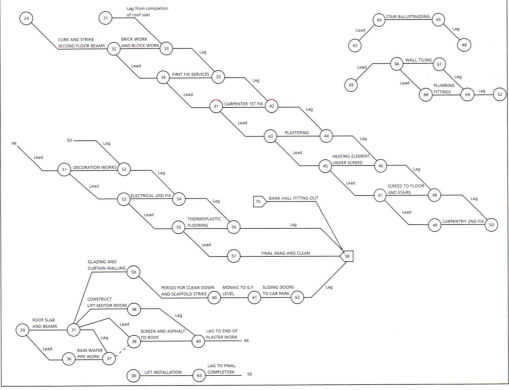

Figure 4.15 *Example network programme*

services and finishes – escape stairs; services and finishes – open plan offices; and external works. For each key activity, the quantity and work targets are given, leading to the calculation of the durations shown as linked bars on the calendar time line section of the programme. The durations bars are related directly to the programme week number and calendar dates shown at the head of the programme. Key project events such as start-on-site, contractor's mobilisation and access for tenant are shown as diamond symbols. Various different activities may be highlighted using a legend of shading, or hatching, illustrated at the base of the programme. Periods of inactivity, such as holidays, are shown as shaded vertical bars.

This bar chart would have been compiled at the pre-contract stage in readiness for producing the contract, or master, programme. Its purpose is to schedule, in draft form, the key construction activities in groups of operations in order to determine anticipated links in activity durations and critical target dates. Once this has been undertaken, a master programme is developed as shown in Figure 4.13. It is evident from the master programme that key activities were compressed into a smaller number of grouped operations for ease and efficiency of scheduling. While short-term planning and programmes would augment the master programme to provide greater depth of detail, the master programme does provide a useful holistic perspective on the project. Furthermore, the master programme provides the basis for progress monitoring, charting and control, for which the concepts, principles and practice are examined in Chapter 8: Progress and Control.

The second example shown is an application of arrow network analysis for a small office development. Figure 4.14 presents the necessary data required for compiling the network. Information for the data spreadsheet shown will have been acquired from the bills of quantitites, method statement and programme calculation sheets, as previously explained. Using the nomenclature for arrow diagram compilation as explained earlier, the earliest event times and latest event times, shown as 'i' and 'j' on the spreadsheet, are inserted along with durations, shown in days, for each construction activity. Dates are calculated automatically by the software programme to produce the completed data spreadsheet illustrated. From this data the network analysis diagram is compiled and plotted using the arrow diagram convention explained previously. The resulting network is shown in Figure 4.15. The final stage in data presentation would be to translate the network into a simple bar chart format for ease of communication during the siteworks. Such a bar chart was described previously and was shown in Figure 4.6.

Summary

A fundamental requirement on all construction projects is the capability to plan, monitor and control the siteworks. At the heart of the planning process is the establishment of a robust construction programme which depicts the particular methods of working and the sequence of activities for the project. This chapter has illustrated the importance of planning, examined the tasks involved in compiling various types of construction programmes and explained the link between the planning function and methods for controlling progress of the works. The concept of planning and the associated concept of progressing are absolutely essential to effective time management on all construction projects.

References

Griffith A., Stephenson P. and Watson P. (2000) *Management Systems for Construction*. Longman, Harlow, ISBN 0–582–31927–7.

Harris F. and McCaffer R. (1995) *Modern Construction Management*. Blackwell, Oxford.

Illingworth J.R. (2000) *Construction Methods and Planning*. Spon, London.

Oxley R. and Poskitt J. (1996) *Management Techniques Applied to the Construction Industry*. Blackwell, Oxford.

Pilcher R. (1992) *Principles of Construction Management*. McGraw-Hill, London.

5 Operations, Cost Planning and Monitoring Expenditure

Introduction

Within this chapter the concept of linking the vital organisational component of 'operations' into the 'corporate decision-making' process is advocated. This is done in order to stress that a 'sustainable competitive advantage' is obtained only if operational functions are linked to strategic decisions and utilise concepts such as Financial Planning and Budgeting Control.

Further, the chapter seeks to establish the importance of cash-flow management. This is one of the most crucial activities undertaken by construction companies. However, the above have not been set out in isolation as they form part of a holistic approach to 'control'. Effective and efficient control is dependent on regular monitoring, and this aspect is demonstrated by way of examples. The key activities of cash-flow analysis, budgetary control and variance analysis incorporated in this chapter, provide a coherent systematic approach for construction managers to engage in the decision-making process.

Operations Strategy

Operations management is concerned with how an organisation delivers its service or product. The operations function should not be a loosely knit aggregation of disparate functions but rather a coherent focused unit. Whether a company is producing goods or services, or a mix of both, there is clearly an operations function. This function should provide the host organisation with a competitive weapon and not be a corporate millstone detracting from its competitive edge.

In a construction company there exists an exceptionally heavy investment of people and resources in the production and processes deployed. For example, a construction organisation employing 250 people would have approximately 210–220 of them active in the operations function. Therefore they consume most of the capital investment and are responsible for the largest amount of fixed and variable costs incurred. Thus it is obvious that the way a business chooses to produce its products or services has a critical effect on the strategies that it can subsequently adopt and the markets in which it can effectively compete. The essential issue is that the choice of processes inevitably creates a range of limitations and trade-offs. These trade-offs impact upon products/services, investment, costs, organisation, controls, and business and market operations. It is the nature of these trade-offs which needs to be clearly understood by senior management if a business is to develop profitably by maintaining/gaining a sustainable competitive advantage and reduce potential conflict between corporate and operational planning. There are various conversion processes available for the production of products or services. The following provides a succinct summary of the options available.

Productive Systems

Project

Organisations such as construction which are in the business of producing or providing one-off, large-scale products or services will normally choose the project process as being the most suitable. Product examples include building contracts; service examples include strategy-based consultancy assignments. The product or service is unique and it is large and made to meet the customer's own requirements. Normally, the necessary resource inputs are taken to the point where the product is to be produced, e.g. the building site. The reason for this is simply that it is not practical to move the product once it has been built or to establish the service anywhere but where it is to be implemented and used. The nature of the product or service dictates the process employed.

The product has therefore to be built on site and the service built around (and oriented towards) the needs of the client's business. The operations task involves not only the effective allocation of resources to the site and then their redistribution to other jobs, but also the co-ordination of the very large number of interrelated activities and resources involved in a way which is both efficient and which also meets the delivery requirements of the customer. The operations task is concerned with both the efficient and effective utilisation of resources.

Jobbing, Unit or One Off

Organisations which choose jobbing, unit or one-off processes as being appropriate will once again be providing products or services of an individual nature and made to the customer's specification. The difference as compared with the process choice of project is that in this case the resources remain on the same site and the products/services flow through them. On completion the products or services (now small enough to be moved) are then transported to the customers to meet their agreed delivery requirements.

As it is a one-off provision, this means that the items will not be repeated – or if they are, the repeat order is both uncertain and over such a long time period that the organisation cannot avail itself of the opportunities associated with repetition.

Batch

This form of process will be chosen when the volume of the product or service has increased and sales of these items are repeated. This means that the business can justify investment in capturing data in deciding how best to complete the tasks required and in the processes necessary to make the product or provide the service. The job itself is divided into a series of appropriate operations. An order quantity is then put through the process which is necessary to complete the first stage. The items then go on to the next operation and so on, until it is completed. An example is component manufacture or fabrication, e.g. roof trusses.

Line

With further increases in volume, additional investment beyond that required for batch processes can be justified in order to dedicate a process to completing one or a small range of products or services. The process is arranged so that the sequence of operations necessary to complete the predetermined range of products or services which the line is designed to accommodate takes place. Each item to be completed then moves from one stage to the next. In product examples, the classic task is a series of assembly operations either to complete sub-assemblies to go into a product or the assembly of the final product itself; the choice of line processes to complete services is relatively infrequent.

Continuous Process

The choice of continuous process is made where the volumes have increased to such an extent that an inflexible, dedicated process is laid down to handle an agreed range of products. The process is designed to run all day and every day, with minimum shutdowns because of the high cost of the starting-up process. This type of process is not used in the provision of a service and would not normally be used in a small business owing to the sales turnover (£) implications involved to justify the high capital investment which goes hand-in-hand with this process choice.

Choice of Process

The choice of project as the appropriate way of providing a product/service is strongly influenced by the product or service itself. As a consequence, there is little possibility of a rationale for a business to move from the project type to one of the other forms of process. However, this is not so with jobbing, batch and line processes. As shown in Figure 5.1, there is a feasible and rational transition from jobbing to low-volume batch to line which is determined by volume changes in demand. As volume increases a different process methodology becomes more economically viable.

Choice between alternatives is, therefore, available to a business and its significance cannot be overemphasised. The importance attached to this decision concerns the fact that each choice has a different set of implications for a business and that a desire to change these is both costly and time-consuming. Consequently, it is essential to take account of the implications associated with each choice as an important element of this critical business decision.

Also the concept of 'learning curve theory' is worth consideration in the decision-making process. Figure 5.2 establishes that the costs of production or service delivery reduce as a result of the increased frequency with which the product or service is provided. Provision of 'x1' has a higher cost because of a reduced quantity; however, 'x2' has a reduced cost because of a higher volume of provision. Thus the more units or times a service is provided, an organisation becomes more proficient in its delivery.

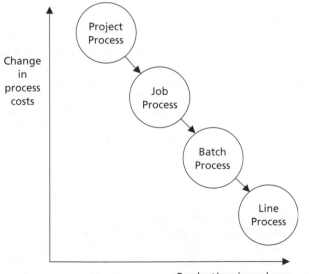

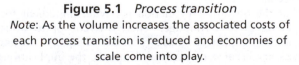

Figure 5.1 *Process transition*
Note: As the volume increases the associated costs of each process transition is reduced and economies of scale come into play.

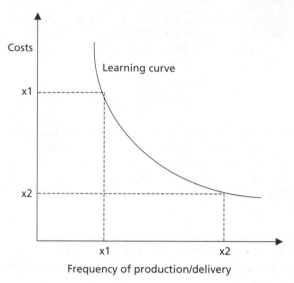

Figure 5.2   *Application of learning curve theory*

Managing Resources

Organisational resources are composed of the 5 'M's:

 (i) Materials
 (ii) Manpower
(iii) Machinery
 (iv) Money
 (v) Management.

All must be available in the required quantity. It is a question of obtaining the right mix so that a large quantity of materials does not lie idle, which leads to locked up capital.

The effective and efficient control of production operations (which may account for up to 80% of resources expenditure within a construction business) is of the utmost importance. In most enterprises it does provide a competitive weapon which should be related to the distinctive competence of the company. The need to control and reduce costs within the operations management function is essential to both the short- and long-term viability of the business. To increase profit we can:

 (i) increase the selling price subject to elasticity of demand
(ii) reduce overheads.

Figure 5.3 provides a pictorial representation of the operations conversion process.

It is vital that the operations department of any construction company is fully understood by the people determining the corporate strategy for the organisation. If the concept of having an 'operations strategy' is not enacted the result could be a mismatch between corporate objectives and the capacity of operations to provide/meet set objectives. Figure 5.4 indicates how the operations strategy should be incorporated into the corporate planning process.

It has been explained in this section that the operational function is a vital ingredient for the attainment/maintenance of a sustainable competitive advantage. Therefore corporate decisions must not impose demands/constraints that lead to a strategic move from the company's distinctive competence. Should a change process be necessary because of external demands, the implications for

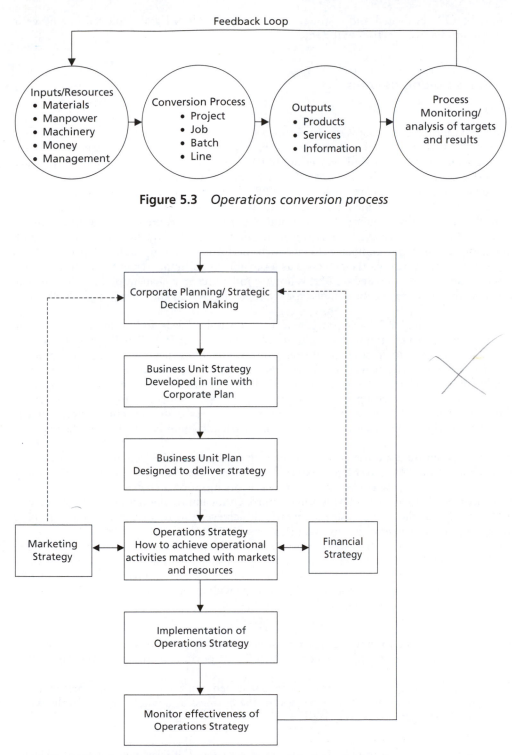

Figure 5.3 *Operations conversion process*

The broken lines depict that the planning process is both top down and bottom up

Figure 5.4 *Corporate planning process model*

operations must be fully evaluated and a strategy for change developed and deployed but never in isolation from the productive function of the enterprise.

Business Process Re-engineering

Introduction

We have already established that organisational performance and the gaining of a sustainable competitive advantage are not mutually exclusive and are issues of great importance to construction companies. By adopting a Business Process Re-Engineering (BPR) philosophy improvements in organisational performance can be obtained. BPR is applicable to the UK construction industry and some of the more innovative construction companies have implemented BPR within their organisations. These firms have obtained the benefits of BPR deployment and there is no fundamental rationale advocating against its applicability to the construction industry. The aims and objectives of BPR are similar to other existing management tools designed to achieve organisational success in respect of quality, cost and time. BPR provides a model for construction companies to view their performance from a different perspective. It highlights the importance of treating the processes as a possible major improvement element by applying the concept of radical process redesign. BPR is not a stand-alone management philosophy, it has to be incorporated with other management tools, for example TQM and the BS EN ISO 9000: 2000 series.

Latham (1994) advocated a 30% improvement in productivity levels and an important element in obtaining this improvement is the re-designing of processes. Interest in BPR will continue to grow as the UK construction industry focuses on process improvements as a methodology for obtaining higher levels of corporate performance. The re-engineering of the construction process(es) requires the development of a co-operative culture between industry professionals in which the aim is to meet the project objectives by co-operation and team building as opposed to a confrontational approach. To achieve this, all team members require an awareness of the benefits appertaining to win–win professional relationships. Everyone, including the client, would benefit by establishing such a relationship.

One of the most important requirements for achieving a successful BPR deployment process is in the acceptance of ideas from both internal and external customers. The main criterion for being a learning organisation is establishing the correct attitude towards organisational change.

Business Process Re-Engineering (BPR) is a new approach for improving business performance by seeking to radically change the way a business operates. It achieves this by focusing on the core business process(es) that deliver value to the customer (Davenport, 1993).

Drucker (1992) considers that *"Re-engineering is new and it has to be done."* Recently, the differences between TQM and BPR have become a 'hot' topic for construction researchers. In general, TQM can be considered as a philosophy for obtaining continuous organisational improvement, whereas BPR can be defined as a management philosophy designed to provide process innovation. The following compares and contrasts the differing philosophies:

- In terms of organisation performance goals, BPR seeks to obtain dramatic improvements whereas TQM pursues an appropriate improvement which is proportionate to the committed effort (Davenport, 1993)
- In terms of organisational focus, BPR is more concerned with the process-orientated tasks. However, TQM aims to improve the effectiveness of the business as a whole (involving the whole organisation and every activity at every level).

Many industries world-wide have found BPR to be an effective and efficient approach to achieving dramatic improvements in production time, costs and quality, yet little attention has been paid to its potential for improvements within the construction industry (Mohamed and Tucker, 1996). BPR is much more than just a cost-cutting or a restructuring technique.

Many successful BPR applications have been reported in areas of business management, public administration and service industries. A typical scenario for BPR applications usually begins with a decision to improve business processes, in line with the vision and goals of the organisation. Following this, business processes are evaluated, critical success factors are determined and a plan for new process deployment is developed. Based on the plan, changes are instigated in the company, leading ultimately to more efficient and effective processes. Such applications, however, limit the definition of BPR to merely introducing changes in the structure of an organisation with a view to accomplishing some identified business objective(s).

An industry, such as construction, represents a major challenge to the implementation of BPR. This is mainly due to the complexity of business relationships and the key role that external factors play in how construction organisations conduct their business activities (Mohamed and Tucker, 1996).

The philosophy of re-engineering has a high positive correlation with innovative learning organisations and is in principle well suited to a project-orientated industry like construction (Betts and Wood-Harper, 1994).

Throughout this section the term BPR refers to re-engineering the entire construction process, i.e. redesigning the pathway by which a conceptual idea is turned into a constructed facility. This definition encapsulates the many phases of the construction process – design, management, procurement, construction, through to the hand-over of the facility to the client or end user.

The Importance of BPR to the UK Construction Industry

The Nature of the Construction Industry

The construction industry is responsible for approximately 6% of the UK's Gross Domestic Product (GDP). Companies engaged in the construction industry have to face many challenges in their operational environment. Clients require more flexibility, shorter project duration times, improved quality levels and lower costs. Business processes in the construction industry are more complex, less structured, subject to rapid change and more difficult to control than in most other industries. In many process(es) a large number of different companies are involved. Therefore the optimal design and management of business process(es) become a critical success factor for the UK construction industry.

The complex structures that connect the wide range of industry partners – architects, engineers, project managers, quantity surveyors, contractors, subcontractors and material suppliers – have limited the ability of the industry to be innovative. The usual existing adversarial relationships between construction organisations, during a project, add greater complexity to the achievement of inter-organisational co-operation, leading to a lack of synergistic advantages.

The construction industry has sometimes resisted changes to the way in which design, tendering, contracting and construction have been conducted. Construction practice has usually been dominated by fragmentation of control and conflicting interests. The former is due to the nature of the industry, where a large number of participants form a temporary group for the construction of a particular project and then disband after project completion, while the latter arises as a result of having different sets of goals and priorities among those participants.

Another remarkable feature of the construction industry is its lack of performance consistency. Project performance is highly dependent on individual circumstances associated with the project. The effectiveness of communication between design and construction teams, for example, can noticeably affect the project performance: "*A variety of other factors such as construction management effectiveness, client sophistication, procurement system adopted, and many more, have different degrees of impact upon performance*" (Walker, 1995); "*Despite the claim that many projects have been completed on schedule and within budget, it is widely believed by construction professionals that estimated schedule and budget targets do not necessarily reflect the actual or required time and cost. This is due to the inherent inefficiency of current practices and mechanisms of the industry which inevitably leads to time wasting*" (Mohamed and Tucker, 1996).

Construction activities in general are divided into sequential activities, which are allocated to different specialists for execution, putting unnecessary constraints on the workflow, increasing the possibility of conflicts and consequently leading to time-consuming corrective actions (Koskela, 1992).

Sir Michael Latham has investigated the procurement and contractual arrangements in the UK construction industry. He emphasises the importance of the construction process as a means of fulfilling the mission statement to reduce average costs. The following are the aspects of the construction process requiring action (Latham, 1994):

- Teamwork on site
- Organisation and management of the construction processes
- Design process, e.g. co-ordinated project information
- Procurement and contractual arrangement
- Tendering process
- Dispute resolution.

The above aspects have been advocated as a means for enhancing the processes of construction, thus providing products that represent value for money to the client. Latham (1994) also suggested that the industry should *"Support the 're-engineering' of the structure and processes of the industries so as to improve their profitability while sharing the benefits with clients and subcontractors throughout the supply chains."*

Efforts to accelerate construction projects by overlapping design and construction activities have been developed and fast-tracking is a typical example, where the overall construction time is reduced by starting construction activities before the design is finalised. This approach does not necessarily lead to an optimum design, because associated construction costs can be higher and the quality is often not as originally specified.

"Instead of embedding outdated processes in silicon and software, we should obliterate them and start over. We should 're-engineer' our businesses – use the power of modern information technology (IT) to radically redesign our business processes in order to achieve dramatic improvements in their performance" (Hammer, 1990).

Attempts to improve performance via concepts adopted from manufacturing, such as quality assurance, automation, prefabrication and standardisation, have been successful in providing some improvement in construction output. However, these concepts alone have failed to radically address the root causes of the existing problematic issues.

Competitive Advantage

In today's competitive operational environment, nothing is constant or predictable, for example market growth, customer demand, product life-cycle, the rate of technological change and the nature of competition are all dynamic. This implies a necessity to revise the culture and style of managing organisations in order to adapt to external influences and hence gain a sustainable competitive advantage.

Successful companies are those that have drastically changed or re-engineered their business processes. Management must re-evaluate their old notions of how business activities should be organised/conducted. They may be required to abandon the organisational and operational principles and procedures they employ and create entirely new ones. Their existing procedures may be based on assumptions related to technology, people and organisational goals which may no longer hold true and hence lack validity in a post-modernist operational environment.

Through the implementation of radical changes to current processes and workflow, the philosophy of BPR is playing an important role in the UK construction industry. It secures a holistic approach to re-engineering, integrating the concepts of concurrent, lean construction and process redesign into construction processes. This leads to the attainment of a competitive advantage by reducing cost increases in such areas as development, construction and overheads, thus enabling UK construction companies to become more competitive. We should not forget that one of Sir Michael Latham's (1994) mission statements is: *"to assist UK construction industries to become and to be seen as internationally competitive."*

Radical Improvement

To meet the requirements of demanding customers, organisations must engage in a radical change to their established ways of working. Radical does not mean a 5% improvement, it means reducing lead time from six weeks to two days or reducing tender prices by 30%.

Hammer and Champy (1993) have stated that "*If we want to reduce the production period from six months to three months, there is no way this can be achieved other than by adopting the concepts of radical process redesign.*"

Process Focus

For two hundred years people have built companies around Adam Smith's brilliant discovery that industrial work should be broken down into its simplest and most basic tasks. In the post-modern business age, corporations should be built around the idea of reunifying those tasks into coherent business processes (Hammer and Champy, 1993).

Re-engineering encourages managers to concentrate on the processes of an organisation rather than its disparate functions. Typically processes cut across business functions – the often skill-based activities such as developing, constructing, finance, sales and marketing around which most organisations are grouped. Hence, the critical challenge here is to focus on the end-to-end processes as seen by the client.

Customer Focus

Today's customers are concerned with quality issues because of the accessibility of information relating to construction products and services. This influences customer requirements in respect of quality and project duration; they tend to require products or services in the shortest time period possible. Establishing customer focused targets means speaking to the customers directly or through surveys in order to ascertain their real needs. (This is now a requirement under BS EN ISO 9001: 2000.) Targets must be sufficiently radical to cause a real change in the customers' perception of value and to establish organisations with a real competitive edge.

With regard to the application of BPR, much of the focus within companies such as Ford, IBM, Hallmark, Bell Atlantic and Capital Holdings, where empirical studies have been documented, has been centred on the issues of customer ordering and sales dealing. These models could be most useful to construction managers given the multiple parties typically involved in construction projects and the observed increasing preoccupation of the research community in tendering and procurement systems (Betts and Lansley, 1993).

Construction companies have to be open to ideas related to organisational improvement; they need to be learning organisations. They have to accept the new management tools and procurement systems that have proved to be most advantageous to the manufacturing industry. These tools and methods of operation can assist in the differentiation of a host organisation from its competitors in the construction environment thus ensuring its long-term survival.

The need for construction organisations to rethink and redesign their operational processes is primarily driven by the following pressures (Lover and Mohamed, 1995):

- Globalisation of the economy
- Greater performance expectations from clients
- Greater competition among domestic organisations
- Continued restructuring of work practices
- Industrial relations
- Enterprise bargaining.

For construction businesses to survive in an ever changing world, they must completely rethink how and why they do what they do. Every construction organisation should recognise the importance of reassessing their operational systems and strategic goals.

Table 5.1 *Characteristics of corporations by category (source: Grover and Kettinger, 1995)*

	Category 1: Industrial Age	Category 2: Information Age
Focus	• cost driven • cost reduction • efficiency	• vision driven • alignment (strategy, people, processes, technology)

Process architecture	• rework existing processes
	  – TQM
	  – incremental improvement
		• new 'clean slate' processes
		• transforming

Organisation architecture	• restructuring • downsizing • outsourcing • functional or flattened • alliances	• cross-functional • team based • 'virtual corporation' • partnering
Information architecture	• new technology for cost reduction and control • workflow computing • co-operative processing • intra-organisational networks • corporate database	• new infrastructure and architecture to exploit new processes • client or server • inter-organisational networks • 'human-centred' information

Information Age

There are two categories of company in respect of organisational focus and process architecture: organisation architecture and information architecture. A category 1 company (Industrial Age) will tend to view change within the existing Industrial Age Model. Category 2 companies (Information Age) will tend to view change as a means of achieving the Information Age Model. A summary and comparison of these two categories, in terms of their positions related to BPR, are shown in Table 5.1.

Toffler (1990) suggests that *"The source of power is moving from wealth in the Industrial Age to knowledge in the Information Age."* Today's organisations should treat the information system as one of the most important elements pertaining to a company's strategic capability.

Emerging Technologies

Koskela (1992) and Hammer and Champy (1993) see information technology as providing an enabling role once processes have been re-engineered to eliminate non-value-adding activities. Areas of application include (Koskela, 1992):

- Making processes more transparent by visualisation and simulation; this is beneficial to all parties, including the client, designer, contractor, product manufacturer and occupier
- Using knowledge-based systems for systematising and standardising operations and as error-proofing devices (such as checking regulations)
- Using knowledge-based systems for providing advice (such as constructability guidance to designers).

Venkatraman (1994) considers that *"The role of IT in shaping tomorrow's business operations is a distinctive one. IT has become a fundamental enabler in creating and maintaining a flexible business network of inter-organisational arrangements – joint ventures, alliances and partnerships, long-term contracts and marketing agreements."*

Technological changes, including ever more rapid telecommunications as well as the cost of capital and the ever changing demands of customers, are leading to such an environment. This makes it necessary for a company to have a constant improvement strategy (Carr and Johansson, 1995).

Benchmarking

When a company expands, the organisation has to consider whether the increased workload can be managed by the current process capability. BPR can play an important role by benchmarking the existing operational processes. It can also aid in the redesigning of work processes by adding or deleting some activity/process in order to achieve the most efficient and effective workflow for a particular task.

It is very important that an organisation understands its current situation by identifying the value-added elements of existing practices and measuring the performance of the business process. It may be necessary for a business to benchmark itself not only against the competition but also against the best-in-class via a generic benchmarking activity.

Latham (1994) has recommended that *"The government should commit itself to being a best practice client. It should provide its staff with the training necessary to achieve this and establish benchmarking arrangements to provide pressure for continuing improvements in performance."*

Concepts of BPR

One-Time Change Frequency

BPR is undertaken as a one-time project that focuses on major magnitude performance improvements. Such large-scale change processes are seldom accomplished quickly or easily. Moreover, it creates initial discomfort for many people involved in the BPR project who will be asked to perform tasks in new ways; remember, most people are inherently resistant to change. Also one must guard against the development of coalitions of resistance.

A Clean-Sheet Starting Point

Unlike continuous improvement teams that build on a foundation already in place, BPR deliberately starts with a clean sheet. It strives to radically redesign the business processes or systems in order to provide dramatic improvements.

Hammer and Champy (1993) introduced the concept of the clean-sheet approach. They emphasise a series of radical actions such as: rejecting conventional wisdom, investing new approaches to delivering business outputs, and, finally, a complete transformation to optimise and improve process oriented systems. *"A clean-sheet approach means getting to the root of things: not making superficial changes or fiddling with what is already in place, but throwing away the old. It also means disregarding all existing structures and procedures and investing completely new ways of accomplishing work."*

However, Davenport (1993) has commented that the concepts of *"throwing away the old"* is not always applicable. He proposes a philosophy of integrating both process innovation and continuous improvement to enhance the existing processes where the radical redesign is not a workable option.

Process Focus

In today's business environment, BPR should commence with an understanding of both organisational processes and the strategic role of the value chain. Gant (1992) notes that BPR is simply *"the redesign of processes to take advantage of the enormous potential of information technology."* Alternatively, Harrington (1991) defines the concept of business process improvement as *"A systematic methodology developed to help an organisation make significant advances in the way in which its business processes operate."*

The following section further elaborates on the concepts of BPR by discussing its special characteristics.

Fundamental Elements of BPR

The defining characteristics of BPR are summarised in Table 5.2.

Table 5.2 *Characteristics of BPR (source: Bogan and English, 1994)*

Time factors	One-time change frequency Long-term time requirements
Risks	Moderate to high
People factors	Operates across organisational units
Primary enablers	Senior management support Process owner concept Best practice identification
Range	Clean-slate starting point Broad and cross-functional scope Improvement change through radical breakthrough
Tools	Process modelling and mapping approach Information technology Best practice benchmarking

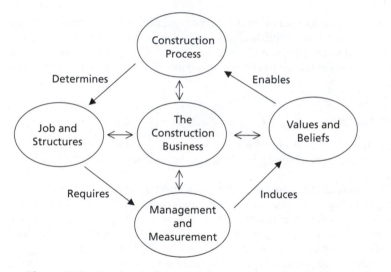

Figure 5.5 *Business diamond for construction business sources (adapted from Hammer and Champy, 1993)*

Champy (1993) defines processes as "*a collection of activities that takes one or more kind of input and creates an output that is of value to the customer.*" Here, the rationale is to improve business processes and streamline the workflow by cost cutting and simplifying workflow. This includes a radical change in respect of tasks and responsibilities designed to enhance the human resource and procurement system of an organisation.

The business diamond (Figure 5.5) is in essence a top-down model which focuses on internal capabilities in order to achieve organisation-wide improvements. It indicates the importance of addressing the business process for construction organisations from a holistic perspective.

Senior Management Support

BPR is a radical approach to performance improvement and senior management's active support and involvement are therefore essential in overcoming the natural resistance to such broad-functional change. They can ensure the integration of both top-down and bottom-up approaches from all staff involved in the BPR project.

Broad and Cross-Functional Scope

Radical change cannot be made by improving the efficiency of each function within a business in isolation. To achieve this radical change the business has to be prepared to work cross-functionally, cutting through functional boundaries and departmental walls (Johansson *et al.*, 1993).

Work tends to flow horizontally across departments, so the functional boundaries will be affected when designing a system to perform work quickly, efficiently and with customer requirements in mind. These changes will affect the business process as a whole.

When rethinking organisational processes, functional boundaries that impede efficiency and effectiveness must be removed, and every step and role in the process should be challenged (Jeremy and Rohit, 1994).

Re-engineering is mainly a managerial implementational change model. It seeks to improve the entire activities of an enterprise. BPR should formulate part of the strategic plans and vision of the host company (Soles, 1995).

People-Centred Culture

Re-engineering must be people-centred, and people should be consulted at every stage of the implementational process. Hammer and Champy (1993) stated that *"teams should try to think as though companies are starting anew."* They also emphasise that organisations should think radically in order to make fundamental changes.

In what has come to be known as classical re-engineering, Hammer and Champy (1993) have suggested that in re-engineering, radical changes yield radical results. Their emphasis is on the importance of sweeping clean, and their tone of language is bold: *"Don't automate the past, obliterate it!"*

"Re-engineering isn't about making marginal or incremental improvements but about achieving quantum leaps in performance" (Hammer and Champy, 1993). Davenport (1993) disagrees with Hammer and Champy when he contends that very few companies can sweep away all their current business operations and start with a clean sheet. He accepts the concept that many businesses have some type of continuous improvement (quality) programmes in place and calls for *"process innovation"* that is a revisionist or pragmatic approach towards business processes. He uses the terms process innovation and process improvement to include various continuous improvement programmes such as TQM. Further, Davenport acknowledges that corporations have many current business processes that are necessary to stay in business. The challenge is to go with the current process improvement programmes and then to select one business process for process innovation.

Technology as Enabler

Technology acts as an enabler that allows organisations to do work in radically different ways. Start with a business process analysis, then employ information technology to fuse the new or amended business processes. Here the rationale is that technology is necessary to save time, but must be subordinate to the business process focus. Hammer and Champy (1993) have stated that *"As an essential enabler in re-engineering, modern IT has an importance to the re-engineering process that is difficult to overstate. But companies need to beware of thinking that technology is the only essential element in re-engineering."*

The Interrelationships of BPR, TQM and JIT

Basically BPR has been used to restructure the production processes and TQM has been adopted to focus on the quality of service. TQM programmes highlighted the role of processes in delivering quality.

Oakland (1989) defines TQM as *"An approach to improving the effectiveness and flexibility of businesses as a whole. It is an essential way of organising and involving the whole organisation and every activity at every level."* However, Fallah and Weinman (1995) define TQM as *"a management system that is based on application of quality principles to all aspects of the business."*

Armstrong (1986) summarised the benefits of TQM implementation as follows:

- Increased customer satisfaction and hence additional sales
- The attainment of a competitive advantage
- The minimisation of waste
- Focusing attention on continuous improvement processes.

Despite the above benefits, Leonard (1982) considers that *"TQM is more concerned with continuous improvement within an agreed framework and does not leave any space for environmental turbulence."* The benefits are more likely to be short term and limited to the immediate context.

Furthermore, TQM concentrates mainly on the improvement of the tasks rather than the processes which make the organisation vulnerable to market changes. Many construction companies currently have developed and used various types of quality and continuous improvement programmes in their organisations. In these quality programmes, attention is focused on small, incremental, systematic and gradual improvements. In contrast, the re-engineering focus is on a holistic and broader scope business process. It is likely to produce major changes in outcomes, but it may take a longer time period to implement.

A relationship does exist between BPR and TQM because many current value-added tasks may well remain in the new design. TQM helps to identify and measure the value-added tasks. Davenport (1993) suggests that both TQM and BPR do exist simultaneously within an organisation, since TQM integrates both behavioural and engineering aspects of quality management, and its scope and objectives distinctly overlap the BPR model.

Davenport has established the concept of integrating both continuous improvement and process innovation by undertaking improvement through innovation. He also notes that *"On the one hand, using innovation-oriented approaches to achieve low levels of change would be overkill. On the other hand, employing improvement-oriented tools to achieve radical change is a recipe for failure"* (Davenport, 1993).

"Ideally, firms would first stabilise a process, then apply innovation for radical change, then settle into continuous improvement so that gains will not be lost" (Davenport, 1993). This clearly shows that the concepts of TQM and BPR are not mutually exclusive. Both TQM and BPR emphasise customer-focus, teamwork and empowerment. The basic difference focuses on the nature of change and the continuum of consequential improvements. TQM is primarily a vehicle for gradual and continuous improvements, and BPR is designed to achieve radical and dramatic changes.

JIT systems are synonymous with processes – the way things are done (processes) and notably not who performs them. TQM programmes, however, have often placed greater emphasis on people and techniques rather than on the process elements. JIT has a similar focus to BPR and places its emphasis on business processes to ensure each and every task is a value-adding activity.

Armstrong (1986) defines JIT as *"A programme designed to enable the right quantities to be purchased or manufactured at the right time without waste."* The prime objective of JIT is the achievement of zero inventory, not just within the confines of a single organisation but ultimately throughout the entire supply chain. It requires far-reaching changes of production systems and techniques and an integrated system between suppliers, customers and the host organisation. A comparison of the key elements related to BPR, TQM and JIT has been summarised by Peppard and Rowland (1995) and is shown in Table 5.3.

Fallah and Weinman (1995) have concluded that *"re-engineering is not a substitute for TQM, it is a means for process redesign within the context of TQM."* In today's globally competitive environment, companies cannot achieve competitive advantage or develop and maintain best-in-class processes simply by relying on continuous improvement. They must take advantage of re-engineering in confluence with their management system, or TQM, to achieve the dramatic results they need to meet the demands of all their stakeholder groups.

Summary

Through the 1990s and into the new millennium, service and product-oriented organisations have realised the importance of having a customer focus, this being evidenced by the number of innovative

Table 5.3 *Business philosophy approach (source: Peppard and Rowland, 1995)*

Element	TQM	JIT	BPR
1. Focus	quality attitude to customers	reduced inventory, raised throughput	processes minimised, non-value-added
2. Improvement scale	continuous incremental	continuous incremental	radical
3. Organisation	common goals across functions	'cells' and team working	process based
4. Customer focus	internal and external satisfaction	initiator of action 'pulls' production	'outcomes' driven
5. Process focus	simplify, improve and measure to control	workflow/throughput efficiency	ideal' or streamlined
6. Techniques	process maps, benchmarking, self-assessment, SPC and diagrams	visibility, kanban, small batches and quick set-up	process maps, benchmarking, self-assessment, IS/IT, creativity

construction companies adopting management tools, such as the BS EN ISO 9000: 2000 series and the EFQM Business Excellence Model, with their emphasis on the quality aspect of organisational provision. However, quality is not the only critical success factor though undoubtedly it is a very important one. Upon what basis will organisations compete when the quality issue has been settled and all organisations operate on a level quality playing field? Construction organisations should realise that in the future customers will require a rapid customised service, one that can respond to clients' demands in terms of both quality and speed of service. These will become the critical competitive issues for the construction industry of the future. In order to fulfil these requirements, organisations should begin to rethink the way they operate their businesses. They should consider the possibility of redesigning business processes from a clean-sheet perspective, hence eliminating non-value-adding activities. BPR can provide the methodological deployment model for attaining a sustainable competitive advantage.

Partnering

Introduction

The principal aim of partnering is to improve corporate performance. While people are working within a traditional framework there are limits to the improvements that can be achieved. Improved performance requires that processes are examined and made more efficient and effective. Partnering provides the means for addressing these two very important performance indicators. It is an approach for addressing the 30% cost reduction advocated by Latham (1994).

Partnering Defined

There are a number of definitions in circulation appertaining to partnering. Some of these are very broad, for example: the drive for a win–win situation based upon trust and mutual objectives. Others are much more detailed but share the same philosophy, for example: *"Partnering can be defined as a voluntary arrangement between two or more partners (it is not a form of contract) to achieve mutual business objectives involving trust and integrity, effective communication, regular review, evaluation and feedback resulting in a win/win outcome"* (CIB, 1997).

Partnering is a long-term commitment between two or more organisations for the purpose of achieving specific set business objectives with the view to maximising the effectiveness of each participant's resources. This relationship should be based on trust, dedication to common goals and an understanding of each other's individual expectations and values.

"There are two different categories of Partnering and within these categories there are a variety of types. The two categories of Partnering are Strategic Partnering and Project Partnering. (Strategic

Partnering is sometimes referred to as 'Multi-project Partnering' or less frequently as 'Second Level Partnering' and Project Partnering as 'Single Project Partnering' or 'First-level Partnering')" (McGeorge and Palmer, 1997). However, Project Partnering can provide a stepping stone towards the achievement of Strategic Partnering.

Partnering is more than simply formalising old-fashioned values, or a nostalgic return to the good old days when a 'gentleman's word was his bond', although moral responsibility and fair dealing are an essential underpinning of any partnership (Hellard, 1995; Schutzel and Unruh, 1996). It is more than a building-procurement technique – although building-procurement techniques can be used to operationalise good practice, bring about cultural change and thus create a more cohesive team (Hinks *et al.*, 1996). The use of partnering in the construction industry has many advocates and many claims of success.

The Origins of Partnering

The origins of partnering, as a construction management concept, are relatively recent, dating from the mid-1980s (Gyles, 1992). This is not to say that partnering did not exist prior to that period and indeed many would subscribe to the view that *"Partnering between contractors and private clients is as old as construction itself."* It has also been claimed that, in the UK, companies such as Bovis have developed a culture and tradition of non-adversarial relationships with particular clients since the 1930s (Hinks *et al.*, 1996).

In effect we focus on *formal* partnering, where there is evidence of an explicit arrangement between the parties. This is not to dispute the existence and importance of *informal* partnering – or as it has been described *"partnering without partnering"* (New South Wales Department of Public Works and Services, 1995). According to the National Economic Development Office (NEDC) report *Partnering: contracting without conflict* (McManamy, 1994), true partnerships in the formal sense only became established in the mid-1980s, the first being that between Shell and partners in 1984. The most frequently cited partnering arrangement of the 1980s is the Du Pont/Fluor Daniel relationship for the Cape Fear Plant project. The partnering agreement between Du Pont and Fluor Daniel was made in 1986 (McManamy, 1994) and was a formalisation of a relationship which had existed since 1975.

Partnering in a Construction Industry Context

Most commentators attribute the emergence of partnering as a force in the construction industry in the late 1980s to the work of the Construction Industry Institute of the United States (CII) and the adoption of partnering by the US Army Corps of Engineers. In the present era, extensive examples of partnering can be found in the USA, with the movement gaining momentum in New Zealand, Australia and the UK. In the latter two countries this gain in momentum is partly as a result of prompting by the Gyles Royal Commission into Productivity in the Building Industry in New South Wales (Stevens, 1993) and the Latham Report (1994) in the UK. Latham, in his foreword to *Trusting the team: the best practice guide to partnering in construction*, states that, *"partnering can change attitudes and improve the performance of the UK construction industry."*

Generations of Partnering

Partnering has developed from its initial inception of first generation through second generation and is currently at the third-generation level. The basic constituent parts of the three generations have been succinctly stated by Bennett and Jayes and are shown in Table 5.4.

First-generation partnering produces a range of benefits which are attainable on an individual or strategic application.

Second-generation partnering requires a strategic decision to co-operate in improving joint performance by a client and a group of consultants, contractors and specialists engaged in an on-going series of projects. Second-generation partnering is underpinned by *'seven pillars'* and those firms which have all seven pillars in place find that cost savings of 40% are not uncommon.

Table 5.4 *The concept of three generations of partnering – new life to the construction industry (Bennett and Jayes, 1998, cited by Watson, 1999)*

First-generation partnering	Second-generation partnering	Third-generation partnering
Centres on principles of: 1. Agreeing mutual objectives to take into account the interests of all the firms involved 2. Making decisions openly and resolving problems in a way that was jointly agreed at the start of a project 3. Aiming at targets that provide continuous measurable improvements in performance from project to project	The so called 'seven pillars' are: 1. **Strategy** – developing the client's objectives and how consultants, contractors and specialists can meet them on the basis of feedback 2. **Membership** – identifying firms that need to be involved to ensure necessary skills are developed and available 3. **Equity** – ensuring everyone is rewarded for their work on the basis of fair prices and fair profits 4. **Integration** – improving the way the firms involved work together by using co-operation and building trust 5. **Benchmarks** – setting measured targets that lead to continuous improvements in performance from project to project 6. **Project processes** – establishing standards and procedures that embody best practice based on process engineering 7. **Feedback** – capturing lessons from project and task forces to guide the development of strategy	Key elements include: 1. Understanding the client's business and aiding its success 2. Construction firms taking full responsibility for producing and marketing products and support services 3. Turning the building process into a cycle of fundamental activities linked by co-operative decision making 4. Accepting equal status for all partners 5. Mobilising development expertise 6. Creating expert teams that specialise in market sectors 7. Producing well-developed supply chains 8. Innovating to produce the best buildings

Third-generation partnering aims to make the construction industry a truly post-modern industry, producing and marketing a range of products and services that clients are eager to invest in. The results in third-generation partnering deliver even greater benefits, with cost savings of 50%.

Figure 5.6 provides a diagrammatic representation of the correlation between partnering deployment and improved organisational performance.

Partnering arrangements are changing attitudes and improving the performance of the UK construction industry, and enable construction firms to meet the demands of their customers. If a balance is to be achieved and maintained, then the partnering agreement should operate as indicated in Figure 5.7.

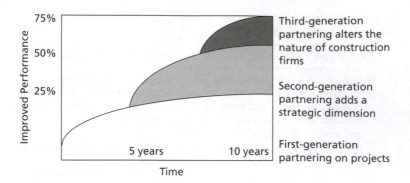

Figure 5.6 *How performance improves with time and growing partnering skills (Bennett and Jayes, 1998)*

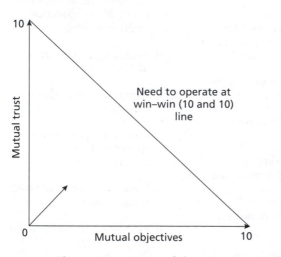

Figure 5.7 *Successful partnering arrangements*

Future Developments

If partnering is to continue developing it is important that all elements involved in the construction supply chain are involved in the process. Currently only the client and the main contractor seem to be in full agreement. However, as with other techniques, such as quality management, the system can only be as good as its weakest link. So subcontractors, material suppliers, plant companies, architects and engineers must all be part of the total team. This has been expressed by the Minister for Construction in a speech at the Annual Conference of the Construction Industry Board, where he called for "*Partnerships between Government, customers, suppliers and the industry.*" Similarly, he intimated that "*the supply chain partnership is the shape of construction in the future.*"

The emergence of the three generations of partnering has resulted in the construction industry making fundamental changes in its attitudes, organisation and technology, thus resulting in improved UK performance and empowering construction firms to meet the varied demands of their clients.

Partnering has received some scepticism, although advantages such as enhanced contract relations, reduced conflict, better team management, earlier contract completion and increased profit clearly outweigh any negative issues. Partnering does increase the efficiency of construction projects, and therefore must be beneficial to the UK construction industry. The partnering approach has been welcomed and accepted by the government, clients, consultants, contractors, suppliers and other industries. For those who can commit to the concept, partnering represents a change of attitude. However, commitment is required by all of the parties concerned.

An Update on the Partnering Paradigm

Latham's Report (1994) recommended that specific advice should be given to public authorities so that they can experiment with partnering arrangements where appropriate long-term relationships can be built up.

Construction studies detailing a partnering approach have shown that costs could be reduced by 2–10% by project partnering and a possible 30% for strategic partnering, including other benefits of improved quality of design, and reduction in the costs of construction and ownership.

Through Latham's (1994) Report, CIB working groups and the Clients' Construction Forum and other parties have been established with the aims of accomplishing the targets set out in the report. Developments from these have led to each publishing its own reports on partnering and working towards non-adversarial relationships. The government took note of clients' wishes, which resulted in the most influential report on the industry since Latham. The government/client-led construction task force headed by Sir John Egan was published in July 1998 entitled *Rethinking Construction*. Egan emphasised many of the ideas recommended in the Latham (1994) Report. In his recommendations he proposed an end to competitive tendering, replacing it with partnering and long-term alliances: "*Clients should use performance measurement and open book accounting to show that 'value for money can be adequately demonstrated' ... too many clients are undiscriminating and still equate price with cost, selecting designers and contractors almost exclusively on the basis of a tender price*" (Egan, 1998).

This section contains the results of a research study conducted via questionnaires on fifty organisations incorporating both clients and contractors in order to obtain their views on the critical issues associated with the partnering concept. One of the crucial issues addressed is whether partnering is applicable to all construction projects independent of size, and this has been tested by asking clients and contractors for their opinions. The primary data collected suggested that both clients and contractors actively involved in partnering arrangements agreed that most contracts, notwithstanding their size, could be incorporated into a partnering agreement.

Egan (1998) called for a £500 million programme of new demonstration projects to take forward the report's reforms, which became the government's 'Best Value' scheme. The above has set the scene for the application of partnering.

With two such vitally influential government sources calling for a move to the partnering approach, it is important to understand the principles and components of successful partnering. The first NEDC report '*Partnering Without Conflict*' (1991), includes components such as:

- all parties must seek win–win solutions
- careful selection of the right partner for the project
- the ability of the client to offer a continuous programme of work
- commitment on all sides
- mutual trust and openness
- an appreciation of the long-term benefits of the relationship.

The essential features as advocated by Bennett and Jayes are indicated in Figure 5.8.

Bennett and Jayes' initial publication, *Trusting the Team* (1995), has since been superseded by *The Seven Pillars of Partnering* (1998), in which Bennett and Jayes suggested that the essential features should include making decisions openly and resolving problems in a way that was jointly agreed at the start of the project, thus producing an updated model as shown in Figure 5.9.

Mutual objectives are a fundamental requirement of partnering (Bennett and Jayes, 1995). The team generally brainstorms its objectives from the project (Table 5.5). There are several important

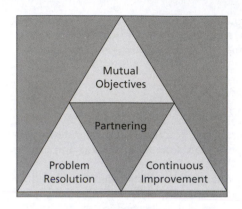

Figure 5.8 *Essential features of partnering (Bennett and Jayes, 1995)*

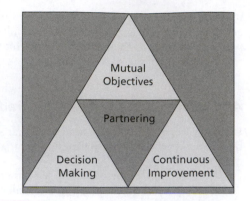

Figure 5.9 *Essential features of partnering 2 (Bennett and Jayes, 1998)*

Table 5.5 *Implementing common mutual objectives (adapted from the Construction Industry Board Mutual Objectives, 1997)*

Objective to be achieved	Implementational strategies
Improved efficiency and information flow	Co-operation
Cost reduction	Continuous improvement
Cost certainty	Early action on problematic issues
Enhanced value	Buildability, value engineering, value management
Reasonable profits	Predictable progress
Reliable product quality	Quality assurance/TQM
Fast construction	No avoidable hold-ups
Certain completion time	Critical path programme
Continuity of workload	Effective programming
Shared risks	Sensibly agreed
Lower legal costs	Dispute resolution procedure
Good public relations	By being proactive
Profit sharing	Prior agreement on sharing savings

success factors for the team to note at this stage. Objectives (CIB, 1998):

- must be agreed and committed to from the outset – the CIOB recognised this as the most essential element of partnering
- must be kept under review through meetings and effective communications
- require long-term goals, such as sustained reasonable profitability rather than making a short-term gain
- benefit from 'open-book accounting' relationships
- must be treated with mutual confidentiality
- work best between businesses with similar cultures.

The Benefits and Pitfalls of Partnering

Latham (1994) and Egan (1998) have set tough targets, with both advocating partnering as one approach to meeting these goals. The benefits of implementing partnering are as follows.

Benefits

A reduction in project costs is a fundamental advantage of partnering and for the client this is possibly the best advantage it has to offer. After all, no one wants to pay more than they have to for a product or service.

A case study provided by the Construction Industry Board (CIB) in *Partnering in the Team* (1997) states a project-specific partnering initiative named 1 St James Square where a £400,000 cost saving occurred on a £19 million project. Further, in a strategic partnering approach adopted by South West Water a 30% cost reduction was achieved.

Further Benefits

- Long-term economies of scale may produce a reduction in costs.
- The quality of the final product is also very important both to the client and the contractors. Reductions in defects and reworking may also be achieved through employing the partnering concept (Bennett and Jayes, 1995).
- Increased contract efficiency is an attainable advantage both for the client and the contractor. This is highlighted in strategic partnering, with the repetition of work leading to reduced staff time as a result of learning curve theory. The generation of less paperwork is an added advantage. A case study on the Pennington waste water treatment scheme emphasised this, where it was noted that much less paperwork was being generated.
- A reduction in completion times is achieved through shortened design times, quicker starts on site and a reduction in project construction time. Of course this is interlinked with a reduction in costs as the less time people spend on a project the less costs they should incur.
- The certainty of continuity of work in a strategic partnering situation is a great advantage for contractors. Contractors have the time to develop a business plan based on secure foundations, thus engaging in strategic planning.
- Enhanced team spirit tends to reduce conflicts and increased job satisfaction flows from a less confrontational attitude, leading to increased intrinsic motivation.
- The design may be improved, such as space maximisation and a more innovative approach due to the early input of the contractor.
- Innovative ideas accompany continuous improvements in partnering arrangements. Innovation occurs more readily in partnered relationships as partners are able to discuss the topic in an open manner and share the risks or rewards without entering into a blame culture (Bennett and Jayes, 1995).
- As the team spirit and adversarial attitudes diminish and problem-solving solutions are achieved at their lowest common denominator, claims and litigation are reduced.
- Continuous improvement is an essential characteristic of strategic partnering and a major advantage.
- Contractors profit in some cases has increased, especially where cost savings on projects are shared. Clients such as Thames Water and South West Water share cost savings 50/50 while others take a 60/40 approach.

Possible Problematic Issues of Implementing Partnering

- Bingham (1998) establishes a possible disadvantage of partnering, because although the contractor may make a profit and ensure continuity of work, it must be wary of too much dependence. Some clients may become bullies and dictate what they want at what price and when. If these prices are too low, when the contractor looks for alternative work it may find that its customers have gone elsewhere.
- The direct costs of partnering have been put forward as a disadvantage. The initial capital cost of a partnering workshop ranges from £2000 to £4000 for a one or two-day course.

Table 5.6 *A comparison of the two philosophies: adversarial and partnering*

Adversarial culture	Partnering culture
Information as power	Open-book want to know
Excluded	All parties included
Some parties lack of communication	Communication at all levels
Why do you want it?	Sure, how can we help you alter it?
Hidden agendas	Straight talking and openness
Setting people up	Setting people straight
Working against	Working with all parties
Preventing	Enabling/empowering
Power over	Power to/enabling
Independence	Interdependence
My work package	Our project

The associated problems that companies may find in a partnering relationship are:

- commitment is an essential success factor within a partnering arrangement and this may not be forthcoming; Hellard (1995) emphasises this by requiring all stakeholders to buy into the concept from the outset
- in strategic partnering situations, stale ideas or even the elimination of risk has resulted in less innovation
- scepticism, lack of trust, job insecurity and people who are unwilling to change are all possible problems for the emerging partnership.

Culture

Partnering requires a culture of co-operation by all concerned. The old adversarial culture cannot sustain a partnership, be it project or strategic in nature. Table 5.6 compares and contrasts the two philosophies.

It is vital that an open and trusting culture exists. What is required is a strategic move to a more enlightened post-modernist approach of conducting business activities. This would ensure that all stakeholders' views and objectives are incorporated into an oganisation business philosophy.

The Adversarial and Partnering Cultures

Contracts

"*Partnering is a reaction to traditional contracting attitudes, characterised by claims and counter claims, disputes and costly arbitration and litigation*" (Parkora and Hastings, 1995).

Latham's Report (1994) considered that modern contracting conditions should contain assumptions for best practice with regard to partnering, and deal with specific issues, some of which are noted below:

- a specific duty on all parties to deal fairly with each other, in an atmosphere of mutual co-operation
- a choice of allocation of risks should be decided for each project, with this risk being given to the parties or party best able to manage, estimate and carry the risk
- firm duties of team work, with a shared financial motivation to pursue this aspect – these should involve a general presumption to achieve 'win–win' solutions to problems which may arise during the course of a project
- take all possible steps to avoid conflicts on site and provide for speedy dispute resolution by a predetermined impartial means
- provide for incentives for exceptional performance.

Some idealists, or perhaps extremists, consider there is no need to have a contract at all. Yet most realists consider the contract to be of importance. Latham (1994) in his report advocates the use of

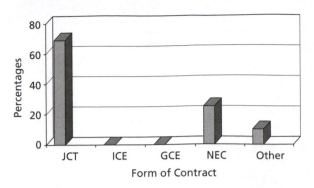

Figure 5.10 *Application of contracts within a partnering agreement – clients' responses to the type of contract used on partnering projects*

the New Engineering for construction contracts of the 1990s. Through the field research undertaken (fifty questionnaires), it can be seen in Figure 5.10 that a varied range of contracts is used under the partnering umbrella. For example, even the JCT and its defensive culture are being used in partnered projects.

Disputes

Alternative dispute resolution (ADR) was devised in the USA to find alternatives to the traditional legal system, which was felt to be costly, damaging to relationships and with limited and narrow remedies, compared to realistic problem solving. It has seen a steady growth in most countries since the 1970s.

ADR schemes are voluntary, non-binding and assisted by a neutral party. These arrangements are particularly useful in partnering agreements because they work best where both parties are looking to settle the dispute in a practical and commercial manner. Through this method, the damage to strategic relationships is kept to a minimum. Decisions can also be made on a strategic basis, ensuring that they are incorporated in future contracts and procedures, which may be creative, using initiative rather than historically based as with the legal system approach.

Private Sector EC Considerations

As public sector partnering has to abide by European legislation, the private sector must also comply. The laws of competition equally affect the private sector with the cornerstone of European Union competition contained in article (1) of the Treaty of Rome. Without covering this subject in detail, it is sufficient to note that trade with member states that prevents, distorts or restricts competition is deemed incompatible with the EU market.

Large partnering arrangements may, at first glance, fall foul of this legislation, with long-term arrangements being essentially anti-competitive. Bennett and Jayes (1995) recommend that advice be sought to ensure the partnering arrangement complies with the complex tests in order to determine whether the arrangement will affect trade between member states, and thus break the law. As a rule-of-thumb, the following points indicate whether EC considerations will affect these agreements:

- partnering arrangements are so widespread that contractors may fix tender prices, because this has the effect of dividing a market or allocating customers or clients
- a third party is disrupting a partnering agreement to gain access to the purchaser's business; notable third party legal actions are claims, damages or injunctions
- some sectors, such as specialist contracts, are already restricted by the small numbers in the field
- the participants exceed 10% of the market, both in product and geographical terms.

In November 1997, the EC published a notice on *Competitiveness of the Construction Industry*, which considered that the industry might benefit from shared risk and co-operation.

Table 5.7 *Clients' response*

Advantages	Highly successful	Successful	Reasonable	Unsuccessful
Cost reduction	29%	71%		
Reduced contract completion times	9%	71%	20%	
Increased contract efficiency	14%	86%		
Reduced arbitration	29%	56%	15%	
Higher standard of workmanship		43%	57%	
Increased chances of future work	N/A	N/A	N/A	N/A
Increased profit	N/A	N/A	N/A	N/A
Increased site safety standards		86%	14%	
More prompt payments	N/A	N/A	N/A	N/A
Problem solving at lowest levels	29%	42%	29%	
Reduced conflict	43%	43%	14%	
Increased job satisfaction	43%	43%	14%	

Table 5.8 *Contractors' response*

Advantages	Highly successful	Successful	Reasonable	Unsuccessful
Cost reduction	67%	33%		
Reduced contract completion times	50%	50%		
Increased contract efficiency	50%	50%		
Reduced arbitration	67%	33%		
Higher standard of workmanship		50%	50%	
Increased chances of future work	33%	67%		
Increased profit		50%	50%	
Increased site safety standards		86%	14%	
More prompt payments		33%	67%	
Problem solving at lowest levels	17%	50%	33%	
Reduced conflict	66%	17%	17%	
Increased job satisfaction	33%	67%		

Field Study Results

Tables 5.7 and 5.8 establish the extent to which clients and contractors view the attainment of the advocated advantages of partnering.

The Advantages of Partnering

The most striking evidence (Tables 5.7 and 5.8) from both clients and contractors is that all the advantages have been realised to some extent. Also both clients and contractors agree that partnering is highly successful in reducing conflicts. This evidence suggests that most of the advantages noted, apart from a higher standard of workmanship, are successfully achieved through partnering.

Table 5.9 *Clients' response*

Disadvantages	Strongly agree	Agree	No opinion	Disagree	Strongly Disagree
Elimination of risk leading to less innovation			29%	71%	
Cosy relationships leading to monopolies		29%	14%	43%	14%
Choosing the wrong partner		43%	14%	43%	
Difficult implementation	15%		14%	71%	
Scepticism		57%		43%	
People unwilling to change	14%	57%	14%	15%	
Lack of trust	14%	72%		14%	

Table 5.10 *Contractors' response*

Disadvantages	Strongly agree	Agree	No opinion	Disagree	Strongly disagree
Elimination of risk leading to less innovation	8%	17%	17%	29%	29%
Cosy relationships leading to monopolies	17%	17%	16%	33%	17%
Choosing the wrong partner		67%	33%		
Difficult implementation		33%		67%	
Scepticism	33%	50%		17%	
People unwilling to change	17%	83%			
Lack of trust	17%	83%			

The disadvantages associated with applying the partnering concept have been tested in a similar way, as indicated in Tables 5.9 and 5.10.

Associated Problems of Partnering

From the results of Tables 5.9 and 5.10 it is apparent that most clients and contractors agree that a reduction in innovation has not generally been experienced. However, choosing the wrong partner is problematic for contractors and this is a crucial factor for attaining a successful partnership.

Both clients and contractors have experienced problems with scepticism, people unwilling to change and a lack of trust. This is all part of a 'culture problem' within the industry at present.

Future Developments

In response to the question *Do you believe that partnering provides the way forward for the UK construction industry?* in the field research, an overwhelming 100% of all contractors and clients believed that it did.

Within the field research both clients and contractors were asked whether they thought the competitive tendering processes would always be used by the construction industry. The results are shown in Figures 5.11 and 5.12.

The results of Figures 5.11 and 5.12 strongly indicate that most clients and contractors believe that competitive tendering processes will always exist in the UK construction industry.

Coulter (1998) indicated that just one-third of clients now believe competitive tendering is the best way of keeping costs down. Therefore partnering may emerge as clients tender for initial schemes.

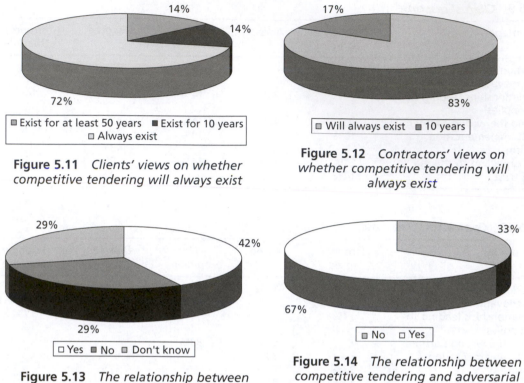

14%

14%

72%

□ Exist for at least 50 years ■ Exist for 10 years
□ Always exist

Figure 5.11 *Clients' views on whether competitive tendering will always exist*

17%

83%

□ Will always exist ■ 10 years

Figure 5.12 *Contractors' views on whether competitive tendering will always exist*

29%

42%

29%

□ Yes ■ No □ Don't know

Figure 5.13 *The relationship between competitive tendering and adversarial attitudes – clients' responses*

33%

67%

□ No □ Yes

Figure 5.14 *The relationship between competitive tendering and adversarial attitudes – contractors' responses*

In a final question, the respondents were asked: *Do you feel adversarial attitudes will always exist while competitive tendering exists in the UK construction industry?* The results are shown in Figures 5.13 and 5.14.

The results support both the Latham (1994) and the Egan (1998) Report that one reason for adversarial attitudes is the competitive tendering process. This attitude, many believe, is strangling the efficiency of the UK construction industry.

Summary

Partnering between clients, contractors and other construction professionals is viewed by many as a panacea for most performance-related problems in the UK construction industry. Partnering can improve the performance of the UK construction industry. The research provided points to an overwhelming range of improved performance indicators. For the client, possibly the most important performance indicators are cost reductions and projects completed within, or ahead, of programme. For the contractor, the most substantial factor is the ability to make a profit. Utilising partnering, a sustainable, and in some cases, a higher profit can be realised.

It has also been recognised that for the partnering arrangement to work, aggressive attitudes must be fully eradicated. This has proved to be difficult as it is often deeply embedded within the industry. As partnering increases, it represents a cultural revolution, or at least a fundamental shift from adversarial methods of working towards a greater collaboration within the UK market.

Clearly partnering is an option to be considered by clients and contractors alike who have the respective ingredients of value and purchasing power and the necessary depth and breadth of financial, technical, staffing resources and cultural commitment. It can be a valuable asset in improving the UK construction industry's performance in this new millennium.

Financial Planning and Control

Control

Control is concerned with the effective and efficient utilisation of resources in the attainment of previously determined objectives contained within a specific plan. This plan may be in the form of a programme of works, e.g. bar chart, network analysis or a financial plan such as a project budget. The plan, being the method to be deployed in order to achieve the predetermined objectives, should be based on the most efficient and effective way of completing the set task(s).

Control is exercised by the feedback of information on actual performance when compared with the predetermined plan and, therefore, planning and control are very closely linked. Control is the activity which measures deviations from planned activities/objectives and further initiates timely, effective and efficient corrective actions.

In order to achieve both efficient and effective control, the Deming Control Cycle should be employed.

Discussion of Plan, Do, Check, Action (PDCA) Control Cycles

In ensuring the success of a project, as well as in attainment of continuous improvements, there is a need to adopt effective and efficient control. The control system needs to include time, cost and quality. The concept of the 'Plan, Do, Check and Act (PDCA) control cycle' can be simply defined as establishing the project objectives and then determining the appropriate methods of reaching the set goals (Plan), implementing the Plan (Do), checking the effect of the implementation by comparing the actual results with the Plan (Check), and taking appropriate corrective action (Act). PDCA is a continuous process not only for the project itself, but can and should provide a holistic approach for the whole corporate enterprise as depicted in Figure 5.15.

This is a continuous improvement spiral. Each time the control cycle is deployed it leads on to the next cycle of deployment and therefore builds upon previous improvements, thus enabling a continuous improvement process.

Platje and Woodman (1998) argue that the Deming Control Cycle has a failing in that it does not indicate that the reset authorised plan is not in the same plane as the action (reset goals) plane: "*There is no activity that implements a new plan after authorisation and yet implementation takes time.*" This suggests that the model is purely a sequential winding up of Plan, Do, Check and Act. These valid comments have been overcome in Figure 5.16 (Watson, 2002) [see pages 144 and 145]. By incorporating

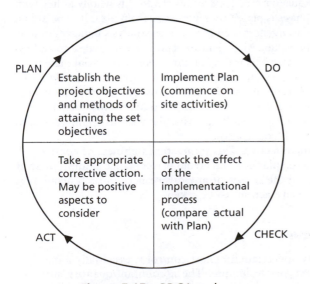

Figure 5.15 *PDCA cycle*

a vertical axis indicating that a more realistic model would be a Plan, Do, Check, Act cycle with the reset authorised plan out of the original planning plane, this reset authorised plan then becomes the plan for the second cycle and so on in an ever-improving control cycle.

The Psychology of Control

Consultation

Managers are ordinary human beings and therefore it is logical to give some consideration to the psychological aspects involved in control. It is a fundamental principle of good control that the person responsible for maintaining control must accept the plan; that is, they must believe it is attainable and agree to be held accountable for those factors within their control in the event of not completing the set tasks/objectives. Budgets and targets, therefore, should not be imposed. Should they be imposed there exists a high probability of failure because the person(s) may:

- be indifferent to failure since they can argue that they cannot be held accountable for something they said could not succeed from its inception
- have a vested interest, in fact, in failure, since success of the plan will prove that their original objections were not valid.

One excellent way of obtaining a person's acceptance of budgets and targets is to allow personnel to set them (possible use of management by objectives). Modification resulting from other people's plans is, of course, inevitable, but if such modifications are tactfully made, acceptance is not usually withheld.

Management by Exception

Once a plan has been made, management's primary task is to correct divergences from that plan. It is unnecessary – indeed it is unwise – for them to concern themselves with matters that are going according to plan. Therefore in order to direct their attention to matters that need their time and efforts, the concept of *management by exception* should be employed. Under this concept only those matters which are not going according to plan are reported to managers. When events are going exactly as planned, no reports are made.

It can be seen that the variance analysis technique (covered later in this chapter) is wholly in line with the concept of management by exception since emphasis is placed on divergences from plan. If there are no divergences there are no variances and if there are divergences then the variance analysis indicates exactly where the divergence occurred and its importance (in value). Managers are therefore encouraged to address such divergences in order of importance rather than spend their time on unnecessary or trivial matters.

In all control work it is very important to appreciate exactly which factor is ultimately to be controlled. In the field of construction managerial economics this factor is usually profit. The person charged with implementing the plan is therefore concerned with profit control. However, because management historically began this form of control by controlling costs, the various techniques used have come to be encompassed under the umbrella of 'cost control'.

Control must be tight for it to be efficient and effective. This means that tightness of control will be attained by, first, making the comparisons between planned and actual performance at appropriately short intervals of time and, secondly, by explaining the causes of any divergences in such detail as is necessary to empower efficient and effective corrective actions to be taken.

Cost Control as a Management Technique

It is important for construction managers to fully appreciate that cost control is essentially a management technique, not an accounting/quantity surveying technique. The accountant/quantity surveyor may certainly be responsible for the work involved in generating the quantitative data that will be part of the managerial decision-making process; however, management is responsible for the planning and action functions.

Cost control can only be attained through managers (or to be specific, persons performing the management function). As previously noted, action to correct divergences is an essential element of effective and efficient control. Such action can only be taken by managers. If the 'actual cost' of operating a process is exceeding the 'planned cost', then only action taken by the person in charge of that process can enable control to be kept. It is obvious then, that the control function is dependent on each manager being advised of both 'planned' and 'actual' costs. Without these, control cannot possibly be really efficient and/or effective.

We have noted that the accountant/quantity surveyor provides 'quantitative' data as part of the managerial decision-making process. This decision-making process is basically choosing between various options or alternatives, and under normal circumstances each will have an associated 'opportunity cost'. This means that the accountant/quantity surveyor can assist managers by showing them the relative economic advantages or disadvantages of one alternative when compared with other valid options. This in turn means that decision-making work should concentrate on analysing the economic differences between alternatives. Figures such as the total cost-per-unit, total sales or total enterprise profits are of little value here. What is required is the difference in profit that will arise as a result of selecting one alternative in preference to another.

Using this basic approach of analysing differences, we may consider what happens if there are differences in activity (break-even charts), if there are minor differences in products (marginal costing) and if there are differences in the timing of receipts and payments (capital investment appraisal). Furthermore, examples of 'cash flow', ' budgetary control' and 'variance analysis' are provided within this section.

However, it is important to understand that the decision-making process can be equally influenced by qualitative data; for example, issues relating to the internal or external politics of the construction company. These are factors which can even lead to decisions being taken that are contrary to the quantitative data provided from a valid analysis.

Marginal Costing and Break-even Analysis

Fixed and Variable Costs

- Dual basis of costs – making correct decisions depends on understanding how costs behave and for short-term decisions this, in turn, depends on appreciating that all costs are essentially either:
    - time based (i.e. change in proportion to time)
    - activity-based (i.e. change in proportion to activity).
- Time-based (fixed) costs are costs that change in direct proportion to the length of time that elapses. Typical of these is rent. Note that such a cost is not affected by activity. No matter whether the enterprise is busy or not, the rent payable is the same. Other examples of time-based costs include rates and debenture interest. Since time-based costs do not vary with activity, then for any given period of time they will remain unchanged regardless of the level of activity. These costs are therefore termed fixed costs. (*Note*: Fixed costs are not costs that never alter, the point being made is that they do not alter as a result of changes in activity level.)
- Activity-based (variable) costs are costs that change in proportion to the level of activity undertaken. The most obvious of these costs is the direct material costs, if, for example, production is increased by 10%. Other such costs are direct wages, sales commission and power. Such costs are termed variable costs and can be defined as costs that vary in direct proportion to activity.
- Mixed-based (semi-variable) costs – there are clearly a number of costs that are neither wholly fixed nor wholly variable. Such costs change with activity but not in direct proportion to it. For example, activity may increase by 10% but a particular cost of this type may only increase by 4% or 5%. Such costs are termed semi-variable and include maintenance or supervision. These costs are really mixed time and activity-based costs. For instance, maintenance can be analysed into time-based maintenance (e.g. weekly, monthly or annual preventative maintenance, the cost of which is independent of activity) and activity-based maintenance.

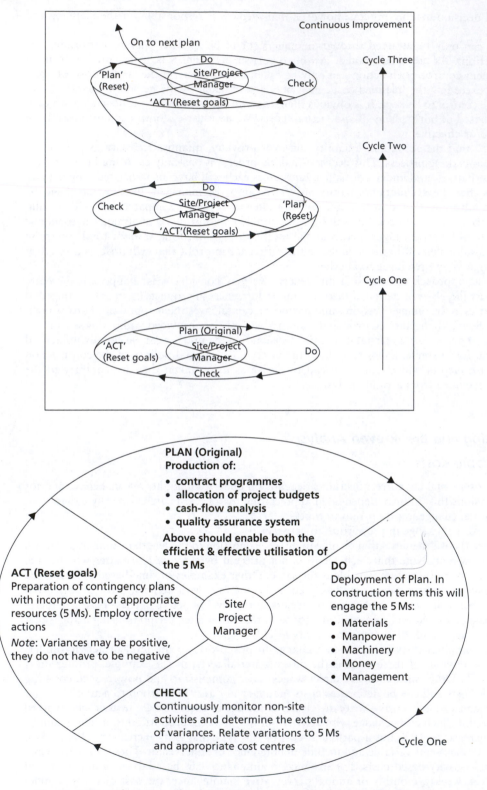

Figure 5.16   *Plan, Do, Check, Action cycle applied to construction (Watson, 2002) [© Dr P Watson]*

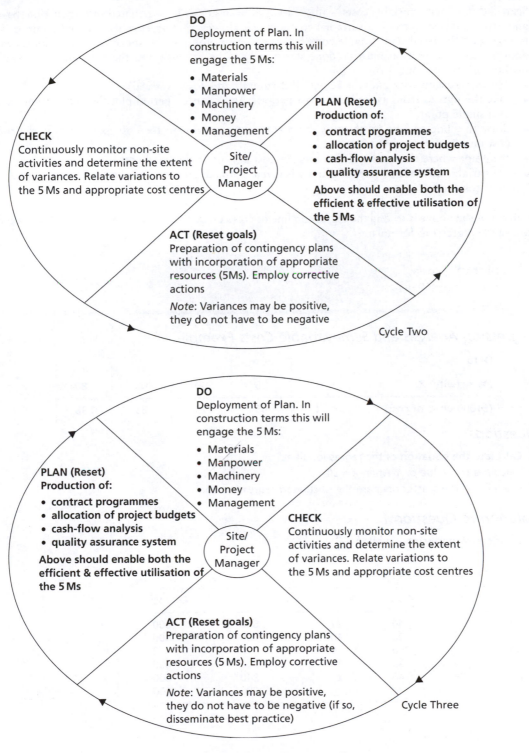

Figure 5.16 *(continued)*

- Segregation of semi-variable costs – all semi-variable costs can be segregated into their time-based and activity-based (fixed and variable) components. However, in practice it is virtually impossible to segregate the fixed and variable costs by simple inspection (as was suggested with maintenance above). Such segregation must be done statistically to be of value and the following is a simple method of segregation:
  - prepare a graph with axes for activity (horizontal) and costs (vertical)
  - take the figures from a number of past periods, and for each period plot the activity and costs as a single point
  - draw the 'line of best fit' through the points and extend it to the cost axis – this is the total cost line
  - the point where the total cost line cuts the cost axis gives the fixed cost
  - the variable cost at any level of activity is given by the difference between the fixed cost and the total cost line.

The object of the above is to find the regression line of costs on activity. A more accurate method would be to use the regression formula:

$$Y = mX + c$$
$$\text{where } Y = \text{costs}$$
$$X = \text{activity}.$$

Regression Analysis and Semi-variable Costs Example

Data

(% Activity) X	0	20	40	60	80
(£1000 units of cost) Y	54	65	75	85	96

Questions

1. Calculate the equation of the regression line.
2. Calculate the value of Y when $X = 50$.
3. Draw in on the scatter diagram the calculated regression line.

Solution to Question 1

Calculation of regression equation

Table 5.11 *Regression calculation*

X	Y	$X - \bar{x}$	$Y - \bar{y}$	$(X - \bar{x})(Y - \bar{y})$	$(X - \bar{x})^2$
0	54	−40	−21	840	1600
20	65	−20	−10	200	400
40	75	0	0	0	0
60	85	20	10	200	400
80	96	40	21	840	1600
$\Sigma200$	$\Sigma375$			$\Sigma2080$	$\Sigma4000$

$$\bar{x} = \frac{\Sigma X}{\Sigma n} = \frac{200}{5} \quad \therefore \bar{x} = \underline{40}$$

$$\bar{y} = \frac{\Sigma Y}{\Sigma n} = \frac{375}{5} \quad \therefore \bar{y} = \underline{75}$$

$$Y = mX + c$$

$$m = \frac{\Sigma(X-\bar{x})(Y-\bar{y})}{\Sigma(X-\bar{x})^2} = \frac{2080}{4000} \therefore m = \underline{0.52}$$

$c = \bar{y} - (m\bar{x})$
$\therefore c = 75 - (0.52 \times 40) = 54.2$ (intercept value)
$54.2 \times 1000 = £54,200$ (fixed costs)

Solution to Question 2

Let $X = 50$
$Y = mX + c = (0.52 \times 50) + 54.2 \therefore Y = 80.2$
$80.2 \times 1000 = £80,200$

Table 5.11 contains the calculations necessary for determining the value of $\bar{x}$ and $\bar{y}$ which are required for producing the regression equation values for m and c.

Figure 5.17 is the graphical (scatter diagram) representation of the results of the regression equation enabling the segregation of the semi-variable costs.

Solution to Question 3

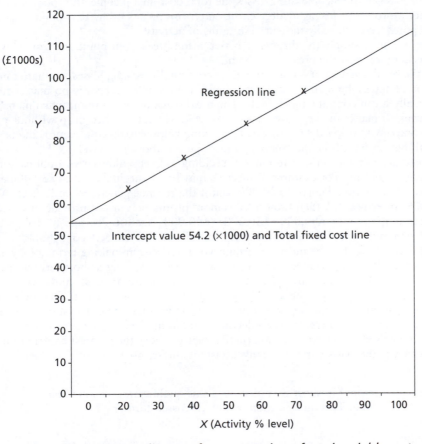

Figure 5.17 *Scatter diagram for segregation of semi-variable costs*

- Total costs = Fixed costs + Variable costs – Since all semi-variable costs can be segregated into fixed and variable components, then all the costs in an enterprise can be analysed into fixed and variable categories. This means that the total cost for any period can be regarded as the sum of costs that are wholly fixed and costs that are wholly variable

  i.e. Total costs = fixed costs + variable costs

- Graphing costs and income – It is a simple matter to plot the foregoing equation on a graph having activity and money axes. On such a graph the fixed costs will show as a horizontal line at a height equal to the fixed costs and the total cost curve as a straight upward-sloping line starting from the fixed cost line at nil activity. (Since there is nil activity, the variable costs must definitely be nil and the total costs therefore equal to the fixed cost.)

  If a third line representing revenue (sales) at the different activity levels is added to the graph, the graph then shows the corresponding profit or loss levels. The sales line will of course start at the origin of the graph, since nil activity leads to nil revenue.

- Features of a break-even chart:
  - *Costs and income* – The total cost, together with the income, can be directly read off the chart for any chosen level of activity.
  - *Profits and losses* – Since the difference between the total cost of operating and the resulting income is profit (or loss), then the gap between the total cost line and the income line at any level of activity measures the profit or loss at that level.
  - *Break-even point* – When the income line crosses the total cost line, income and total costs are equal and neither a profit nor loss is made, i.e. the enterprise breaks even. This point is called the 'break-even point' and is usually measured in terms of activity.
  - *Margin of safety* – This is simply the difference between the break-even point and any activity selected for consideration past the break-even point.

- The importance of the break-even point – If a company is continually making losses, no matter how small, then its life is definitely limited. On the other hand, if it is making profits no matter how small, then theoretically it can continue indefinitely. The break-even point is important to management therefore, because it marks the very lowest level to which an activity can drop without putting the life of the company in jeopardy. Occasionally working below break-even point is of course not necessarily fatal, but on the whole the company must operate above this level.

- Contribution – Costs are not the only criterion for deciding between alternative options – the income from each is also relevant. For example, Project 'A' may have marginal costs of £30,000 and an alternative Project 'B' a marginal cost of £20,000. But if the revenue generated on Project 'A' is £80,000 and £60,000 for Project 'B', then Project 'A' is more profitable, since it provides an excess of income over marginal costs of £80,000–£30,000 = £50,000, against Project 'B' £60,000– £20,000 = £40,000. This difference between income and marginal cost is termed 'contribution'. The concept of contribution is probably the most important word in decision-making terminology and it must be fully understood. Its importance warrants restating: when making a short-term decision between a number of alternatives the only relevant figures are the marginal costs and incomes of each alternative, as the fixed costs are the same whichever alternative is selected. The difference between the marginal cost and income, i.e. the contribution of an alternative, measures therefore the net gain to the business that will result from selecting that alternative. Consequently the profits of the business are maximised by selecting the alternative that provides the greatest contribution.

  This principle, which is the basis for very many decisions, must, however, be qualified in the following two respects:

  (a) Fixed costs must not alter and in short-term decisions this qualification usually holds.
  (b) No key factors are involved. This qualification is now discussed further.

- Explanation of $\frac{P}{V}$ ratio

  The $\frac{P}{V}$ ratio is $\frac{\text{contribution}}{\text{sales revenue}}$, and can be used to assess the profitability of a contract for production.

Example

Let Contribution   = £35,000
and Sales revenue = £100,000

$$\therefore \frac{P}{V} = \frac{£35,000}{£100,000} \times 100 = 35\%$$

or for every £1 of sales revenue, 35p of contribution is obtained.

- Limiting or key factor – There is always something that limits an organisation from achieving an unlimited profit. When comparing two products, management should look for a limiting factor. For example, there is no point in producing 200 units if the company can only sell 100 units. The limiting or key factor here is sales, in other cases it may be machines, men or materials.

Example

Let the limiting factor be wages at £10,000

$$\therefore \text{Limiting factor ratio} = \frac{\text{Contribution}}{\text{Limiting factor}} = \frac{£35,000}{£10,000} = 3.5$$

or for every £1 of wages, £3.50 of contribution is generated.

- The application of marginal costing provides a yardstick in the following situations:
  - in deciding prices during a recession etc.
  - to compare the results for different contracts
  - in assessing profits following increases or decreases in sales
  - to assess if it is suitable to sell below total costs or even the marginal cost perhaps for a limited period in order to retain the services of the skilled labour during a recession or to maintain production when competition is at its fiercest. There comes a time, however, when costs increase because of inflation, and losses will then become evident.

Marginal Costing and Break-even Analysis Example

The following data is used to provide an example of marginal costing and break-even analysis in practice.

Data: A medium-sized building company located in the North East of England specialises in the construction of speculative housing projects. It has acquired land suitable for the construction of 25 detached houses and has been approached by a developer who has negotiated the sale of the properties for £90,000 per house with a contract duration period of 12 months. Should the company accept the offer? The marginal costs will be £30,000 per house and the fixed overheads are £450,000 per annum.

Question 1

After what period of time will the company start to break even? (graphically)

Question 2

At what percentage level of activity will the company break even? (graphically)

Question 3

What effect would a revised contract period of 2 years have on the break-even point (BEP) of the company if the sale price remains the same (£90,000) with no increase in marginal costs (in time)?

Question 4

Check your answer to Question 2 above via the calculation method in units.

Question 5

Determine the BEP using actual profits (on profit graph).

Question 6

Check your answer to Question 5 above via the calculation method.

Solution to Question 1

Total revenue = revenue per unit × number of units = £90,000 × 25 = £2,250,000
Total marginal costs = marginal cost per unit × number of units = £30,000 × 25 = £750,000
Total costs = total fixed costs + total marginal costs = £450,000 + £750,000 = £1,200,000

See Figure 5.18 for the plotting of the above values to obtain the BEP in time.

Solution to Question 2

Figure 5.19 provides the BEP in % of activity level using the above data.

Solution to Question 3

Total fixed costs = 2 (years) × £450,000 = £900,000
Total costs = total fixed costs + total marginal costs = £900,000 + £750,000 = £1,650,000
These figures are now plotted on Figure 5.20 to obtain the BEP for an extended contract time of two years.

Solution to Question 4 (activity level)

$$BEP = \frac{\text{Total fixed costs}}{\text{Total revenue} - \text{Total marginal costs}} = \frac{£450,000}{£2,250,000 - £750,000}$$

∴ BEP = 0.3% × 100 = 30% of activity level (as indicated on Figure 5.19)

Question 5 relates to a profit graph and Figure 5.21 provides an example of this approach to determining break-even points. Profit cannot be read directly off a break-even chart. It is necessary, therefore, to deduct the total cost reading from the total income reading. However, a profit graph overcomes this problem by plotting the profit directly against activity. This means that at nil activity a loss equal to the fixed cost will be incurred. On Figure 5.21, total fixed costs are set at £90,000 and profit at £60,000 at 100% activity level.

Solution to Question 5

Total revenue = £2,250,000
Total contribution = Total revenue − Total marginal costs
         = £2,250,000 − £750,000 = £1,500,000 contribution
Contribution − Total fixed costs = Profit
         = £1,500,000 − £450,000 = £1,050,000 profit

Solution to Question1

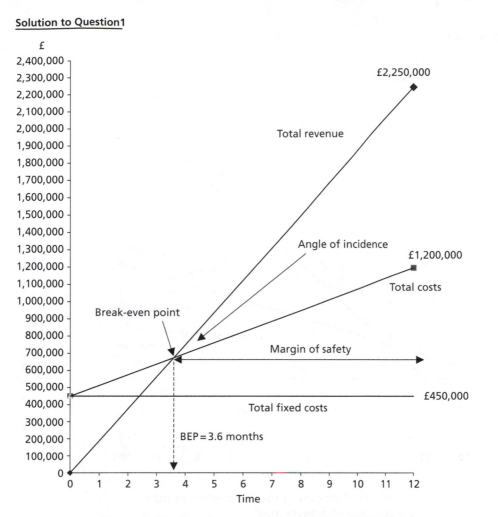

Figure 5.18 *Break-even chart – for example in time*

These figures are now plotted on Figure 5.22 (profit graph), and the break-even point can be read off as per the break-even chart.

Solution to Question 6

$$BEP = \frac{Total\ fixed\ costs}{Revenue\ per\ unit - Marginal\ cost\ per\ unit}$$

$$= \frac{£450,000}{£90,000 - £30,000}$$

BEP = 7.5 units (as indicated on Figure 5.22)

Solution to Question 2

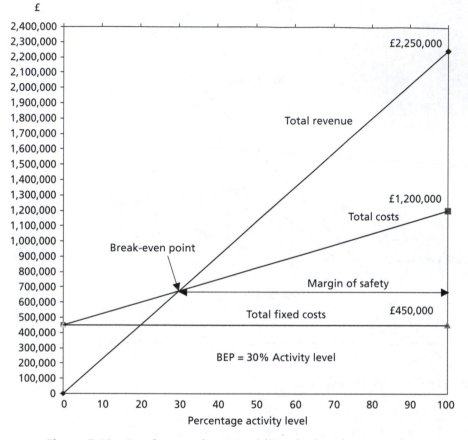

Figure 5.19 *Break-even chart providing the break-even point in percentage of activity level*

Solution to Question 3

(£100,000)

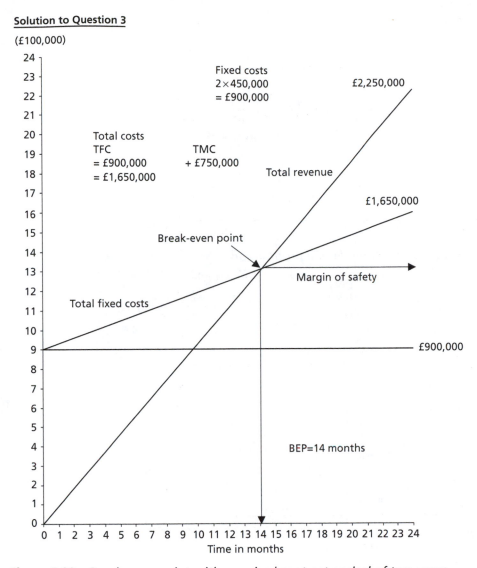

Figure 5.20 *Break-even point with a revised contract period of two years*

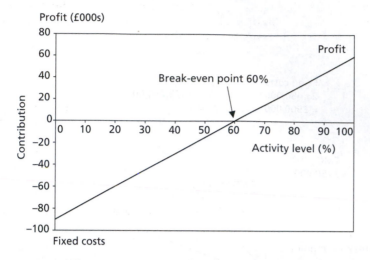

Figure 5.21 *Example of a profit graph*

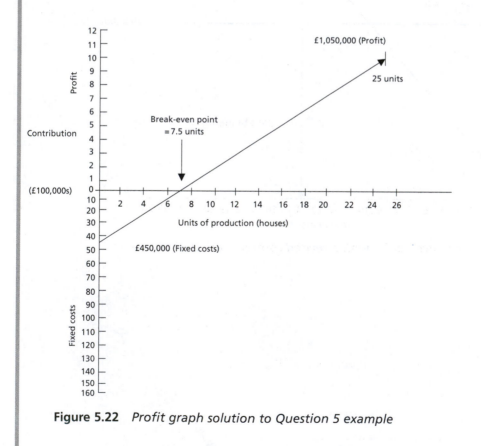

Figure 5.22 *Profit graph solution to Question 5 example*

As can be seen from Figures 5.19 and 5.22, the break-even chart and profit graph are useful tools for relating costs and profits to activity, but they do have certain limitations that must be borne in mind. First, break-even charts are only valid within the actual limits of the activity on which they are based (statistically speaking, you cannot extrapolate). Secondly, fixed costs may vary at different activity levels.

Capital Expenditure

In the previous section consideration was given to short-term decision-making aids. However, capital expenditure involves payments with long-term implications, which in turn can significantly affect the cash position of a construction company in the short term.

Capital expenditure is that cash which is paid out in order to purchase the necessary fixed assets of a company. Decisions have to be taken to invest in long-lived assets, and these are important decisions for the following reasons:

- a company will probably be involved in expending large sums of money at a particular moment in time
- a company might be committing itself to utilising those assets for some considerable time in the future
- sometimes the strategic position of a company is shifted in some way by the decision, if only slightly, and this could impact upon its competitive advantage.

Capital expenditure decisions are important to all stakeholders as they can impact on a firm's earning power, liquidity and its potential for growth. There is a stage at which such projects must be appraised and for this there exists a variety of available methods. They all involve the estimation of cash inflows and outflows associated with the capital expenditure, on the grounds that such investments can only be appraised on a cash-flow basis. A construction company needs to carefully consider the effect of such expenditure on the total cash-flow situation and the form of financing required when the company needs to acquire funds for the investment purposes.

Investment Appraisal Techniques

The following methods are the most frequently used by organisations in the investment appraisal process:

Payback period technique – In this technique the *time that must elapse* before the net cash flows from a project result in the entire initial capital outlay being repaid is found. This time figure is called the *payback period*, and users of the technique look for short payback periods when making capital project decisions. The main objections to this technique are:

- it does not take into consideration the receipt of cash flows received once the payback period has been reached
- it ignores the timing of receipts – and hence the time value of money.

The main advantage claimed for the application of this technique is that if there is considerable uncertainty about the future, then an early replacement of the initial outlay by the end of a period in which events are *relatively* predictable means that the chance of at least regaining one's money is reasonably high. In the converse situation, a loss is a serious possibility and may represent too grave a risk, even though a distinctly improved profit situation would result should events run as planned.

Average annual percentage return technique – In this technique, projects are evaluated according to the average return earned per annum as a percentage of the outlay.

- *Limitation* – This technique ignores the timing of receipts and is also subject to distortion if the life pattern is at all unusual.

The main advantage is that profitability is measured and presented in the form of a percentage which can be compared with alternative investment project opportunities.

Net Present Value

The essence of this technique is the fact that £1 in a year's time has not the same value as £1 now. There exists more than one reason for this (e.g. inflation) but the only reason this technique recognises is that £1 now can be invested so that by the end of the year its value will have grown by the amount of interest (gained at a specified interest rate). To pay out £1 now, therefore, for a promise of £1 in two or three year's time is (quantitatively) unprofitable. Consequently, when investing money now in long-term projects, it is not only the amount of money that will be earned later that is relevant but also when the amount is received, i.e. the timing of the subsequent cash flows is important.

This technique involves selecting a moment in time and expressing the value of all cash flows in respect of that moment. The moment selected is usually the start of the project investment period. In other words, if the project is to begin now, then we shall convert all future cash flows into the amounts we would pay now for the promises of those future flows. Such amounts are termed *present values* since they are the value of those flows at this present moment. Note, incidentally, we are considering *cash* flows, not cost values, since only cash can be invested to earn interest.

The present value of the future £ – Just what is the promise of £1 in one year's time worth today? Clearly, the amount would equal the sum of money we would need to invest *now* to end the year with a total of principal and interest of £1. This would obviously depend on the interest rate.

- *Illustration* – Assume that the rate is 100% per annum. If £1 is invested now, then in one year this will have increased to £2. This means that we would be prepared to exchange £1 now for the promise of £2 in one year's time, so if the promise was only £1 in one year we would only pay £0.5 now. *Therefore, the present value of £1 in one year at 100% is £0.5.* Continuing the illustration, if £1 is invested now then in two years it will amount to £4. (Note that compound interest is always used in this technique.)

The main disadvantage of this approach is that it is not as simplistic in its application as the two previous methods. However, this is also its strength as the quality of data generated from its application is enhanced and therefore so is the output of the quality of the decision-making process. Under the net present value method the cash flows of a project are discounted at the selected rate and the sum of all the present values gives the net present value (NPV) of the project. It is 'net' because the outlay is deducted from the discounted positive net cash flows resulting from the investment. If the NPV is positive then the interpretation is that the return on cash invested in the project is above that of the selected rate. Conversely, if it is negative the return is below the selected rate. The NPV analysis can be further utilised in determining the internal rate of return (IRR) on a given project.

Under this method (internal rate of return), a trial-and-error procedure is adopted to ascertain what interest rate must be employed in order to make the NPV of the project zero. This means arbitrarily choosing a rate, discounting the project at this rate and then examining the NPV to see if it is zero. If not, a new rate must be taken and the procedure repeated. (If the NPV is positive, it means that the rate chosen was too low, and vice versa.) Ultimately, the rate that gives a NPV figure of zero is found and this is termed the discounted yield. It is then compared with the predetermined minimum rate which is often called 'the cut off rate' and a decision is then taken on the basis of this comparison.

Application of the Three Appraisal Techniques

The following provides an example of the application of all the above techniques of investment appraisal. The net cash flows for three project investment opportunities have been provided in order to demonstrate how the various methods can be used in ranking a range of capital investment opportunities (see Table 5.12).

Table 5.12 *Data for capital expenditure problems*

End of year	Project 'A'	Project 'B'	Project 'C'
0	−£30,000	−£30,000	−£40,000
1	£10,000	15,000	12,000
2	15,000	15,000	12,000
3	20,000	10,000	15,000
4	20,000	5,000	20,000
5	5,000	20,000	20,000
	70,000	65,000	79,000

Questions

1. Calculate the payback period (PBP) for the projects.
2. Calculate the average annual percentage rate of return (AA%RR) for the projects.
3. Calculate both the net present value (NPV) and the internal rate of return (IRR) for the projects.
4. Rank the alternative projects based on the data produced from the appraisals conducted in Questions 1–3 above.

Project 'A' will be taken and subjected to appraisal via

*   Payback period
*   Average annual percentage rate of return (AA%RR)
*   Net present value (NPV)
*   Internal rate of return (IRR)

in order to provide a detailed explanation of how to employ the methods. Then the data for the remaining projects 'B' and 'C' is provided in order to 'rank' the three projects. Project 'A' data for appraisal purposes is provided in Table 5.13.

Table 5.13 *Data for appraisal techniques example*

End of year	Project cash flow
0	−£30,000
1	10,000
2	15,000
3	20,000
4	20,000
5	5,000
	70,000

Payback Period, Method One (for Project 'A')

The amount to be recovered is £30,000 (Table 5.13) and after two years (Table 5.14) shows we have recovered £25,000. Therefore a shortfall of £5,000 exists and the next income increment is £20,000 (again indicated in Table 5.14). Therefore

$$\frac{\text{shortfall}}{\text{next increment}} \times 12 = \text{time in months to recover shortfall}$$

$$= \frac{5,000}{20,000} \times 12 = 3 \text{ months}$$

Payback period for project 'A' = 2 years 3 months

Table 5.14 *Payback period appraisal example*

Year	Cash flow	Cumulative cash flow	
1	10,000	10,000	
2	15,000	25,000	Capital investment
3	20,000	45,000	← £30,000
4	20,000	65,000	
5	5,000	70,000	

Average Annual Percentage Rate of Return (AA%RR), Method Two Example

Data from Table 5.13 is used in this example.

Step 1: Calculate the average annual rate of return (AARR)

$$AARR = \frac{\text{Total return (see Table 5.13)}}{\text{Investment period}}$$

$$= \frac{70,000}{5}$$

$$= £14,000$$

Step 2: Calculate the AA%RR

$$AA\%RR = \frac{AARR}{\text{Capital invested}} \times 100$$

$$= \frac{14,000}{30,000 \text{ (see Table 5.13)}} \times 100$$

$$= 46.67\%$$

Net Present Value (NPV), Method Three Example

This method is based on the inverse of the compound interest formula.

Compound interest $= (1 + i)n$

Discounted cash flow $= \dfrac{1}{(1 + i)n}$

Let the discount rate (cut-off-rate) be 15%. Table 5.15 indicates that we do receive sufficient funds to warrant the expenditure economically. The return on capital expenditure is +£17,200.

Table 5.15 *NPV data at 15% (from net present value tables, Table 5.18)*

Year	Cash flow	NPV factor 15%	NPV
0	£30,000	1.000	−30,000
1	10,000	0.87	8,700
2	15,000	0.76	11,400
3	20,000	0.66	13,200
4	20,000	0.57	11,400
5	5,000	0.50	2,500

NPV = Total NPV − Capital invested (47,200 − 30,000) = +£17,200

From the previous example (NPV), it has been established that we are not obtaining a return equal to the 15% cut-off rate. We can use the IRR method to determine the actual NPV rate of return. The return is considerably higher than 15%, so let us try 30% first. Table 5.16 establishes that at 30% NPV factor the project makes a profit of £4,100.

Therefore we need to increase the NPV factor in order to reduce the end product, so let us now try 40% NPV factor. The results of this calculation are shown in Table 5.17, and the return is not as high as 40%; actual figure is −£1,900. The true NPV can be obtained by interpolation as follows.

Table 5.16 *NPV calculation tabulated with a NPV factor of 30% (from Table 5.18)*

Year	Cash flow	NPV factor 30%	NPV
0	£30,000	1.00	−30,000
1	10,000	0.77	7,700
2	15,000	0.59	8,850
3	20,000	0.46	9,200
4	20,000	0.35	7,000
5	5,000	0.27	1,350

NPV = +£4,100

Table 5.17 *NPV calculation tabulated with a NPV factor of 40% (from Table 5.18)*

Year	Cash flow	NPV factor 40%	NPV
0	£30,000	1.00	−30,000
1	10,000	0.71	7,100
2	15,000	0.51	7,650
3	20,000	0.36	7,200
4	20,000	0.26	5,200
5	5,000	0.19	950

NPV = −£1,900

Table 5.18 *Net Present Value – tables for examples*

Factor rate / Years	1%	2%	3%	4%	5%	6%	7%	8%	9%	10%	11%	12%
1	0.99	0.98	0.97	0.96	0.95	0.94	0.93	0.93	0.92	0.91	0.90	0.89
2	0.98	0.96	0.94	0.92	0.91	0.89	0.87	0.86	0.84	0.83	0.81	0.80
3	0.97	0.94	0.92	0.89	0.86	0.84	0.82	0.79	0.77	0.75	0.73	0.71
4	0.96	0.92	0.89	0.85	0.82	0.79	0.76	0.74	0.71	0.68	0.66	0.64
5	0.95	0.91	0.86	0.82	0.78	0.75	0.71	0.68	0.65	0.62	0.59	0.57
6	0.94	0.89	0.84	0.79	0.75	0.70	0.67	0.63	0.60	0.56	0.53	0.51
7	0.93	0.87	0.81	0.76	0.71	0.67	0.62	0.58	0.55	0.51	0.48	0.45
8	0.92	0.85	0.79	0.73	0.68	0.63	0.58	0.54	0.50	0.47	0.43	0.40
9	0.91	0.84	0.77	0.70	0.64	0.59	0.54	0.50	0.46	0.42	0.39	0.36
10	0.91	0.82	0.74	0.68	0.61	0.56	0.51	0.46	0.42	0.39	0.35	0.32
11	0.90	0.80	0.72	0.65	0.58	0.53	0.48	0.43	0.39	0.35	0.32	0.29
12	0.89	0.79	0.70	0.62	0.56	0.50	0.44	0.40	0.36	0.32	0.29	0.26
13	0.88	0.77	0.68	0.60	0.53	0.47	0.41	0.37	0.33	0.29	0.26	0.23
14	0.87	0.76	0.66	0.58	0.51	0.44	0.39	0.34	0.30	0.26	0.23	0.20
15	0.86	0.74	0.64	0.56	0.48	0.42	0.36	0.32	0.27	0.24	0.21	0.18

Factor rate / Years	13%	14%	15%	16%	17%	18%	19%	20%	30%	40%	50%
1	0.88	0.88	0.87	0.86	0.85	0.85	0.84	0.83	0.77	0.71	0.67
2	0.78	0.77	0.76	0.74	0.73	0.72	0.71	0.69	0.59	0.51	0.44
3	0.69	0.67	0.66	0.64	0.62	0.61	0.59	0.58	0.46	0.36	0.30
4	0.61	0.59	0.57	0.55	0.53	0.52	0.50	0.48	0.35	0.26	0.20
5	0.54	0.52	0.50	0.48	0.46	0.44	0.41	0.40	0.27	0.19	0.13
6	0.48	0.46	0.43	0.41	0.39	0.37	0.35	0.33	0.21	0.13	0.09
7	0.43	0.40	0.38	0.35	0.33	0.31	0.30	0.28	0.16	0.09	0.06
8	0.38	0.35	0.33	0.31	0.28	0.27	0.25	0.23	0.12	0.07	0.04
9	0.33	0.31	0.28	0.26	0.24	0.23	0.21	0.19	0.09	0.05	0.03
10	0.29	0.27	0.25	0.23	0.21	0.19	0.18	0.16	0.07	0.03	0.02
11	0.26	0.24	0.21	0.20	0.18	0.16	0.15	0.13	0.06	0.02	0.01
12	0.23	0.21	0.19	0.17	0.15	0.14	0.12	0.11	0.04	0.02	0.008
13	0.20	0.18	0.16	0.15	0.13	0.12	0.10	0.09	0.03	0.013	0.005
14	0.18	0.16	0.14	0.13	0.11	0.10	0.09	0.08	0.03	0.009	0.003
15	0.16	0.14	0.12	0.11	0.09	0.08	0.07	0.06	0.02	0.006	0.002

The IRR is located in-between 30% and 40%. Let us now employ interpolation to obtain the IRR.

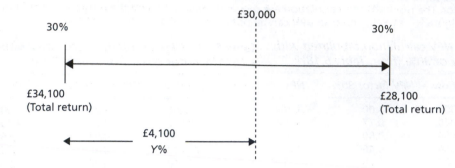

Therefore in monetary terms the distance from 30% to 40% = £34,100 − £28,100

$$= \underline{£6,000}$$

In percentage terms this is 40% − 30% = <u>10%</u>

Therefore $\dfrac{6000}{10} = 600 = \dfrac{4100}{Y\%}$

$$\therefore Y\% = \dfrac{4100}{600}$$

$$= 6.83$$

$$\text{IRR} = 30 + 6.83$$

$$= 36.83\%$$

The techniques have been applied to Project 'A' only. However, their main advantage is as a decision-making aid when evaluating between various options. Therefore they are now applied to Projects 'B' and 'C' from the data provided in Table 5.12.

Table 5.19 *Payback period for Project 'B'*

Year	Cash flow	Cumulative cash flow	
1	15,000	15,000	Capital investment
2	15,000	30,000	← £30,000
3	10,000	40,000	
4	5,000	45,000	
5	20,000	65,000	

Solution = Payback period = 2 years

Table 5.20 *Payback period for Project 'C'*

Year	Cash flow	Cumulative cash flow	
1	12,000	12,000	Capital investment
2	12,000	12,000	← £40,000
3	15,000	15,000	
4	20,000	20,000	
5	20,000	20,000	

$$\therefore \left[\dfrac{\text{shortfall}}{\text{next increment}} \right] \times 12 = \text{time in months to recover shortfall}$$

$$= \dfrac{1,000}{20,000} \times 12 = 0.6 \text{ month}$$

Payback period = 3 years 0.6 month, rounded up to 3 years 1 month

Table 5.19 clearly indicates that the payback period for Project 'B' is 2 years. Table 5.20, with associated calculations, provides a payback period for Project 'C' of 3 years 1 month. Therefore a ranking of all three projects based on the payback period is

Ranking 1st   Project 'B'   2 years
Ranking 2nd   Project 'A'   2 years 3 months
Ranking 3rd   Project 'C'   3 years 1 month

Average Annual Percentage Rate of Return (AA%RR)

(Data from Table 5.12 again.)

Table 5.21   *AA%RR for Project 'B'*

Year	Cash flow
1	15,000
2	15,000
3	10,000
4	5,000
5	20,000

Stage I

$$AARR = \frac{\text{Total return}}{\text{Investment period}} = \frac{£65,000}{5} = £13,000$$

Stage II

$$AA\%RR = \frac{AARR}{\text{Capital investment}} \times 100 = \frac{13,000}{30,000} \times 1000 = 43.33\% \text{ for Project 'B'}$$

Table 5.22   *AA%RR for Project 'C'*

Year	Cash flow
1	12,000
2	12,000
3	15,000
4	20,000
5	20,000

Stage I

$$AARR = \frac{\text{Total return}}{\text{Investment period}} = \frac{£79,000}{5} = £15,800$$

Stage II

$$AA\%RR = \frac{AARR}{\text{Capital investment}} \times 100 = \frac{15,800}{40,000} \times 1000 = 39.5\% \text{ for Project 'C'}$$

Table 5.21 with associated calculations provides an AA%RR of 43.33%. Table 5.22 with calculations indicates an AA%RR of 39.5%. Therefore a ranking of the three projects based on AA%RR is

Ranking 1st   Project 'A'   46.67%
Ranking 2nd   Project 'B'   43.33%
Ranking 3rd   Project 'C'   39.5%

Net Present Value (NPV) example for Projects 'B' and 'C' based upon the data from Table 5.12:

Table 5.23 *NPV calculations for Project 'B' at 15% factor*

Year	Cash flow	NPV factor 15%	NPV
0	£30,000	1.00	−30,000
1	15,000	0.87	13,050
2	15,000	0.76	11,400
3	10,000	0.66	6,600
4	5,000	0.57	2,850
5	20,000	0.50	10,000

NPV = +£13,900

Table 5.24 *NPV calculations for Project 'C' at 15% factor*

Year	Cash flow	NPV factor 15%	NPV
0	£40,000	1.00	−40,000
1	12,000	0.87	10,440
2	12,000	0.76	9,120
3	15,000	0.66	9,900
4	20,000	0.57	11,400
5	20,000	0.50	10,000

NPV = +£10,860

Table 5.23 establishes the NPV of +£13,900 for Project 'B'. Table 5.24 provides an NPV of +£10,860 for Project 'C'. Therefore the ranking of Projects 'A', 'B' and 'C' based on NPV calculations is

Ranking 1st	Project 'A'	+£17,200
Ranking 2nd	Project 'B'	+£13,900
Ranking 3rd	Project 'C'	+£10,860

The internal rate of return (IRR) example for Projects 'B' and 'C' is based on the data from Table 5.12. NPV factors are taken as 30% and 40% from Table 5.18.

Table 5.25 *IRR for Project 'B' at 30% factor*

Year	Cash flow	NPV factor 30%	NPV
0	£30,000	1.00	−30,000
1	15,000	0.77	11,550
2	15,000	0.59	8,850
3	10,000	0.46	4,600
4	5,000	0.35	1,750
5	20,000	0.27	5,400

= +£2,150

Table 5.26 *IRR for Project 'B' at 40% factor*

Year	Cash flow	NPV factor 30%	NPV
0	£30,000	1.00	−30,000
1	15,000	0.71	10,650
2	15,000	0.51	7,650
3	10,000	0.36	3,600
4	5,000	0.26	1,300
5	20,000	0.19	3,800

= −£3,000

Interpolation Project 'B'

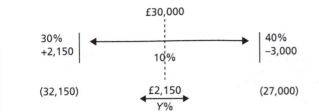

$$\frac{5150}{10} = \frac{2150}{Y\%} \therefore 515 = \frac{2150}{Y\%}$$

$$\therefore Y\% = \frac{2150}{515} = 4.175\%$$

30% + 4.18% = 34.18% IRR for Project 'B'

Table 5.27 *IRR for Project 'C' at 30% factor*

Year	Cash flow	NPV factor 30%	NPV
0	£40,000	1.00	−40,000
1	12,000	0.77	9,240
2	12,000	0.59	7,080
3	15,000	0.46	6,900
4	20,000	0.35	7,000
5	20,000	0.27	5,400
			= −£4,380

Table 5.28 *IRR for Project 'C' at 40% factor*

Year	Cash flow	NPV factor 20%	NPV
0	£40,000	1.00	−40,000
1	12,000	0.83	9,960
2	12,000	0.69	8,280
3	15,000	0.58	8,700
4	20,000	0.48	9,600
5	20,000	0.40	8,000
			= +£4,540

Interpolation Project 'C'

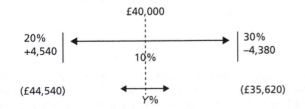

$$\frac{8920}{10} = \frac{4540}{Y\%} \quad \therefore 892 = \frac{4540}{Y\%}$$

$$\phi Y\% = \frac{4540}{892} = 5.09\%$$

$$20\% + 5.09\% = 25.09\%$$

The data provided from Tables 5.25 and 5.26 shows, with associated calculations, an IRR of 34.18% for Project 'B'. Tables 5.27 and 5.28 clearly establish (again with associated calculations) an IRR of 25.09% for Project 'C'. The ranking of all three projects, based on IRR, is as follows:

Ranking 1st   Project 'A'   36.83% IRR
Ranking 2nd   Project 'B'   34.18% IRR
Ranking 3rd   Project 'C'   25.09% IRR

Table 5.29 *Ranking matrix for appraisal methods and projects in examples*

Appraisal technique utilised	Ranking 'Projects'	Result of appraisal calculations
Payback period	'B'	2 years
	'A'	2 years 3 months
	'C'	3 years 1 month
Average annual percentage rate of return	'A'	46.67%
	'B'	43.33%
	'C'	30.50%
NPV	'A'	+£17,200
	'B'	+£13,900
	'C'	+£10,860
IRR	'A'	36.83%
	'B'	34.18%
	'C'	25.09%

Table 5.29 notes Project 'A' as the first choice in all investment appraisal techniques other than by the payback period method. The assessment ranking matrix is a useful methodology to be adopted by the person undertaking the analysis as it provides the outcomes of the appraisal for senior management (without all the calculations). The calculations used in the appraisal process should be made available to the Construction Manager by the analyser, if this information is requested.

Summary

The techniques provided within this section for appraisal purposes are useful decision-making aids. The techniques are essentially tools for problem solving and these tools can be combined. For example, the payback period method could be combined with the net present value approach. This would result in determination of the payback period after the funds flow has been discounted. Hence the data would be more valuable to the decision maker(s).

Budgets, Cash Flows and Variance Analysis

A budget is a plan which has been converted into quantitative and monetary form, a plan that has been agreed by all those in the management of the project to be sensible and achievable. Budgetary control is applied at both individual project level and in a holistic way to the whole company. Budgets are targets and *"the challenge for the construction industry is to use the targets to improve performance"* (Bennett, 2000).

There are many advantages which might be obtained from the use of budgetary control. At the budget-setting stage it is hoped to achieve the discipline of getting all managers to give thought to the capabilities of the business and to achieve co-ordination of their efforts by obtaining a proper fit or match to budgets. Budgetary control also involves a monitoring stage. It is used to check against actual performance and the differences which occur, normally called variances, are regarded as being divergences from the plan which need to be addressed by corrective actions, especially if they are variances having the effect of reducing the profit or worsening the cash-flow situation. The cycle is depicted in Figure 5.16.

From a construction perspective the financial budget is closely related to the contract programme for a project. The financial budget converts all resources utilised in the conversion process to a monetary basis for control purposes.

A budget period is the period of time for which a budget is prepared and deployed. Budget periods depend very much on the planning horizons relating to construction projects where there are budgets for various cost centres which may be the elements of the structure or resources such as plant, materials, labour and site overheads.

Cash Flow

Cash flow is a phrase meaning the actual movement of cash in and out of an organisation. Cash flow in (or positive cash flow) is cash received and cash flow out (negative cash flow) is cash paid out. The difference between these two is termed the 'net cash flow'.

Cash-flow analysis is a decision-making/analytical technique that utilises the differences between the cash inflows and outflows. For construction companies cash-flow analysis provides a relatively simple method for monitoring the cash flows related to a contract programme linked to allocated budgets. This process is used at the pre-tender stage and empowers an organisation to be proactive in the planning and controlling processes.

Most of the problems of cash-flow control occur in the production/operational aspects of a company, especially on construction-related projects. This makes the budget allocation developed for construction programmes, cash-flow analysis followed by variance analysis, a key activity for effective and efficient control.

Variance Analysis

Once a site manager has received the contract programme and associated budgets, and cash-flow plans and site work are commenced, we enter the Do and Check phase of the Plan, Do, Check, Act control cycle.

The establishment of variances can only be identified after comparing planned budget allocations with actual costs incurred on site. Such a comparison results in the provision of a variance analysis table. The production of the variance analysis table essentially involves taking each cost centre in turn, computing the allowance in view of the actual performances, and then comparing this allowance with the figure attained. This data is then used in the 'Act' phase of the Plan, Do, Check, Act cycle (Figure 5.16) and may result in a change in the Plan: "*In setting targets for individual projects, strategic teams need to take account of the nature of the work being undertaken*" (Bennett, 2000). An example has been provided incorporating the development of a cash flow for a project and the incorporation of a budgetary control exercise.

Cash Flow and Budgetary Control Example Incorporating Variance Analysis

A contractor has entered into a contract to build an extension to a hospital. The terms of the contract are that the amounts to be paid will be assessed from valuations taken at the end of each month, and payment will be made one month later. The retention percentage has been agreed at 5% of the monthly valuation figures, with half being paid on practical completion and the remaining amount six months later.

The programme for the works is shown in Figure 5.23 and the budgeted costs for the work elements and the value of each element are provided in Table 5.30. From the data provided in Figure 5.23 and Table 5.30 we shall:

- calculate the cash flow for the project
- plot cumulative cash in and cash out against time.

The contract commenced on 1 August 2002 and the project duration was 8 months as shown in Figure 5.23.

The contract budget is shown in Table 5.30.

Tables 5.31, 5.32, 5.33 and 5.34 are tabulated representations of the allocation of costs and profits for the contract based on the data provided in Table 5.30. The budgeted costs and profit allocations are based on weekly units related to Figure 5.23.

Figure 5.24 provides the cost and profit elements loaded onto the contract programme. The individual monthly totals have been calculated by adding vertically the cost and profit elements encapsulated within each month. For example, November 2002 has a cost associated with the activity 'Substructure to 1st floor' of £38,499 and a cost associated with '1st floor to roof' of £63,524.50. Therefore, the total cost for November is £38,499 + £63,524.50, giving a total cost of £102,023.50.

The allocation of profit for November is £7,700 for 'Substructure to 1st floor' and £17,325.50 for '1st floor to roof', giving a total profit allocation for November of £7,700 + £17,325.50 = £25,025.50.

In order to determine the cash flow for the hospital extension project, the production of a tabulation sheet is recommended. Table 5.35 provides an example format related to the contract under consideration. The profit and cost components have been abstracted from Figure 5.24. The headings for the columns on Table 5.35 (Cash flow – tabulated analysis sheet) indicate how to work through the sheet. They enable the contract conditions to be built into the analysis. For example the 'Amount due' is the monthly valuation figure comprising costs and profit allocation. The retention row is 5% of the 'Amount due' figures.

Activities \ Month	Aug 2002	Sept 2002	Oct 2002	Nov 2002	Dec 2002	Jan 2003	Feb 2003	Mar 2003
Substructure	▬▬▬	▬						
Substructure to lst floor			▬▬	▬				
1st floor to roof				▬▬				
Services					▬			
Finishes						▬▬▬	▬	
External works							▬▬	▬

Figure 5.23 *Contract/programme for extension of hospital project*

Table 5.30 *The contract budgeted costs and value of works*

Budgeted cost	Work element	Value of work
231,000	substructure	288,200
115,499	substructure to 1st floor	138,600
127,049	1st floor to roof	161,700
57,750	services	98,175
103,949	internal finishes	150,150
254,100	external works	317,625
£889,347		£1,154,450

Table 5.31 *Budget allocation per weekly increment*

Work elements of structure	Duration in weeks	Budgeted costs (£)	Budgeted costs ÷ Weekly duration
Substructure	8	231,000	$\dfrac{231,000}{8} = 28,875$
Substructure to 1st floor	6	115,499	$\dfrac{115,499}{6} = 19,250$
1st floor to roof	4	127,049	$\dfrac{127,049}{4} = 31,762$
Services	2	57,750	$\dfrac{57,750}{2} = 28,875$
Internal finishes	8	103,949	$\dfrac{103,949}{8} = 12,994$
External works	10	254,100	$\dfrac{254,100}{10} = 25,410$
Totals	38	889,347	

Table 5.32 *Profit allocation per weekly increment*

Work elements of structure	Duration in weeks	Value of work element (£)	Value − budget = profit (£)	Profit ÷ Weekly duration
Substructure	8	288,200	288,200 − 231,000 = 57,200	$\dfrac{57,200}{8} = 7,150$
Substructure to 1st floor	6	138,600	138,600 − 115,499 = 23,101	$\dfrac{23,101}{6} = 3,850$
1st floor to roof	4	161,700	161,700 − 127,049 = 34,651	$\dfrac{34,651}{4} = 8,663$
Services	2	98,175	98,175 − 57,750 = 40,425	$\dfrac{40,425}{2} = 20,212.50$
Internal finishes	8	150,150	150,150 − 103,949 = 46,201	$\dfrac{46,201}{8} = 5,775$
External works	10	317,625	317,625 − 254,100 = 63,525	$\dfrac{63,525}{10} = 6,352.50$
Totals	38	1,154,450	Profit total 265,103	

Table 5.33 *Allocation of budgeted costs onto contract programme calculation*

Weekly budgeted total and activity	Activity duration in weeks per month	Month	Allocation per month
Substructure 28,875	4 4	August 2002 September 2002	115,500 115,500
Substructure to 1st floor 19,250	4 2	October 2002 November 2002	77,000 38,499
1st floor to roof 31,762	2 2	November 2002 December 2002	63,524.50 63,524.50
Services 28,875	2	December 2002	57,750
Internal finishes 12,994	4 4	January 2003 February 2003	51,974.50 51,974.50
External works 25,410	2 4 4	January 2003 February 2003 March 2003	50,820 101,640 101,640
Totals	38		889,347

Table 5.34 *Allocation of profit onto contract programme calculation*

Weekly profit total and activity	Activity duration in weeks per month	Month	Allocation per month
Substructure 7,150	4 4	August 2002 September 2002	28,600 28,600
Substructure to 1st Floor 3,850	4 2	October 2002 November 2002	15,400 7,700
1st floor to roof 8,663	2 2	November 2002 December 2002	17,325.50 17,325.50
Services 20,213	2	December 2002	40,425
Internal finishes 5,775	4 4	January 2003 February 2003	23,100 23,100
External works 6,353	2 4 4	January 2003 February 2003 March 2003	12,705 25,410 25,410
Totals	38		£265,101

Month / Activity	August 2002	September 2002	October 2002	November 2002	December 2002	January 2003	February 2003	March 2003	Totals
Substructure	115,500	115,500							231,000
	28,600	28,600							57,200
Substructure to 1st floor			77,000	38,499					115,499
			15,400	7,700					23,101
1st floor to roof				63,524.50	63,524.50				127,049
				17,325.50	17,325.50				34,651
Services					57,750				57,750
					40,425				40,425
Internal finishes						51,974.50	51,974.50		103,949
						23,100	23,100		46,200
External works						50,820	101,640	101,640	254,100
						12,705	25,410	25,410	63,525
Monthly profit	28,600	28,600	15,400	25,025.50	57,750.50	35,805	48,510	25,410	265,101
Monthly costs	115,500	115,500	77,000	102,023.50	121,274.50	102,794.50	153,614.50	101,640	889,347

Figure 5.24 Contract programme for hospital project

Table 5.35 Cash flow – tabulated analysis sheet for contract example

Month	Costs (£)	Profit (£)	Amount due	Retention 5%	Amount due less retention	Cumulative amount due	Time-adjusted cumulative amount due (inflow)	Cumulative costs	Time-adjusted cumulative costs (outflow)	Cash flow
1	115,500	28,600	144,100	7,205	136,895	136,895	136,895	115,500	115,500	+21,395
2	115,500	28,600	144,100	7,205	136,895	273,790	273,790	231,000	231,000	+42,790
3	77,000	15,400	92,400	4,620	87,780	361,570	361,570	308,000	308,000	+53,570
4	102,023.50	25,025.50	127,049	6,352.50	120,697	482,267	482,267	410,023.50	410,023.50	+72,243.50
5	121,274.50	57,750.50	179,025	8,951.30	170,073.70	652,340.70	652,340.70	531,298	531,298	+121,042.70
6	102,794.50	35,805	138,599.50	6,930	131,670	784,010.70	784,010.70	634,092.50	634,092.50	+149,918.20
7	153,614.50	48,510	202,124.50	10,106.30	192,018.20	976,028.90	976,028.90	787,707	787,707	+188,321.90
8	101,640	25,410	127,050	6,352.50	120,697.50	1,125,587.70 (+1/2 R)	1,125,587.70	889,347	889,347	+236,240.70
9				Total 57,722.60*						
10				1/2R = 28,861.30						
11										
12										
13										
14								889,347		
15						1,154,450 (+1/2 R)	1,154,450		889,347	265,103
16										

* Plus half retention of £28,861.30 due on practical completion and the end of defects liability period (six months)

Notes

1. 'Time-adjusted cumulative amount due' column figures provide the stepped lines on the 'S' curve graph.
2. 'Time-adjusted cumulative costs' figures provide the 'S' curve line on the 'S' curve graph. See Figure 5.25.

Standard 'S' Curve Graph for Hospital Contract Example

In order to be able to plot the data (see Figure 5.25) and determine a cash-flow profile we have to use cumulative figures. Therefore the 'cumulative amount due' column is a running total of the 'amount due' column. An allowance needs to be made for the fact that the contract conditions state that payments carry a month's delay, so we have to adjust the cumulative amount due payment by one month. Cumulative costs are then determined and they also have to be delayed by a month in payment; this is the 'time-adjusted cumulative costs' column.

The cash-flow row is calculated by subtracting the 'time-adjusted cumulative costs' column from the 'time-adjusted cumulative amount due' column. In this example the cash-flow column clearly shows a cash surplus from month 2 onwards (see Table 5.35). The cash-flow figures can then be plotted to produce a cash-flow graph (see Figure 5.26).

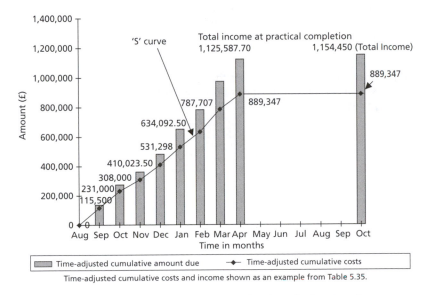

Time-adjusted cumulative costs and income shown as an example from Table 5.35.

Figure 5.25 *Standard 'S' curve graph for hospital extension contract*

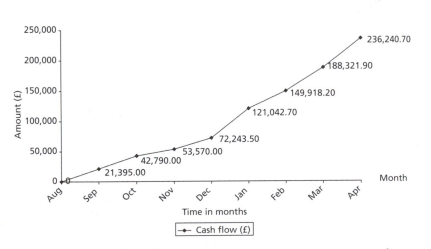

Figure 5.26 *Cash-flow graph for hospital contract example*

The pictorial representations depicted in Figures 5.25 and 5.26 are part of the pre-tender analysis phase of the contract. They have been produced using the contract programme (Figure 5.23) and the allocated budgets from Table 5.35. One must remember that the rationale for producing the cash-flow analysis is to provide data for the Plan, Do, Check, Act control cycle (see Figure 5.16). The contract programme, budgets and cash-flow graphs are all part of the 'Plan' function. Thus, these have to be compared with the results obtained from the actual work done on site.

The 'Do'

As an example and leading on to the 'Act', let us take the cash-flow graph of Figure 5.26 and develop this further with the production of Figure 5.27. Included in Figure 5.27 is a further plot produced from the actual cash flows incurred on commencement of work activities. As can be seen from Figure 5.27, the contract is showing a higher positive cash flow than the projected cash-flow. However, we must remember not to use one piece of data in isolation, and the data for the actual cash-flow figures would need to be related back to the basis of our 'plans', which was the contract programme, and it may be that we are ahead of contract programme and this is the reason for the difference between planned and actual cash flows. Also, on completion of the contract we may arrive back at the total cumulative cash-flow figure of £236,240.70.

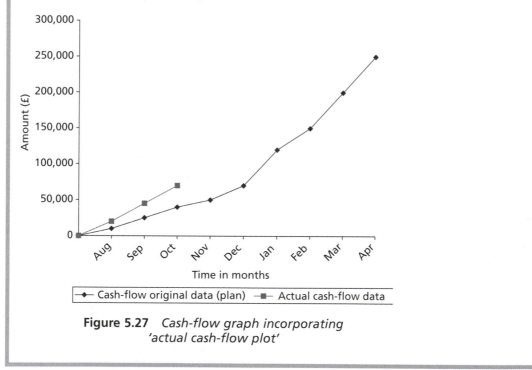

Figure 5.27 *Cash-flow graph incorporating 'actual cash-flow plot'*

Holistic Aspects of Cash-flow Graph

The usefulness of the cash-flow graph in representing the total cash-flow requirements for an organisation should not be forgotten. It is possible to plot the cash flow for more than one contract on the same graph. If this is done then the cumulative impact of cash flows can be readily noted and fed back into the corporate decision-making process. Figure 5.28 incorporates the example contract data cash flow and a fictitious contract so that the cumulative profile can be demonstrated.

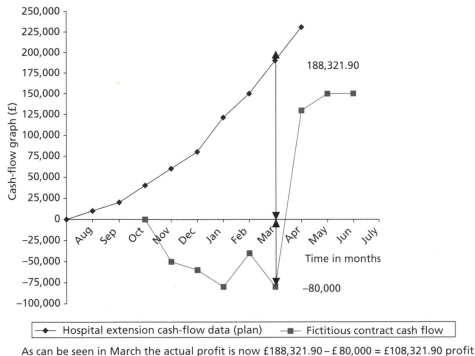

As can be seen in March the actual profit is now £188,321.90 – £80,000 = £108,321.90 profit

Figure 5.28 *Cash-flow graph incorporating fictitious contract cash flow*

Figure 5.28 can be further utilised when new contracts are introduced and their cumulative effects can be determined. This is an important aspect of cash-flow analysis for construction companies because it enables them to be proactive in managing their finances and avoids the dangers of insufficient liquidity.

Thus far, the hospital example has provided the data for the 'Plan', 'Do' components of Figure 5.16. Now we can consider the 'Check' and 'Act'. To do this we must make the comparison between our 'Plan' and the 'Do' (actual results). This will lead us on to the budgetary control and variance analysis later in this chapter.

Calculation of Interest Payments for the Hospital Contract

Stages in the calculation process:

1. From the tabulated analysis sheet (Table 5.35) establish:
   - total adjusted cumulative costs, see Figure 5.29
   - the difference between costs and revenue (cash flow)
   - then produce Table 5.36.
2. Plot the cash-flow sawtooth graph (Figure 5.30). This includes the negative and positive values from Table 5.36.
3. Produce Table 5.37 (funding required) from Table 5.36 and Figure 5.30.
4. Calculate interest payments as per Table 5.38.

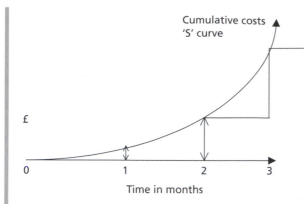

Figure 5.29 *Allocation of cumulative costs for the hospital project; example for interest calculation*

Month 3 will be the total value at month 3 − total of month 2.
Months 1 and 2 are full cash requirements.
Note: The costs are cumulative.
Do not subtract any value from month 1 or month 2;
commence at month 3. This process will provide the maximum cost requirements.

Table 5.36 *Calculation of interest payment for hospital example*

No.	Month	Time-adjusted cumulative costs*	Difference between costs and revenue†	Plotting the negative value
1	August 02	£50,000.00		−£50,000.00
2	September 02	£115,500.00	£21,395.00	−£65,500.00
3	October 02	£231,000.00	£42,790.00	−£115,500.00
4	November 02	£308,000.00	£53,570.00	−£77,000.00
5	December 02	£410,023.50	£72,244.00	−£102,024.17
6	January 03	£531,298.00	£121,043.25	−£121,274.50
7	February 03	£634,092.50	£149,918.75	−£102,794.50
8	March 03	£787,707.00	£188,323.00	−£153,614.50
9	April 03	£889,347.00	£236,241.75	−£101,640.00
10	May 03			
11	June 03			
12	July 03			
13	August 03			
14	September 03	£1,154,450.00	£265,103.00	

† For plotting positive values.
* £50,000.00 in month 1 from cash-flow graph, i.e. the cumulative costs read off the 'S' curve graph (Figure 5.27).
Time-adjusted cumulative costs and cash flow are from the tabulated analysis sheet (Table 5.35).
Plotting the negative values: these figures are obtained by subtracting the time-adjusted cumulative costs from each other.

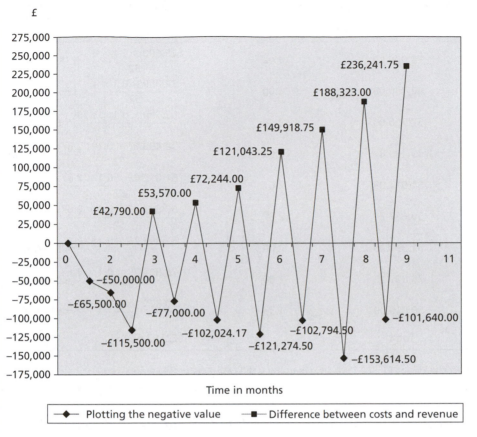

Figure 5.30 *Cash-flow sawtooth graph – the negative and positive cash flows for the hospital contract*

Table 5.37 *Calculation of the funding required for hospital project*

Month	Start of month	End of month	Average	Funding period (weeks)
1	0.00	50,000.00	25,000.00	4.00
2	50,000.00	115,500.00	82,750.00	4.00
3	0.00	115,500.00	57,750.00	3.33
4	0.00	77,000.00	38,499.67	2.67
5	0.00	102,024.17	51,012.09	2.67
6	0.00	121,274.50	60,637.25	2.67
7	0.00	102,794.50	51,397.25	2.00
8	0.00	153,614.50	76,807.25	2.00
9	0.00	101,640.00	50,820.00	1.33

Note: £50,000 in Month 1 from cash-flow graph.

Table 5.38 *Calculation of the average monetary sum lock-up for the hospital contract*

Month	£	Weeks	Interest at 10% per annum
1	25,000.00	4.00	$\dfrac{25{,}000 \times 0.1 \times 4}{52} = £192.31$
2	82,750.00	4.00	$\dfrac{82{,}750 \times 0.1 \times 4}{52} = £636.54$
3	57,750.00	3.33	$\dfrac{57{,}750 \times 0.1 \times 3.33}{52} = £369.82$
4	39,499.67	2.67	$\dfrac{38{,}499.67 \times 0.1 \times 2.67}{52} = £197.68$
5	51,012.09	2.67	$\dfrac{51{,}012.09 \times 0.1 \times 2.67}{52} = £261.93$
6	60,637.25	2.67	$\dfrac{60{,}637.25 \times 0.1 \times 2.67}{52} = £311.30$
7	51,397.25	2.00	$\dfrac{51{,}397.25 \times 0.1 \times 2}{52} = £197.68$
8	76,807.25	2.00	$\dfrac{76{,}807.25 \times 0.1 \times 2}{52} = £295.42$
9	50,820.00	1.33	$\dfrac{50{,}820 \times 0.1 \times 1.33}{52} = £129.99$
	Total		£2592.67

Note: Total interest charges of £2592.67 will be due utilising the cash-flow profile established.

Budgetary Control and Variance Analysis (related to the hospital contract)

We have previously established the importance of the budgetary function but it may be useful to remember that budgeted operating costs are our most realistic expectation of what the actual costs will eventually be. Once we have generated some actual costs (by conducting on-site activities) these may be compared and contrasted with our budgeted costs (at budgeted output levels). This will serve as an overall budget check on the efficiency and effectiveness of the project: *variance analysis compares and contrasts forecasts with actual results in order to establish differences.*

A breakdown of the contract's budgeted costs is shown in Table 5.39.

Table 5.39 *Budget breakdown for extension to hospital contract*

Activities	Labour	Materials	Plant	Site overheads	Totals
Substructure	127,050.00	92,400.00	10,395.00	1,155.00	231,000.00
Substructure to 1st floor	46,200.00	62,947.00	5,775.00	577.00	115,499.00
1st floor to roof	57,750.00	62,947.00	5,775.00	577.00	127,049.00
Services	15,708.00	40,425.00	1,155.00	462.00	57,750.00
Finishes	58,558.00	42,735.00	1,732.00	924.00	103,949.00
External works	92,400.00	68,145.00	92,400.00	1,155.00	254,100.00
Total	397,666.00	369,599.00	117,232.00	4,850.00	889,347.00

We require some actual costs so that a comparison ('Check') can be undertaken with the budgeted costs ('Plan'). These are provided in Table 5.40 for the month ending January 2003. Table 5.40 also incorporates the percentage completions for each element of the structure at this point in time.

The first stage of analysis requires the production of a table which incorporates the percentage completions. This is Table 5.41 and contains the adjustment of Table 5.40 values in accordance with the percentage of work actually completed on site.

Once the percentage adjustments have been made and the new set of budgeted totals determined, the variance analysis table can be produced. This requires the adjusted budgeted costs from Table 5.41 to be compared with the stated actual costs from Table 5.40 ('Check').

The actual costs are subtracted from adjusted budgeted costs. The variance figures may be utilised in the calculation of percentage variances on the contract. Tables 5.42 and 5.43 incorporate this analysis.

It should be noted that the 'Plan', 'Do', 'Check' and 'Act' cycle is not only a process for identifying areas where corrective actions are necessary to maintain control. In fact the process should also identify where performance is better than planned. Table 5.42, for example, has established that the cost centre of 'Plant' has in fact generated a positive value of 4.96% or £2,079.25. Management should establish the reason for this and it could be used to disseminate 'best practice' on other sites. A note of caution is required, as it should be remembered that the results need to be related to the contract programme as is the case with the cash-flow analysis.

Figure 5.31 provides a graphical representation of the variance analysis in percentages.

Table 5.40 *Incorporating 'actual costs' and percentages completed as at January 2003*

Cost centre	Actual costs (£)
Labour	300,000
Materials	287,595
Plant	39,847
Site overheads	3,344

Elements of structure	Percentages completed
Substructure	100%
Substructure to 1st floor	100%
1st floor to roof	95%
Services	80%
Finishes	50%
External works	20%

Table 5.41 *Percentage completions incorporated with values adjusted*

Activities	Work done	Labour	Materials	Plant	Site overheads	Totals
Substructure	100%	127,050.00	92,400.00	10,395.00	1,155.00	231,000.00
Substructure to 1st floor	100%	46,200.00	62,947.00	5,775.00	577.00	115,499.00
1st floor to roof	95%	54,862.50	59,799.65	5,486.25	548.15	120,696.55
Services	80%	12,566.40	32,340.00	924.00	369.60	46,200.00
Finishes	50%	29,279.00	21,367.50	866.00	462.00	51,974.50
External works	20%	18,480.00	13,629.00	18,480.00	231.00	50,820.00
Totals		288,437.90	282,483.15	41,926.25	3,342.75	616,190.05

Table 5.42 *Variance analysis for the hospital contract*

Cost centre	Budgeted cost	Actual cost	Variance	Individual variance/total adjusted budget for cost centre (%)
Labour	288,437.90	300,300.00	−11,862.10	−11,862.10 / 288,437.90 × 100% = −4.11%
Materials	282,483.15	287,595.00	−5,111.85	−5,111.85 / 282,483.15 × 100% = −1.81%
Plant	41,926.25	39,847.00	+2,079.25	2,079.25 / 41,926.25 × 100% = +4.96%
Site overheads	3,342.75	3,344.00	−1.25	−1.25 / 3,342.75 × 100% = −0.04%
Total	616,190.05	631,086.00	−14,895.95	

Table 5.43 *Total variances as a percentage of total budgeted cost*

$$\text{Percentage variance} = \frac{\text{Total variance}}{\text{Total budgeted costs}} \times 100 = \frac{-14,895.95}{616,190.05} \times 100 = -2.42\% \text{ total}$$

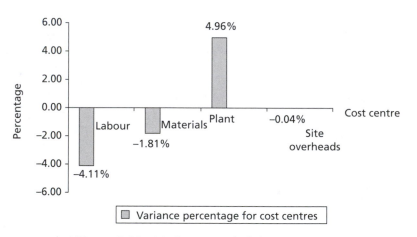

Figure 5.31 *Variance analysis in percentages*

Figure 5.32 shows the pictorial representation of the variance analysis in monetary terms. Tables 5.44 and 5.45 indicate the predicted variance in percentages (Table 5.44) and in monetary terms (Table 5.45) based on the current variance between budgeted and actual costs.

The total variance on the contract has been established along with individual cost-centre variances. However, the value of such analysis is dependent on both the timeliness and the level of detail encapsulated in the analysis. The above two points relate to Figure 5.16 which shows the control cycle ('Plan', 'Do', 'Check' and 'Act'). To be both efficient and effective in the operation of the control function (control loop) we require regular comparisons ('Check') and 'short cycle times' for the information to circulate around the loop. The quality of the information in circulation needs to be sufficient to ensure effective corrective actions. For example, knowing the total variance on the hospital contract is −2.42% is not as useful as knowing the individual variances for the cost centres as depicted in Table 5.42. The decision-making process is enhanced by the quality of data that goes into it.

The whole point of conducting the cash-flow analysis, budgetary control and variance analysis is to enable the management function of 'control' to be exercised by construction managers. The control activity can only be undertaken before completion of set tasks and it should be remembered that it is impossible to have effective retrospective control. The utilisation of Figure 5.16 cannot be overemphasised.

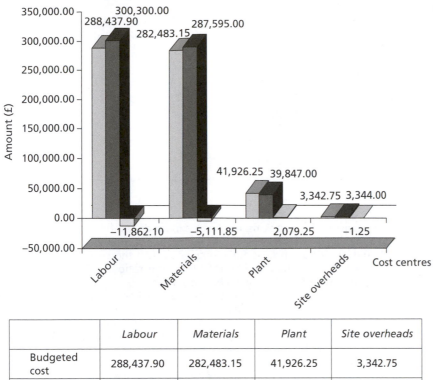

	Labour	Materials	Plant	Site overheads
Budgeted cost	288,437.90	282,483.15	41,926.25	3,342.75
Actual cost	300,300.00	287,595.00	39,847.00	3,344.00
Variance	−11,862.10	−5,111.85	2,079.25	−1.25

Figure 5.32 *Variance analysis in monetary terms*

Table 5.44 *Prediction of total variance at the end of the contract*

	Variance % (£)	Budgeted (£)	Predicted actual (£)	Variance value (£)
Labour	−4.11%	397,666.00	414,010.07	−16,377.07
Materials	−1.81%	369,599.00	376,288.74	−6,689.74
Plant	4.96%	117,232.00	111,417.29	+5,814.71
Site overheads	−0.04%	4,850.00	4,851.94	−1.94
	Total	889,347.00	906,568.05	−17,221.05

Total variance as a percentage of total budgeted costs

$$\text{Percentage variance} = \frac{\text{Total variance}}{\text{Total budgeted costs}} \times 100$$

$$= \frac{-17,221.05}{889,347.00} \times 100 = -1.94\%$$

Table 5.45 *Prediction of profit at the end of the project, based on actual variance*

	Budgeted (£)	Predicted actual (£)
Contract sum	1,154,450.00	1,154,450.00
Cost	889,347.00	906,568.05
Profit	£265,103.00	£247,881.95
	29.8%	27.3%

Summary

Within this chapter the integrative nature of 'operations' has been expounded. It has been advocated that 'operations' should provide a means for the attainment and maintenance of a 'sustainable corporate competitive advantage'. Should operations not be fully integrated into the corporate decision-making process, the result will be a 'corporate millstone' effect. Both the rationale for and models enabling the full inclusion of the operational function have been provided.

Further, in this chapter a range of various decision-making techniques has been outlined and examples included. These have been incorporated because it is important to realise that the concept of control is a fundamental one and impacts upon the profitability of construction organisations. Cost control is a management function and for construction managers to perform this task efficiently and effectively they require timely, high-quality information to be encapsulated within their decision-making process. Some of the most tragic errors in the construction industry are made as a result of a failure to predetermine the cash flows, monitor them and deploy any necessary actions consequent on the establishment of variations from planned activities/costs.

Thus the chapter provides a framework for performing the critical activities that form the 'Plan', 'Do', 'Check', 'Act' cycle applied to a construction project depicted in Figure 5.16.

The concepts presented within the chapter are corroborated by Bennett (2000): *"Feedback is essential for control and for continuous improvements in performance. Feedback is knowledge to a system about the effects of its actions on its environment."*

References

Armstrong M. (1986) *A Hand Book of Management Technique*. Kogan Page, London, ISBN 1–85091–077–4.

Bennett J. (2000) *Construction the Third Way*. Butterworth-Heinemann, London, ISBN 07–5063093–0.

Bennett J. and Jayes S. (1995) *Trusting the Team: The Best Practice Guide to Partnering in Construction*. Reading Construction Forum. Thomas Telford, Reading.

Bennett J. and Jayes S. (1998) *The Seven Pillars of Partnering – A Guide to Second Generation Partnering*. Reading Construction Forum. Thomas Telford, Reading.

Betts M. and Lansley P. (1993) A Review of the First Ten Years, *Construction Management and Economics*, **11**, 221–45. Spon, London, ISSN 0144–6193.

Betts M. and Wood-Harper T. (1994) Re-engineering Construction – A New Management Research Agenda, *Construction Management and Economics*, **12**, 551–6. Spon, London.

Bingham T. (1998) Spot the Balls, *Building*, 21 August, p 31.

Bogan C.E. and English M.J. (1994) *Benchmarking for Best Practices – Winning Through Innovative Adaption*. McGraw-Hill Inc., London.

Carr D.K. and Johansson H.J. (1995) *Best Practice in Re-engineering – What Works and What Doesn't in the Re-engineering Process*. McGraw-Hill Inc., USA, ISBN 0–07–011224–X.

Champy J. (1993) In Re-engineering, Organisational Change Must Start from Day One, *CSC Insights*, No. 4.

CIB (1997) *Partnering in the Team*, Working Group 12. Thomas Telford, London, ISBN 0–7277–2551–3.

CIB Fact Sheet (1998) Take your Partners by the Hand, *Construction News*, 18 June.

Coulter S. (1998) Most Clients yet to try Partnering, *Building*, 5 June, p 17.

Davenport T.H. (1993) *Process Innovation – Re-engineering Work through Information Technology*. Harvard Business Press, Cambridge, MA, ISBN 0–87584–366–2.

Drucker P.F. (1992) *Tasks, Responsibilities and Management Practices*. Butterworth-Heinemann, London.

Egan Sir E. (1998) *Rethinking Construction*. A Task Force Report, HMSO, London.

Fallah M.H. and Weinman J.B. (1995) *Will Re-engineering Replace TQM?* Engineering Management Conference, ISBN 0–7803–2799–3.

Gant J.G. (1992) Work Management – The Next Step in Imaging, *Chief Information Officer Journal*, Fall, pp 60–4.

Grover V. and Kettinger W.J. (1995) *Business Process Change – Concepts, Methods and Technologies*. Idea Group Publishing, London, ISBN 1–878289–29–2.

Gyles R.V. (1992) *Royal Commission into Productivity in the Building Industry in New South Wales*. Government of New South Wales, Sydney.

Hammer M. (1990) Re-engineering Work – Don't Automate, Obliterate, *Harvard Business Review*, July–August, pp 104–12.

Hammer M. and Champy J. (1993) *Re-engineering the Corporation – A Manifesto for Business Revolution*. Nicholas Brealey Publishing, ISBN 1–85788–029–3.

Harrington H.J. (1991) *Business Process Improvement – The Breakthrough Strategy for Total Quality, Productivity and Competitiveness*. McGraw Hill Inc., New York, ISBN 0–07–026768–5.

Hellard R.B. (1995) *Project Partnering: Principle and Practice*. Thomas Telford, London, ISBN 0–7277–2043–0.

Hinks A.J., Allen S. and Cooper R.D. (1996) Adversaries or Partners? Developing Best Practice for Construction Industry Relationships, *The Organisation and Management of Construction: Shaping Theory and Practice*. (eds D.A. Langford and A Retik). *Proceedings of the CIB W65 International Symposium*, Glasgow, Scotland, 28 August–3 September, 1, pp. 220–8.

Jeremy S.G. and Rohit T. (1994) *Re-engineering the Business*. BBC Enterprises, London, ISBN 101–431–535–2.

Johansson H.J., McHugh P., Pendlebury A.J. and Wheeler W.A. (1993) *Business Process Re-engineering – Breakpoint Strategies for Market Dominance*. Wiley, England, ISBN 0–471–93883–1.

Koskela L. (1992) Application of the New Production Philosophy to Construction, *CIFE, Technical Report* No. 72. Stanford University, USA.

Latham M., Sir (1994) *Constructing the Team – Final Report of the Government/Industry Review of Procurement and Contractual Arrangements in the UK Construction Industry*. HMSO, Department of the Environment, London, ISBN 0–11–752994–X.

Leonard F.S. (1982) The Incline of Quality, *Harvard Business Review*, 59.

Lover P. and Mohamad S. (1995) Construction Process Re-engineering, *Building Economist*, No. 4.

McGeorge D. and Palmer A. (1997) *Construction Management – New Directions*. Blackwell Science, Oxford, ISBN 0–632–04258–3.

McManamy R. (1994) CII Benchmarks Savings, *ENR*, 15 August.

Mohamed S. and Tucker S. (1996) Options for Applying BPR in the Australian Construction Industry, *International Journal of Project Management*, 14(6), 279–385. Elsevier Science Ltd and IPMA, Great Britain, ISSN 0263–7863.

New South Wales Department of Public Works and Services (1995) *Contractor Accreditation Scheme to Encourage Reform and Best Practice in the Construction Industry*. Government of New South Wales, Sydney.

Oakland J.S. (1989) *Total Quality Management*, 2nd edn. Heinemann, London.

Parkora J. and Hastings C. (1995) *Building Partnerships: Teamworking and Alliances in the Construction Industry*. Construction Paper No. 54.

Peppard J. and Rowland P. (1995) *The Essence of Business Process Re-engineering*. Prentice-Hall International, London, ISBN 0–13–310707–8.

Platje A. and Woodman S. (1998) *International Journal of Project Management*, 16(4), 201–8.

Schutzel H.J. and Unruh V.P. (1996) *Successful Partnering: Fundamentals for Project Owners and Contractors*. Wiley, New York.

Soles S. (1995) *Work Re-engineering and Workflow – Comparative Methods*. Future Strategies Inc., USA, ISBN 0–9640233–2–6.

Stevens D. (1993) Partnering and Value Management, *Building Economist*, 5–7 September.

Toffler A. (1990) *Power Shift*. Bantam Books, New York.

Venkatraman N. (1994) IT Enabled Business Transformation – From Automation to Business Scope Redefinition, *Sloan Management Review*, No. 5, Spring, pp 15–29.

Walker D.H.T. (1995) *An Investigation into Construction Time Performance*. Spon, London. *Construction Management and Economics*, **13**, 263–74, ISSN 0144–6193.

Watson K. (1999) Is Partnering Starting to Mean … All Things to All People? *Contract Journal*, 24 March, p 8.

Watson P. (2002) Developing an Efficient and Effective Control System, *Journal of the Association of Building Engineers*, **77**(3), 28–9.

6 Site Establishment

Introduction

An essential component of pre-tender planning, pre-contract planning and start on site is 'site establishment'. An efficient and effective site establishment provides the foundation for a successful project by configuring, structuring and organising those temporary facilities needed to support the works on site. While small construction works might require little temporary site establishment, larger projects may require extensive site infrastructure. This needs to be costed, built into the tender price and actually delivered on site to provide the management and resource capacity to carry out the works and to provide appropriate welfare facilities for persons working on site. Legislation must be adhered to when considering site organisation and welfare facilities, the prominent legislation being The Construction (Health, Safety and Welfare) Regulations 1996. In addition, local bye-laws and particular working restrictions affecting the site will need to be complied with. This chapter introduces those key aspects of site establishment which are common to large construction projects and, although not prescriptive, provide appropriate considerations.

Site Establishment

Site establishment may be usefully considered under four main headings, or groups of components:

(1) Preliminaries
(2) Site organisation
(3) Site layout
(4) Welfare provision.

Preliminaries

The preliminaries section of the tender documents fulfils a number of important functions. At a holistic level, the preliminaries section provides a management overview of the project, including a description of the project and the site, and details of the conditions of contract to be used. The preliminaries will also include a considerable amount of more detailed information about specific issues. Information covered in the preliminaries section normally includes:

- General details of the project
- Contractual matters
- Specific requirements of the employer (client)
- Principal contractor's general cost items
- Information about works to be carried out by persons other than the principal contractor.

General Details of the Project

This section will contain project information, including the name, nature and location of the project, the names and addresses of the employer (client), lead consultant, quantity surveyor, structural engineer, mechanical engineer, electrical engineer etc., and lists of the relevant drawings. Only the drawings from which the bills have been prepared are specifically required to be listed in this section.

In addition, information will be given about the site and the works:

(i) A description of the work including the types of work involved and details of any unusual features or conditions likely to be encountered on the site.

(ii) Details of site boundaries, either by description and/or by reference to site plans or map references.

(iii) Details of the existing buildings on or adjacent to the site where their presence may have an influence on cost because of their proximity causing restricted access or limitations in the use of plant and equipment.

(iv) Any drainage, water, gas and other mains services known to exist on or over the site, or any services that exist outside the site boundary and which may be affected by site operations.

Most of the information given in this section is not easily costed but will serve to give the contractor some guidance regarding the degree of simplicity or difficulties with the site works.

Contractual Matters

(i) Form of contract to be used, together with the schedule of clause headings of the standard conditions.

(ii) Special conditions to be imposed by the employer, or any amendments to standard conditions.

(iii) The insertions to be made in the appendix to the conditions of contract, where applicable.

(iv) Employer's insurance responsibility.

(v) Any performance guarantee bonds required.

Specific Requirements of the Employer (Client)

The specific employer's requirements on a project may relate to:

(i) Conditions of contract additional to those stated in the tender documentation.

(ii) General requirements for the provision, content, use and interpretation of documents, whether by the architect, consultants, principal contractor or subcontractors.

(iii) Requirements for the general management, supervision and administration of the works, with particular reference to budgeted programme. Matters which may be covered are:
  • Responsibility for co-ordination, supervision and administration of the works, including all subcontracts
  • Notification of commencement and completion of the works
  • Relevant insurances
  • Recording of weather conditions
  • The master programme and its monitoring
  • Arrangements for site meetings
  • Conditions relating to materials on site
  • Provision of site labour and plant records
  • Notice to the quantity surveyor before covering up work for site measurement and payment
  • Dayworks and variations.

(iv) General requirements for the quality of materials and workmanship and arrangements for supervision and inspection. Completion of the work and making good of defects. This might include items such as:
  • Protection of unfixed materials
  • Requirements in relation to samples for testing

- Setting out the works, appearance and fit of components, critical dimensions and record drawings
- Proposals for the rectification of defective work or materials
- Clearing the works, removal of rubbish
- Final finishing, including touching up of paintwork, easing, adjusting and lubricating ironmongery and other moving parts, and leaving the works secure on completion.

(v) Requirements for safeguarding the site, the works, unfixed materials, existing buildings and their contents, adjoining property and mains services, and for preventing danger and nuisance; also requirements for health, safety and environmental matters such as listed buildings, tree preservation, orders and aspects of environmental significance.

(vi) Employer's requirements which specifically and directly limit the contractor's/subcontractors' methods of operation, their sequence and timing.

(vii) Facilities, temporary works and temporary services specifically required during the construction of the works.

(viii) Facilities and services required to be provided by the principal contractor/subcontractors at completion or thereafter to help the employer operate and maintain the finished building.

Principal Contractor's General Cost Items

Items covered will include:

(i) The principal contractor's site organisation: the size of the principal contractor's site organisation will vary depending on factors such as the size of the contract, the type and complexity of the work to be undertaken, availability of staff and the contractor's staffing policy. The number of people employed on site by the contractor in supervisory and administrative roles will also tend to vary during the course of the work to suit the needs of the construction process. In assessing the cost of site staff, the estimator must therefore carefully study the requirements of the construction programme.

(ii) Site accommodation for the use of the principal contractor: the cost of supplying and maintaining suitable office accommodation for the principal contractor and the client's representatives will vary depending on many factors, the main ones being the number of the people to be accommodated, the size of offices required, the type of temporary buildings to be used, the location of the offices on the site and the amount of space available. The estimator will have to make a detailed build-up of the cost of the site offices for each contract, taking into account both fixed and time-related expenditure. Fixed costs would include capital cost and depreciation, or alternatively the hire rate, for office buildings and furniture, transport to and from the site, labour and plant costs for loading and unloading, erection and dismantling. Time-related costs would include the cost of weekly cleaning and attendance, lighting and heating costs and any rates of taxes payable to the local authority.

(iii) Services and facilities, including power, lighting, water, telephones, storage of materials and attendance on subcontractors. Temporary services will require very careful consideration. The principal contractor will need to consider availability of public supplies or whether generators, water browsers etc. will be required.

(iv) Small items of site equipment and associated consumables will be included in this section.

(v) Temporary works, including general scaffolding, fencing, hoarding and temporary access roads.

Information About Works to be Carried Out by Persons Other Than the Principal Contractor

This section will include information on site activities to be conducted by other parties, for example, subcontractors nominated by the client.

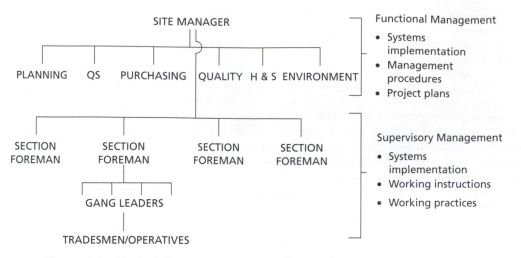

Figure 6.1 *Typical site management and supervisory organisation on site*

Site Organisation

The form of site organisation adopted for any construction project is determined by many factors, ranging from the constitution of the principal contracting organisation to the individual characteristics of the particular contract. It is absolutely essential that project sites should adopt commonly accepted principles of good organisation and construction management practice.

Chapter 2 introduced common conventions for structure and organisation of human resource inputs presented in the form of organisation charts. It is not intended to repeat the discussion presented earlier, except to reiterate the key elements of operation – clearly defined and structured channels of:

- authority
- responsibility
- communication
- control.

Although organisation differs from project to project, the functions of the site team personnel are generally the same from project to project. The site functional disciplines also reflect the functional specialist departments operated at corporate level. Together, the corporate organisation and the temporary project organisation must be structured in such a way as to translate effectively the company's operating systems into supervisory management, procedures and practices on site.

There must be a clear line of authority from the site manager to all functional managers and first-line supervisors. Everyone on site needs to be clear and understand the duties for which they are responsible and to whom they are accountable. In addition, structure must also clearly depict lines of general communication and put in place the routes and duties for site control. Chapter 2 presented various organisational structures illustrating the relationship between the corporate and project organisations. Figure 6.1 reinforces the earlier discussion.

The Principal Contractor's Responsibilities On Site

A principal contractor is responsible for all activity on the construction site. In almost all principal contracting companies, this responsibility would be delegated by the corporate organisation to a construction, or site, manager with single-point responsibility. Those responsibilities can be summarised as follows:

- Liaison with the client and consultants, i.e. architect, clerk of works, quantity surveyors, consultants etc.
- Checking all drawings, details and specifications

- Providing a detailed programme of site works
- Determining the methods and sequences of construction and choosing the plant/equipment required
- Setting out all siteworks
- Application of method study and works study to determine optimum performance of site tasks
- Operation and maintenance of plant and equipment
- Site layout including the positioning and erection of offices, stores, welfare facilities and temporary infrastructure
- Administrative and clerical work involved in the management on site of:
  - receipt, checking, storage and distribution on site of materials
  - receipt, maintenance and records of plant/equipment
  - the assignment and performance of labour
  - the utilisation and performance of plant
  - cost control
  - the provision of control information to the corporate organisation of the principal contracting company
- Maintaining a detailed site works diary
- Compliance with all health, welfare and safety regulations and site requirements
- Compliance with environmental legislation on and around the project site and local community
- Effective storage of all materials used on site
- Ensuring the quality and workmanship throughout the works, including that of contractors.

Site Layout

In association with the pre-tender and pre-contract planning stages, there will be supporting planning schemes for the layout and organisation of the physical site infrastructure. While it is inappropriate to specify any one approach to site layout, as this would vary according to many site factors, the construction methods adopted and the programme to be followed, some main factors for consultation include:

- size of site, proposed construction and site space
- volume of building, or structure, relative to site area
- amount of work above and below ground level
- location
- general site topography and environment
- levelness of the site
- access and egress abilities
- materials and plant to be used
- contract period and note of progress needed.

In carrying out site activities, the works must not produce excessive environmental effects and all works must be conducted safely with respect to those on and around the site.

In setting out the site, the key aspects that need to be considered are:

- access and traffic control
- storage
- communications and site control
- administrative and general site facilities/office
- site enclosure – fencing and hoarding
- quality control
- safety and security
- signposting
- effect of design on equipment and layout.

Access and Traffic Control

A construction site is affected by the need for traffic control in two key ways:

- To remove surplus excavated material from site.
- As most sites have restricted storage space, the materials required for the job have to be brought in throughout most of the construction period. This means a constant flow of transport, bringing materials that are required, within a short space of time, at different points on the site. Materials might need to be closely phased and delivered when assembled (just-in-time management).

To ensure that the contract does not suffer owing to lack of forward planning, the following points should be considered and a plan of the site should be studied in relation to the nature and volume of excavation and the methods to be used to excavate and remove surplus materials:

  (i) How equipment is to be brought in
 (ii) How surplus material is to be taken out
(iii) Whether any other operations or work anywhere on the site affect (i) or (ii).

A decision will then be required on the following:

- Traffic routes on and off the site, taking care to ensure that there is no danger to other traffic if the site is adjacent to a public highway.
- If the excavation is deep, whether or not ramps and bridges will be used, and if so, what type and where will they be situated.
- If roads are intrinsic to the project, whether or not to do the excavation and laying of the sub-base as the first site operation so that they can be used to service the temporary infrastructure. If this is done, making good towards the end of the job will have to be carried out and prior to completion when there is no danger of excessive wear or damage by transport. If roads are not included in the contract and the conditions are such that trouble is anticipated because of soft or difficult ground, provision should be made in the estimate for temporary roads. These can be either:
  – temporary roads of various types, depending on ground conditions and the sort of traffic anticipated
  – semi-permanent roads if the job is to be of long duration.

It is important to ensure that the traffic routes enable safe control to be exercised over all vehicles. This can only be done if the traffic is well regulated and controlled by good signposting and strategically placed observation points. If traffic is likely to contravene the terms of any Road Traffic Acts, the authorities should be approached before any nuisance is committed.

Storage

Storage on site will be needed for the following:

  (i) materials
 (ii) tools and equipment
(iii) spare parts for mechanical plant
(iv) manufactured goods for incorporation in the works.

Materials represent a major component of cost of most construction projects and therefore sensible precautions and use will be needed. It is important to:

- Reduce double-handling of materials to a minimum
- Protect materials from damage by weather

- Prevent damage arising through carelessness or proximity to building operations without suitable protection
- Prevent losses arising through careless stacking and handling.

Careful thought should be given to the location of materials in relation to their ultimate position in the job. Reducing the distances that the materials have to be transported on site will make a very positive contribution to a reduction in overall costs. A building itself can very often be used for storage, particularly after it has been made weatherproof, so use of existing buildings or parts of the constructed building might be considered. Timber and steel reinforcement should be stored in a dry safe place and in a way which makes it easy to extract the required pieces, i.e. size and/or length. Special stores are needed for some items, for example fuel oils are determined by the requirements of legislation.

Communications and Site Control

The physical extent of a site has a considerable bearing on the need for adequate communications. On a small site, there is very little need for anything special to facilitate direct contract between the manager, the supervisor and the operatives, mainly because the lines of communication are short and effective. On a larger site, in carrying out duties of control by observation and checks, the site manager will spend a large proportion of time out of the site office. This poses the problem of contact between the office staff and the manager, and also applies in the reverse manner when the manager requires speedy contact with the supervisors or operators on the site. There are several methods and systems available to facilitate communications and the choice will depend on circumstances and conditions, for example bells, loudspeakers, telephone, short-wave walkie-talkies, mobile telephones, light systems etc. might be considered. A common method of general communications, particularly in the sphere of discipline and welfare, is a simple noticeboard, and this can be extremely effective, provided that its use is controlled by a responsible person.

There is another aspect of communication that is complicated by the conditions normally found on a building site, namely the passing of written matter from the operatives and supervisors up to site management. It is therefore advisable to limit the returns of reports required to be passed upwards. Those that are necessary, i.e. timesheets, day worksheets etc., should be simple to read and complete.

It is an important part of a site manager's job to ensure that all heads of functional disciplines should be kept in the picture and well-informed with regard to site activities, programmes and progress. It is in this respect that a principle of limited span of control and clearly defined chain of command is of extreme importance and will facilitate the smooth running of a large contract.

Administrative and General Site Facilities

It is well recognised that physical conditions in construction can sometimes be quite poor, with inadequate facilities being provided on *ad-hoc* or short-term works. If the labour force is to be stabilised and the standard in general raised, it is imperative that improvements are made in this respect. Particular facilities needed are:

- Canteens
- Drying and changing rooms
- Shelter from inclement weather and for rest
- Washing arrangements and adequate toilet facilities.

Details of facilities needed are given in the section on 'Welfare Provision' later in this chapter.

Site Office

There are essentially two types of layout for site offices:

(i) Provision of a separate office for the site manager with the rest of the staff housed in one large open-plan office. This has the advantage of making communications easy and is obviously economical from the point of view of cost and the utilisation of space. The major disadvantage is that it may tend to distract people and to encourage the wasting of time.

(ii) Provision of offices divided into small units with each specialist function housed in its own office – for example quantity surveyors, site engineers, planning engineers, supervisory staff etc.

The buildings themselves should ideally be portable, particularly if they have to be moved at some stage during the contract or are on a job of short duration. Mobile offices are very useful on small jobs or as sub-offices on larger sites. When members of the staff have to spend long periods of time in site offices, it is obviously false economy to provide inadequate accommodation, particularly in the winter period. Where the size of the contract warrants it, special provision for site meetings would have to be made. In most cases, the site manager's office with the provision of a suitable table and seating should be large enough to accommodate all those likely to attend these meetings. The bill of quantities normally stipulates the requirements with regard to the provision of offices for the architect, consulting engineer, clerks of works etc.

Site Enclosure – Fencing and Hoardings

The principal contractor is often obliged to erect fences and hoardings either by virtue of local regulations or because of a specific clause in the bill of quantities to this effect. In any event, such works provide a secure means of bordering-off the site. Local authorities usually have regulations governing:

- the type, height and lighting of hoardings
- in what circumstances a covered way and/or handrail is required for protection of pedestrians.

Information on these points should always be obtained prior to commencing operations. From the principal contractor's point of view, the provision of adequate fences and hoardings can be an asset in helping to reduce unauthorised entrance to sites and consequent pilferage and losses.

Quality Control

There are four aspects of control of quality which have to be catered for on a construction site:

(i) Materials delivered to site: provision should be made for the inspection and checking of all materials delivered to the site to ascertain whether or not they conform to the specification laid down in the contract documents. All materials which do not meet the requirements should be rejected.

(ii) Components delivered to site: all components should be checked to see whether they:
- meet the specification requirements as to materials
- are the correct dimensions
- have been manufactured in a sound and workmanlike manner.

(iii) Materials produced or components manufactured on site: it is the site manager's responsibility to introduce inspection procedures and control systems to ensure compliance with specification. This may involve, for example, the setting up of a complete testing laboratory.

(iv) Workmanship: the good name of the contractor is dependent on his producing, consistently, a well-built job. It is therefore advisable, wherever possible, for the contractor to introduce a system of independent inspection of quality, in order to safeguard its good name and to make sure that standards of workmanship are consistently high.

Safety and Security

Adequate safety measures have a marked effect on productivity in the building industry, because accidents are the cause of a large proportion of unproductive time. There are numerous regulations governing safety measures on construction sites but, without the whole-hearted support of management and positive efforts to educate and discipline the supervisory staff and operatives, they can be ineffective. It is the ultimate responsibility of site management to ensure the enforcement of all safety measures. Large contracting companies have found that it is well worth employing safety specialists to educate and advise site management on the implementation of safety policy.

The provision of adequate fences or hoardings will contribute towards making a site secure. In certain areas it may also eliminate the need for watchmen to be employed outside normal working hours. One must bear in mind that, wherever there is any possibility of damage to the building under construction, a deterrent is well worth the time and money because making good can be very costly.

Even should it be considered unnecessary to fence in the site as a whole, it is usually prudent to provide an adequate stores compound. In addition to a suitable compound, there must be a system of efficient and effective stores control.

Three rules which should be rigidly applied, whatever the circumstances, are:

(i) that no unauthorised person should be allowed on the site
(ii) that all visitors should report first to the site office
(iii) that all subcontractors and suppliers must comply with regulations in force on a particular site.

Signposting

This does not normally present a problem in urban or built-up areas, but when sites are remote, external signposting will save time, particularly if the site is likely to be difficult to find.

On large sites, internal signposting will tend to assist in avoiding unnecessary movement on the site, and waste of time by people and transport looking for particular officials or unloading.

Effect of Design on Equipment and Layout

Items of plant work best in specific conditions and are limited by various factors. If plant is to be fully and economically utilised, one has to consider:

- whether conditions and circumstances dictate the type of plant to be used, by virtue of limitations on size, reach, power etc.
- whether its position in relation to the work expected of it will enable the anticipated output to be achieved.

In general, a major influence is the relationship between design and mechanisation of construction. Unfortunately the degree of co-operation between the designer and the constructor has, in the past, sometimes left a lot to be desired, with the result that very often the contractor had little choice in the type of equipment to be used.

Each of the major phases of construction needs to be considered. Information will be contained in the project method statements:

- Excavation: the method of excavation and the means of transporting the surplus material are important when considering the layout of offices, entrances, exits etc. It may, in some circumstances, be advisable to make do with temporary accommodation and site facilities until completion of the excavating and concreting of the basement slab, and then move to a permanent site.
- If it is inevitable that the site installations be moved, then the final location and transfer should be considered and planned in the initial stage of the job.
- Concreting: the main problems here are the location of mixers in relation to the size of the site and the proposed method of distributing the concrete. The method of distribution is also an important factor, i.e. by dumper, monorail, pneumatic pump etc., and will in each case warrant consideration in relation to layout, because tracks, pipe routes and the like will limit free passage on the site.
- Lifting and handling: the main factors here are the use of cranes and hoists which will very often fix the location of the following:
  – mixing plant
  – unloading points for lorries
  – pre-casting yard.

These decisions snowball in that they limit the choice for location of other site facilities. Where cranes are considered, the limitations imposed can be clearly seen by plotting on a site plan the position of the crane and its maximum and minimum radii, within which one must position all the unloading and lifting points, otherwise the cranes cannot be fully utilised.

Case Study – Site Layout

Figures 6.2–6.11 illustrate various aspects of site layout and temporary works involved with a major city-centre office development. The figures include services positions, lighting changes to the vicinity, site enclosure, traffic routing, fencing, site delivery and off-loading, access points and scaffolding positioning. It is pertinent to note that the layout is modified over time as works proceed. The figures are presented unabridged except for the deletion of all identifying names and locations in the interests of confidentiality. These figures will be seen later in this book as part of the principal contractor's construction-phase health and safety plan.

Welfare Provision

The principal contractor has a legal obligation to ensure that appropriate welfare facilities are provided on site. Such requirements are specified within The Construction (Health, Safety and Welfare) Regulations 1996 (HSE, 1996). The Health and Safety Executive (HSE) provides authoritative information to assist principal contractors to comply with current legislation. HSE Information Sheet No. 18: *Provision of Welfare Facilities at Fixed Construction Sites* (HSE, 1998) is one such excellent reference, providing good practice advice.

The following list identifies those important aspects requiring consideration by the principal contractor when providing appropriate welfare provision. The list source referenced is from HSE Information Sheet No. 18 and acknowledged as such:

Planning

- Make sure welfare arrangements are clearly addressed in the health and safety plan, where The Construction (Design and Management) Regulations 1994 (CDM) apply.
- Consider welfare facilities, their location on site and regular maintenance during the planning and preparation stage of any project, whether or not CDM apply.
- Arrange for equipment to be available, provided to sites and connected to services before construction work (including demolition) starts or when additional numbers of workers start on site.
- Make sure the facilities reflect the site size, nature of the work and numbers of people who will use them. If a large number of people are working on site or the work being carried out is particularly dirty or involves a health risk (e.g. pouring concrete), you will need more washing facilities (which may include showers), toilets etc.

General Welfare Requirements

- Ensure that all toilet, washing, changing, personal storage and rest areas are accessible and have adequate heating, lighting and ventilation.
- Facilities may need to be provided at more than one location to make sure workers have easy access.
- Make sure someone is responsible for keeping the facilities clean and tidy. How often the facilities will need cleaning will depend on the number of people on site and on how quickly they become dirty. Basic daily cleaning may not always be enough.

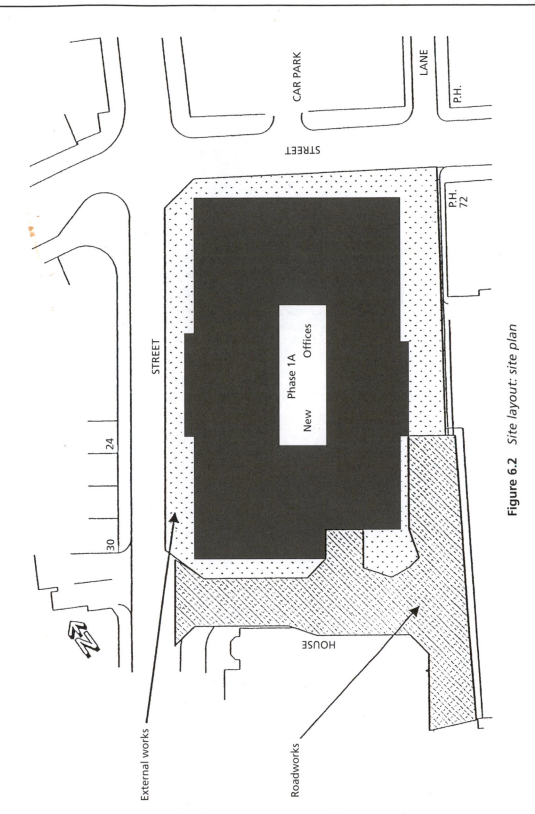

Figure 6.2 *Site layout: site plan*

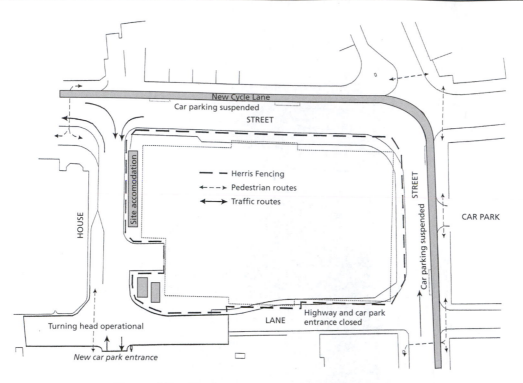

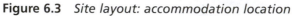

Figure 6.3 *Site layout: accommodation location*

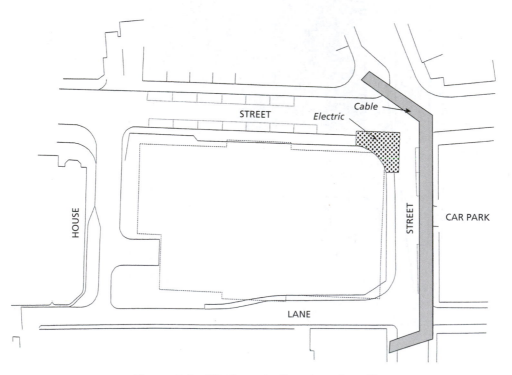

Figure 6.4 *Site layout: diversions for piling*

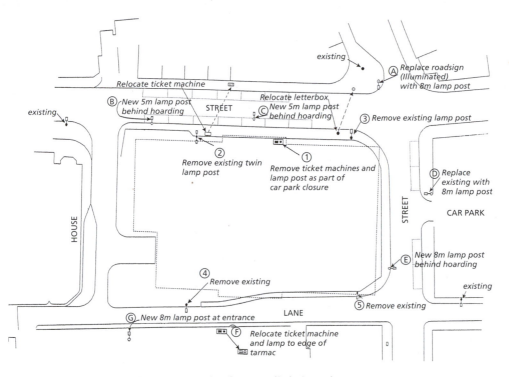

Figure 6.5 *Site layout: lighting changes*

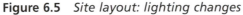

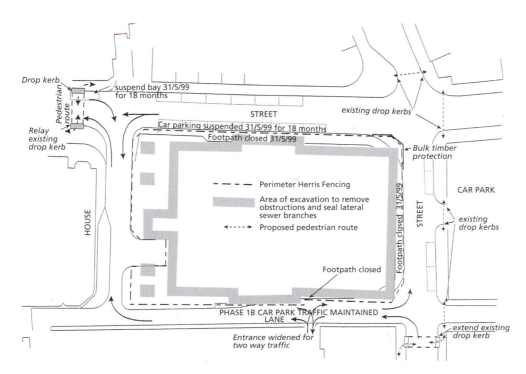

Figure 6.6 *Site layout: sewer sealing for foundations*

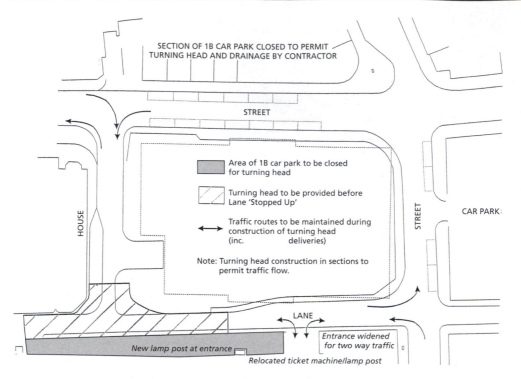

Figure 6.7 *Site layout: traffic routing*

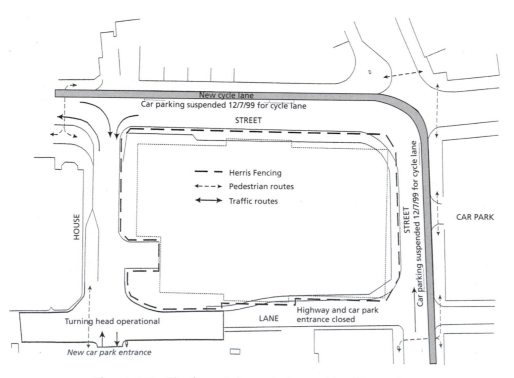

Figure 6.8 *Site layout: boundaries and traffic routing*

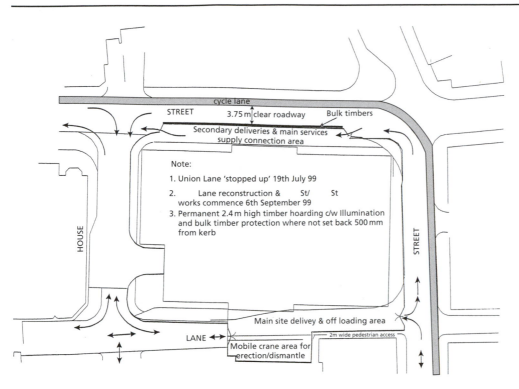

Figure 6.9 *Site layout: highway works*

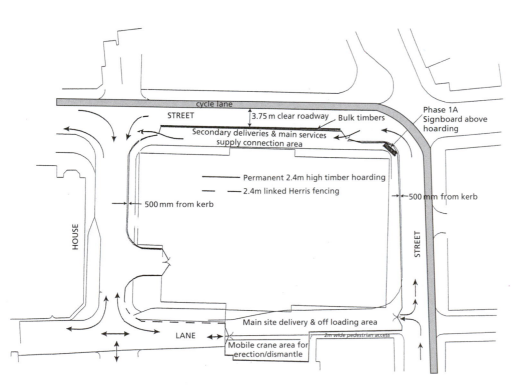

Figure 6.10 *Site layout: hoardings*

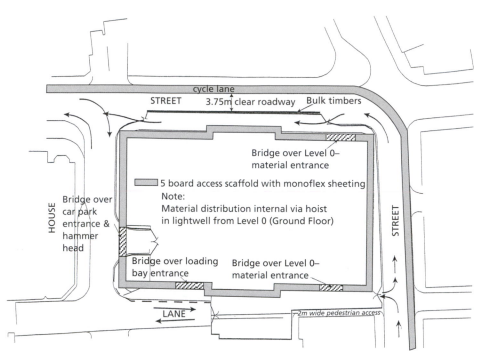

Figure 6.11 *Site layout: scaffolding*

Toilets

- Make sure that an adequate number of toilets are provided at all times.
- Men and women may use the same toilet, provided it is in a lockable room and partitioned from any urinals which may also have been provided. Otherwise separate toilets will be needed.
- Wherever possible, connect toilets to a mains drainage system and ensure that they are water-flushing. If you cannot do this, use facilities with built-in supply and drainage tanks.
- Where portable no-mains toilets (chemical) are used, make sure they are serviced regularly in accordance with the suppliers' recommendations. The frequency of servicing will depend on the amount of use. Make sure the toilet is always accessible for servicing. Suppliers of portable facilities will be able to advise you.
- Units used by female workers should have effective means for disposal of sanitary waste.
- Make sure adequate supplies of toilet paper are always available.

Washing Facilities

- Put washing facilities next to both toilets and changing areas and make sure they include:
  (i) basin(s) or sink(s) large enough for people to wash their face, hands and forearms
  (ii) a supply of hot and cold or warm running water
  (iii) soap and towels (either cloth or paper) or dryers.
- If mains water is not available, use clean water supplied from a tank.
- You may need more washing facilities, including showers, where the work is particularly dirty or when workers are exposed to especially hazardous substances, e.g. development of contaminated land, or demolition of old industrial buildings which are contaminated with toxic substances etc. These will need to be separate from the main facilities.
- You may need specialist facilities for certain activities, e.g. working with lead, asbestos, tunnelling under compressed air etc.

- Men and women can share basins used for washing hands, face and arms.
- A shower may be used by both men and women so long as it is in a separate, lockable room which can be used by one person at a time.

Storing and Changing Clothing

- Every site should have arrangements for storing:
  (i) clothing not worn on site (e.g. jackets, training shoes etc.)
  (ii) protective clothing needed for site work (e.g. wellington boots, overalls, reflective jackets etc.)
- Separate lockers might be needed although, on smaller sites, the site office may be a suitable storage area, provided it is kept secure.
- Where there is a risk of protective site clothing contaminating everyday clothing, store items separately.
- Men and women should be able to change separately.
- Make sure that wet site clothing can be dried.
- Many fires have been caused by placing too much clothing to dry on electrical heaters, making the heater overheat. If electrical heaters are used, ensure they are properly ventilated and, if possible, fitted with a high-temperature cut-out device.

Rest Facilities

- Provide facilities for taking breaks and meal breaks. The facilities should provide shelter from the wind and rain and be heated as necessary.
- The rest facilities should have:
  (i) tables and chairs
  (ii) a kettle or urn for boiling water
  (iii) a means for warming-up food (for example, a gas or electrical heating ring, or microwave oven).
- Non-smokers should be able to use the facilities without suffering discomfort from tobacco smoke. Provide ventilation or, if this is not possible, you may need to provide separate areas for smokers and non-smokers, or ban smoking in the presence of non-smokers.
- On small sites, the site office or hut can make a suitable rest area, especially if it is one of the common portable units.
- Do not store plant, equipment or materials in rest areas.

Drinking Water

- Make sure there is a supply of wholesome drinking water readily available. Where possible it should be supplied direct from the mains.
- If water is stored, protect it from possible contamination and make sure it is changed often enough to prevent it from becoming stale or contaminated.
- Clearly mark the drinking water supply to prevent it being confused with water which is not fit to drink or hazardous liquids.
- Provide cups or other drinking vessels at the water tap, unless the water is supplied in an upward jet which can be drunk easily (for example, a drinking fountain).

Heating

- Inadequately ventilated LPG cookers and heaters can produce carbon monoxide. Gas may escape from leaking cylinders which have not been properly turned off. You can eliminate these risks by using properly maintained electrical equipment instead.
- If this is not possible, reduce the risk by:
  (i) using and storing the cylinders in safe, well-ventilated places outside the accommodation (including overnight)

(ii) providing adequate combustion ventilation (provide fixed grills at high and low level)
(iii) checking that cylinders are properly turned off when not in use – turn off the tap at the appliance and isolate the cylinder
(iv) using wall or ceiling mounted carbon monoxide detectors.

Summary

Site establishment, incorporating contract preliminaries, site organisations, site layout and welfare provisions, is essential to a successful construction project. It provides the sound basis for configuring and structuring the project organisation, layout and facilities needed to operate the site works effectively. Principal contractors have overall responsibility for ensuring that legal requirements for welfare are met. Site establishment provides an essential vehicle for fulfilling such legislation.

References

Health and Safety Executive (HSE) (1996) *The Construction (Health, Safety and Welfare) Regulations 1996.* HMSO, London.
Health and Safety Executive (HSE) (1998) Information Sheet No. 18, *Provision of Welfare Facilities at Fixed Construction Sites.* HMSO, London.

7 Plant and Materials

Introduction

The premise of this chapter is not to engage in the basic technologies of materials or the varied types of plant available to the principal contractor, but rather it is concerned with the management process of these two valuable resources. Plant, if poorly managed on construction projects, suffers from low utilisation rates, and materials are renowned for their high wastage figures. Therefore the principal contractor needs to concentrate efforts on the removal of non-value-adding activities.

Plant Management

Considerable care is necessary when purchasing plant and equipment of any kind. The person charged with making the purchase needs to thoroughly investigate the various aspects identified below. The successful and profitable operation of plant is dependent not only on the price paid but also on the necessity to conduct a critical evaluation of its suitability. This is an activity that should not be left solely to a plant department but should involve consultation with the operational and user personnel. The plant department should not function in isolation or this will impact upon operational effectiveness and efficiency.

If due consideration is given during consultations between plant department and operations the benefits available include:

- simple installations (will link with existing plant if required)
- minimum commissioning time (reduced time span before operating at maximum efficiency)
- rapid training of operators (learning curve theory)
- high utilisation with the minimum of 'down time'
- longer life of plant (doing the job for which it was designed)
- lower operational costs
- ease of maintenance with the minimum investment in spare parts.

Purchasers can protect their interests, safeguard production capacity and ensure an adequate return on investment by paying attention to:

- maintainability
- reliability
- installation and commissioning
- product support
- costs.

Definitions

Maintainability

The designers of all industrial plant should ensure that their designs provide:

- a minimum maintenance requirement
- rapid fault diagnosis and repair
- low maintenance and repair costs.

Equipment which conforms to these three objectives has maintainability.

Reliability

This is demonstrated by the length of time between breakdowns in operational service. The best manufacturers of plant have quality and reliability departments which ensure that the materials, manufacturing processes and the final product are all of a high standard of reliability.

Installation and Commissioning

The signs of an unsatisfactory purchase usually become obvious during the installation and commissioning period. This is the time when plant is delivered and put into operation.

Product Support

This includes providing the following:

- operating, installation and maintenance manuals
- maintenance schedules
- training aids
- special tools
- special test gear
- technical assistance.

The most effective product support and customer service operations are seen in those companies which rent and service equipment.

Cost

The total cost is the cost of the equipment plus the cost of delivery, installation, servicing and product support (life-cycle cost).

Benefits

Some of the benefits to be obtained through careful procurement are explained below.

Commissioning

This can be a long process if the purchaser selects the wrong vendor or incorrect equipment. A competent vendor arrives on site with proof in the specification that the plant meets set objectives. Installation, testing and operator training should not take long if the plant attains the required performance and if the technical information for training is available.

Minimum Down Time

This is the time in which the plant is not available for production because of a need for maintenance or repair. Down time can be caused by a breakdown of plant or by a shutdown period for preventive maintenance.

Long Life

The purchasers must assure themselves of the ability of the plant to attain long service. This is closely linked with 'reliability'. Much can be determined from an advance study of the plant by the plant engineer, who must consider whether the plant is designed for long service with a continued availability of spare parts from the manufacturer or whether the vendor's policy is to make frequent changes of design with built-in redundancy. Reference can be made to other users of plant from the selected vendor.

Low Operating Costs

A vendor should be asked for predictions of operating costs and its approach to this problem will indicate whether it has adequately assessed this before selling its product. Operating costs in each year could include:

- interest on capital cost
- cost of operating labour
- cost of maintenance
- depreciation of plant
- cost of down time
- cost of space (infrastructure).

Cost of Maintenance

This is not a simple calculation and among the many factors to be taken into account are:

- the quality of the technical information and training provided
- the standing investment in spare stock parts
- the consumption of spares and lubricants
- the numbers and grades of maintenance staff
- the maintainability of the design itself
- the reliability of the design.

Minimum Investment in Tools and Spare Parts

Sometimes it can be costly to change vendors solely to satisfy lowest tender requirements. If previous plant is satisfactory and meets all the requirements, it does not always benefit the user to change to another supplier to save a small percentage on the purchase price. Not only does a new type of plant require further investment in special tools and spare parts but also maintenance staff may have to learn new techniques.

Maintainability

Plant and equipment should be designed so that maintenance is as easy as possible and that much routine maintenance is 'designed out'. Designed out maintenance (DOM) is a well-known practice. A typical example is the use of sealed bearings to remove the need for lubrication. However, some maintenance will always be required, as faults will occur and repairs will always be necessary. The purchasing officer can ensure that this is kept to a minimum by asking questions such as (or by using the plant engineer to ask them):

- Has a reliability and maintainability study been undertaken for the plant? Such a study will produce MTBF (Mean Time Between Failure) and MTTR (Mean Time To Repair) values. Critical parts can then be identified.
- Have arrangements been made for fast fault diagnosis when there is a breakdown? The time taken to locate a fault is added to the repair time to produce the total down time. Among the points to

look for are:
– undue dependencies (a whole process depending on one small part which could easily fail)
– too few test positions
– requirement for special test gear which is not provided with the plant.

Fault diagnosis is fast and simple if a designer provides facilities in its design. Another important aspect of maintenance is performance monitoring. This may range from simple running tests for accuracy on a machine tool to a vibration test position provided on a generator. In many installations performance checks are used to provide a warning when maintenance is required.

Quality Assurance

Best practice plant vendors have quality and reliability departments. Their task is to ensure that the incoming materials for the product, at the various stages of manufacture and final product as delivered to the user, are of high quality. The purchasing officer should ask whether quality assurance (QA) specifications are available and if they are QA certificated.

Terotechnology as a Management Technique

The concept of terotechnology can be summarised in many ways and other names have been used which convey the underlying concept better than does the word 'terotechnology', for example:

- resource management
- whole-life costing
- costs in use
- total cost of ownership
- cradle to grave management
- physical assets management
- life-long care.

The development of the principle of terotechnology is a reflection of the fact that, when deciding on the purchase of any physical asset, the user often has no coherent plan for obtaining a holistic, efficient and effective operation of the asset at an optimum cost.

Systems Approach

The application of terotechnology does depend very much on a systems approach. If applied correctly, the entire process of specifying, procuring, commissioning, installing, maintaining and replacing capital equipment will be organised in a co-ordinated and disciplined fashion. This constitutes a significant breakthrough from the 'traditional' approach to manufacturing technology and should lead to impressive improvements in management efficiency and equipment reliability. In order to implement the principles involved, it is necessary that a comprehensive and relatively sophisticated system of information be established. Data detailing the costs of breakdown, service defects and reliability needs to be collected and analysed so that it can be used for decision making.

The information required on reliability, costs and performance should already be available from several sources in an organisation. If it is not then a feedback system or a means of retrieval and collation will have to be established. This ought to be as simple as possible because it will have to be paid for and thereby contribute to the total cost of ownership.

Multi-disciplinary Involvement

There can be no one 'most suitable' form of organisational structure to promote terotechnology. The practice of terotechnology involves the bringing together of experts from a variety of functional areas, such as purchasing, maintenance, design, production, accountancy and marketing, and there is no unique way of achieving this. In addition, there is no reason to suggest that the existing organisational structure need be disturbed in any way if a 'project team' approach is adopted. This involves the setting up of a team to manage the entire capital equipment programme.

Life-cycle Costing

The life-cycle costs consist of:

(1) *Costs of use* – which are the costs of materials, energy and manpower used in producing goods or services, and the costs of loss of output due to malfunction or failure.
(2) *Costs of ownership* – which are the costs of acquisition, installation, commissioning, maintenance, modification and repair, and of monitoring condition and performance.
(3) *Costs of administration* – which include costs of data acquisition, recording and analysis.

The intention of employing life-cycle costing is to ensure that a maximum return on a particular asset is achieved. This implies that an asset's life-cycle costs must be kept to a minimum. What appears often not to be appreciated, however, is the extent to which the initial purchase price, as only one aspect of the total cost of ownership of an asset, is often far exceeded by subsequent costs during the remainder of the life-cycle.

Terotechnology is simply one of many management concepts which can contribute to the overall efficiency and effectiveness of a company. To a large extent it is 'commonsense' (it may indeed be simply more a style of management than anything else) but it does require careful thought and planning if the full benefits of its application are to be realised. What it probably does not need to any great extent are new techniques or disciplines. It is essentially a way of grouping familiar activities together and involves combining well-tried techniques in a different way so as to improve the management of physical assets in accordance with the objectives of an enterprise. For the practical application of terotechnology in a construction operational environment it would seem that two factors must be in evidence: one is a systems approach, the other is multi-disciplinary involvement.

It is evident that the procurement of plant and equipment is a vital activity that can add to or detract from the performance of an organisation. Therefore construction firms must ensure that it is recognised within its corporate and operational strategic thinking and planning processes.

Materials Management

Construction operations consume materials in order to provide the client with a completed structure. However, the supply of materials tends to be a problematic activity for contractors. In order to safeguard against having insufficient stocks of materials to complete production operations, contractors may engage in the activity of maintaining 'buffer stocks'.

Buffer Stocks

Buffer or safety stocks are maintained as insurance against unpredictable events that can disrupt operations on construction sites. For example, suppliers can be late delivering material or the material may be defective when it arrives. Equipment can break down at the most inconvenient time, employees can be absent and intermediate processes may produce defective products. For these and many other reasons, buffer or safety stocks are held to balance the costs of being out of stock with the costs of carrying the additional inventory. An alternative to buffer stocks is the elimination of the disrupting element. For example, Just-In-Time practices would promote selecting and developing highly reliable suppliers and thus reduce the need for material buffers.

Materials have been established previously as one of the resources of organisations (the 5 Ms). Therefore the efficient and effective utilisation of this resource is a vital consideration for construction organisations. Within this section Just-In-Time (JIT) is advocated as a means for ensuring that materials (and plant) are considered under the umbrella of waste reduction.

Just-In-Time Concept

Introduction

JIT has now become established in a wide range of industries such as food, drink and processing as well as the more traditional manufacturing, engineering and construction industries. JIT also exists under a number of new 'buzz' words such as 'quick response' and 'short cycle manufacturing'. However, one thing has not changed and that is its fundamental philosophy.

JIT is now being extended in some leading companies to embrace 'kaizen' which is Japanese for continuous improvement and is very much focused in instilling continuous improvement involving every employee. It is essentially a company-wide Total Quality Management concept, with particular emphasis on delivering specific results by involving all employees in problem solving, and making customer satisfaction paramount.

History

JIT has its origins in the Toyota company in Japan. In the 1960s, Toyota worked hard on developing a whole range of new approaches to manufacturing. The development of the noted approaches was hastened by the 'oil crisis' of the 1960s. By 1972 these new approaches had begun to attract wide attention in Japan and in the mid 1970s other Japanese companies began to experiment with and adopt them. At this stage, and for some time later, these were not known as JIT but the approach was called 'Toyota Manufacturing System'. By the end of the 1970s the Toyota Manufacturing System had begun to attract attention in the West. One of the many elements of this system was a pull scheduling technique using 'kanban' (Japanese for card). The system first became known in the West as the 'Kanban System'. This was rather misleading as kanban is only a small part of the total system and is very difficult to operate independently of other critical activities.

Just-In-Time management is not one technique or even a set of techniques for manufacturing but is an overall approach or philosophy which embraces both old and new techniques.

The Philosophy

The philosophy underpinning JIT is founded on the elimination of waste. A suitable definition of waste in this context is 'Anything other than the minimum amount of equipment, materials, parts, space and worker's time, which are absolutely essential to add value to the product'. In other words, we should be trying to use the minimum amount of resources in the most efficient and effective way to meet customer demands. The definition refers to added value and once the concept of looking at the operation activities in terms of value added is understood, the opportunities that JIT offers can be obtained. Investigating the business and observing where value is added and, more importantly, where costly non-value-added activities occur, is a key to understanding JIT and the elimination of waste.

Another way of looking at the whole issue is to say that JIT is focused on improving the return on capital employed by impacting on cost inventories (materials) and fixed assets (plant). Usually the amount of waste within an operation can be measured by examining the stock levels. High inventory levels tend to hide quality problems. The JIT philosophy attempts to expose these problematic issues and develop advocated solutions for addressing them.

Management of Change

JIT requires all employees to be involved. It is a philosophy of continuous improvement and therefore requires commitment and participation by everyone. In short, companies need to manage 'a change process' at all levels.

JIT will require a significant change in employee and management relationships. Usually incentive schemes require a reappraisal, and piecework-based incentive schemes are not suited to a JIT environment. Therefore a move to group incentive schemes – value-added or gain-sharing schemes – is appropriate. Employees may need to become more flexible and multi-skilled and this may involve cross-training. The application of Anthropocentric Production Systems, discussed in Chapter 3, could provide a useful model for the implementation of JIT. It would certainly address the managerial and flexibility issues.

JIT Implementation Strategy

The development of a JIT implementation plan can only be commenced when business and customer needs have been properly understood and defined. Once this task has been completed, the appropriate resources can be determined along with any systems and organisational requirements.

When identifying opportunities for implementing JIT, a certain amount of data collection is necessary covering costs, product/process and information flows. The data should highlight material delays and disruption, opportunities for lead-time reduction, stock analysis and conflicting objectives.

From this information a suitable pilot project(s) can be defined. Pilot projects enable companies to gain experience with JIT principles and techniques and, if selected correctly, can be undertaken with minimum disruption to existing production. This is an incremental approach as opposed to a 'big bang' deployment strategy. Normally pilot projects are selected around a particular process, operation, product or group of products. If proved to be successful, a pilot project can provide a motivational drive for other areas of the company.

On average it takes up to three months to obtain a properly prepared and selected pilot project, and another two to three months to implement it, so benefits are usually achieved quickly. In addition, JIT projects are normally not capital intensive but do require a substantial input of time with its associated opportunity costs. It is therefore crucial that sufficient time and resources are made available for the training needs of all personnel.

Implementation Issues

Since the concept of JIT may include elements such as business process redesign, radical change and dramatic improvement, senior management needs to reconsider the strategic capability of its organisations in order to verify that all relevant aspects have been addressed. This is necessary in order to obtain co-operative participation from everyone in the host organisation. One must remember that consultation should precede implementation. JIT generally requires less specialisation and more generalisation from employees. This approach means that re-engineering must be undertaken in conjunction with the strategic capability of the enterprise. This is a fundamental concept for oganisations to appreciate, as the operational activities of the firm should provide (as previously stated) a competitive weapon.

Analyses

Organisations should not adopt new management tools or technologies without first conducting a full internal and external analysis incorporating, for example, a SWOT and/or PEST analysis focusing on maintaining/gaining a competitive edge. JIT is a management philosophy which addresses the importance for an organisation to reconsider the way it performs its tasks, in order to ensure that the correct methodology is utilised on its operational tasks within the required time frame. This will provide dramatic benefits for a company by eliminating any wasteful non-value-adding activities, empowering the business to produce its product/service at a lower cost, faster and at a higher quality level. Thus, a company can differentiate itself from its competitors and gain a sustainable competitive advantage based on strategies of least cost, differentiation and focus.

Improving Customer Satisfaction

Customer satisfaction is a core element of JIT and the advocated advantages of deployment focus on the improvements in customer satisfaction and include:

- A faster service in design and construction – the elapsed time between initial customer contact and the completion of a project is reduced, with some steps being eliminated entirely. Most importantly, the reduction in the number of different people involved reduces delays associated with multiple involvement.

- Improved product/service quality – refining the processes for product design, construction and inspection can produce improved product/service quality.
- Enhanced innovation – improvements in operations offer the potential to enable organisations to innovate in new ways more quickly and continuously.
- One-call response to enquiries – because customer service and marketing representatives have on-line access to work management and design information they are able to respond quickly to a customer's enquiries on the status of a construction project.
- A reduction in the time required to conduct a design process – there are savings in pre-construction time appertaining to briefing, designing, tendering and the preparation of specifications, and in the management of the construction process. This allows a construction organisation to undertake more projects with reduced contracting out during times of high demand for construction services.
- Improved marketing and customer service – filing and handling paperwork are reduced substantially by the new systems. Also, less time is required to respond to customer information requests, and this results in the company's representatives spending more time in direct contact with customers and potential customers.

The above benefits are derived from employing the concepts encapsulated within JIT.

The Utilisation of Human Resources

JIT is concerned with processes, people and technology; it is therefore important to ensure the optimum utilisation of all human resources. Working within a JIT approach, employees work in empowered teams with the ability and autonomy to solve customers' problems. They are also involved with a project as it develops from design to completion and are therefore more likely to experience greater intrinsic motivation and hence job satisfaction. Employees need to be educated to understand how the success of teams within the company influences the company's overall success (Harrington, 1991).

In general, the major difference between a winning and a losing organisation is that the winning organisation knows best how to complete organisational tasks. Usually larger firms have specific and documented ways of conducting their business operations, with very precise job descriptions. Because of the nature and complexity of construction projects, employees must be very clear about their work tasks and areas of responsibility. Tasks should be kept as simple as possible and be incorporated within simple processes: this will ensure the optimum work flow relating to each process and thus the optimum contribution to the value chain: "*In order to meet the contemporary demands of quality, service, flexibility, and low cost, processes must be kept simple*" (Hammer and Champy, 1993).

The JIT approach views inspection as a non-value-adding activity which ideally should be eliminated from the construction process. This calls for a sea change from the inspection paradigm, i.e. a move from inspecting quality of products (a dubious activity at best) to designing quality into products and services.

Redesigning Information Systems

A successful JIT application process has a high positive correlation with the information systems used by the firm. It is of prime importance when redesigning the information systems to ensure that the least time and cost possible to perform business activities are achieved. Furthermore, the quality of information can be improved, thus adding value to the information system by enabling greater control of operational activities by employing dynamic closed feedback loops (see Figure 5.15).

Much research and development and the practical applications of IT within the construction industry have focused on the automation of traditional and conventional ways of working. These traditional methods of working are now being widely criticised by various authors (Koskela, 1992).

Miozzo *et al.* (1997) claimed that drawing on new process-based management and organisational methods in manufacturing could provide a strategy for improving both the efficiency and effectiveness of organisations employed within the UK construction industry, and hence should be explored in earnest. This would enable the identification of construction-specific process improvements which could then be developed to provide a strategy for implementing changes in order to achieve a consistent process innovation strategy.

JIT is not likely to provide the improvements desired without the application of new techniques. Today, advanced computer tools provide opportunities for solving many of the existing problems facing the construction industry. Three-dimensional (3D) modelling, simulation and virtual reality are used to support and enhance the design and construction processes. IT can effectively promote the integration of activities within the construction industry – directed towards removing non-value-adding activities, especially those related to plant and materials.

Improvements in the Value-Added Tasks

In any company, a number of inefficiencies exist that, once identified, can usually be easily corrected. JIT is not concerned with correcting inefficient operations; instead, it is concerned with fundamentally changing the processes so that the results are dramatically improved. In many cases, inefficient operations can be eliminated or replaced by changing work practices. The following notes explore how to take advantage of opportunities for attaining improvements related to value-added tasks:

- Independent sequential processes – these processes provide opportunities for designing parallel systems which can dramatically reduce cycle times and the resulting operational and competitive costs. This approach would be a means of improving the utilisation rates of plant.
- Redundant processes – it is necessary to conduct a thorough analysis of all processes employed by a firm: this will often reveal redundant processes. Duplicated processes are relatively easy to spot; however, other redundancies can be very subtle and their identification is often the result of some concentrated detective work on the part of the business process analyst. Changing the process may remove the need for costly plant.
- Visible competitive processes – "*Anytime your customer sees you doing something better, you win*" (Champy, 1993). The message is, of course, to identify those parts of the processes that are visible to customers and subject them to re-engineering. Even a modest performance improvement to internal operations can have a dramatic impact on the business's bottom line if customers believe that efforts have been instigated on their behalf.
- Time-sensitive information – define the timeliness of the information needed and ensure that it can be achieved. Make sure the information used to support the processes is, in fact, timely enough to make the processes worthwhile: this is a critical issue for feedback control loops and materials call-up.
- Data entry replacement – eliminate data entry if at all possible; it is expensive and error prone, and most replacement technologies easily pay for themselves.

The Advantages of Applying JIT to Construction Activities

JIT challenges the traditional approaches undertaken by construction companies when engaged on construction projects, and the critical success areas are design, procurement and construction. Latham (1994) has suggested that "*Clients will benefit by improving the industry's performance and teamwork and thereby achieve better value for money. For example the Stanhope, British Airports Authority plc and large retail chains, have emerged as leaders of the construction process, introducing new methods and techniques of procurement systems.*" Within the construction industry the term 'Construction Process Re-engineering' (CPR) has evolved and, for example, advocates that attention should be placed on the integration between design and construction in order to eliminate the problem of non-value-adding activities. CPR does provide an opportunity for obtaining improvements in the key resources of plant and materials.

Technological advancement, market expansion, global competition and a renewed demand for quality and productivity have highlighted the critical issue of integration of the various stages involved in the construction process (Schimming, 1993).

It is acknowledged by both industry researchers and practitioners that there are many wasteful activities incorporated in the design and construction process. The majority of these activities consume time and effort without adding value, thus undermining the efficiency and effectiveness of the whole process (Mohamed and Tucker, 1996). Plant and materials certainly fall into this category. Therefore JIT should be adopted in order to address this problem and thus fully utilise these key resources.

Problematic Issues of Implementation

Certainly there are difficulties associated with any JIT project, for example, as employees are asked to learn new working processes and associate with unfamiliar people and departments, a natural reaction is fear and trepidation. Davenport (1993) noted three powerful reasons why problematic issues exist in process innovation: the cost is too high in funds and human effort; the psychic shock and upheaval are too great; and the co-ordination of all the change projects is too formidable.

Soles (1995) commented on implementation: "*It is complex and long term, it requires sufficient finances and commitment.*" JIT implementation involves the following factors – though listed separately they are not mutually exclusive and most organisations experience a combination.

Problematic Issues

Insufficient commitment by senior management: senior management must instil in all employees of the company a desire to improve competitiveness. JIT requires total commitment, which must be extended to all employees at all levels and in all departments. Therefore senior management must be fully committed to the deployment process. This can be evidenced by senior management providing all the required resources for the initiative. Senior management must drive the deployment process and lead from the front, expounding enthusiasm at all times. JIT deployment will fail without clear and consistent project leadership. Management leadership has to overcome considerable emotional and political problems that will arise from such a change (Hammer and Champy, 1993).

Inappropriate Corporate Culture

JIT requires a corporate culture based on trust and a desire to identify problems in order to eliminate them, thus improving production processes and eliminating non-value-adding activities. The concept of 'empowerment' is a vital part of the JIT philosophy. However, if a climate of distrust exists between senior management and the rest of the company the implementational process is doomed to fail. This will result in the demotivation of staff with its associated opportunity costs.

No Formal Implementation Strategy

The deployment process should be planned. JIT is a project and therefore requires planning as a project: to treat it as a company bolt-on activity will lead to failure. JIT is a means of improving the competitiveness, effectiveness and flexibility of an entire organisation. Achieving these noted advantages requires organisations to plan and organise every operational activity at all levels of the firm. This process must be part of the strategic implementational development and must not be treated in isolation. Ibbs *et al.* (1995) support the argument that "*Many re-engineering projects are doomed to failure from the moment they are initiated, because vital pieces of the jigsaw are missing.*"

Lack of Effective Communication

The life blood of any company is communication, and the importance of this organisational activity cannot be overemphasised. Do not forget the concept of 'internal' and 'external' customers with its requirement for effective communication mechanisms. If employees are to become part of the organisational decision-making process they need a means of expressing their views to senior management.

Control within any organisation is dependent on the communication systems function. The specific control of plant and materials is definitely dependent on a clear and effective communication system.

Narrowly Based Training

The key to a successful JIT application is having staff who are competent in executing its allocated tasks. If employees are empowered to plan and perform work activities it is vital that they possess all the necessary skills and competencies. A primary function for a construction-related enterprise seeking to gain a competitive advantage is to implement "*training and education in teamwork*" (Hellard, 1995). As an example, if staff are to participate in group discussions, training in group dynamics and public speaking would be advantageous.

Necessity to Concentrate on Organisational Strengths

Senior management should not lose sight of the fact that sustained competitive advantages are obtained by implementing strategies which exploit their strengths through responding to environmental opportunities, while neutralising external threats and avoiding internal weaknesses. The following two standard corporate planning techniques can be utilised: (1) a Strengths, Weaknesses, Opportunities and Threats analysis (SWOT), and (2) a Political/Legal, Economic, Social/Cultural and Technological analysis (PEST).

Key Elements for the Successful Application of JIT

- Senior management must possess a full understanding of the philosophy and requirements of JIT.
- A common vision is required by all employees of the organisation; this may be accomplished by adopting awareness sessions, customer surveys, benchmarking and common vision workshops. Staff must be made aware of the advantages, i.e. the elimination of non-value-adding activities, thus leading to increased competitive advantage.
- Provision of the necessary resources, which include humanistic as well as financial requirements, to educate and train for quality improvements.
- The development of a fully resourced implementational strategy.
- Senior management must conduct frequent reviews of the deployment process in order to monitor and maintain progress. This could be improved by employing the 'Plan', 'Do', 'Check', 'Action' cycle (PDCA) – see Figure 5.15.
- Designing procedural systems relevant to work practices, and concentration of organisational effort, should be focused on the adoption of improved operational processes. In the case of JIT this needs to be extended to suppliers.

Construction-related organisations should begin to rethink the way in which they operate their businesses. They need to consider the possibility of redesigning operational processes from a clean-sheet perspective (incorporating stakeholders). The analysis should focus on the elimination of non-value-adding activities. JIT can provide a methodological model for conducting such an analysis and thus contribute to the attainment of a sustainable competitive advantage.

Problematic Issues Concerned with Implementation

Issues associated with JIT implementation were listed in a postal questionnaire and respondents given the opportunity to indicate their impact. The sample consisted of twenty construction organisations. The factors and their relative importance to the respondents are detailed in Table 7.1, with the respondents' ratings for each factor in terms of importance given as percentages.

Table 7.1 *Factors associated with successful JIT implementation*

	Identified as 'very important' (%)	Identified as 'quite important' (%)	Identified as 'not important' (%)	Factor positively confirmed?
Commitment from senior management	85	15	0	Yes
Leadership at all levels	85	11	4	Yes
Staff participation at all levels	85	12	3	Yes
Clearly defined aims and objectives understood by all	85	7	8	Yes
Use of well-established value management techniques	44	44	12	Partly
Use of relevant technology	33	37	30	Partly
Application of teamwork by all those involved	78	19	3	Yes
Flexibility in organisational structure	48	22	30	Partly
Training and education	52	33	15	Partly

Note: Greater than 75% 'very important' fully confirms advocated success factor, and greater than 60% 'very/quite important' partly confirms advocated success factor.

JIT could improve efficiency in the construction process. Specifically, the following potential benefits of JIT are advocated (Wafa and Yasin, 1998):

- eliminating waste in production and materials utilised
- improving communication internally (within the organisation) and externally (between the organisation and its external stakeholders)
- reducing purchasing costs, which is a major issue for most construction organisations
- reducing lead time, improving production quality, increasing productivity and enhancing customer responsiveness
- fostering organisational discipline and managerial involvement
- integrating the different functional areas in the organisation – this especially bridges the gap between production and accounting.

Conclusions

Although there are significant differences between manufacturing and construction, with appropriate modification and the development of an implementational strategy an increase in productivity in the construction industry could be achieved through the utilisation of JIT:

- JIT is a manufacturing philosophy which can improve a company's performance
- the concept is based on making the process more customer orientated, simple and co-ordinated
- to achieve efficient and effective utilisation of resources, attention needs to be concentrated on the non-value-added aspects of a construction business and two key areas for potential improvement are plant and materials
- there is a need to establish a more co-operative relationship with suppliers, built on mutual trust and strategic relationships
- a well-developed and resourced implementational strategy is vital for success, with senior management support.

Holding excess inventory can bring severe cash-flow problems. In designing a JIT system, the aim is therefore not only to lower work in progress levels and manufacturing lead times but also to identify possible problematic areas and plan appropriate actions.

Supply Chain Management

Supply chain management (SCM) is a concept that has flourished in manufacturing, originating from Just-In-Time (JIT) and logistics; it represents an autonomous managerial concept. SCM endeavours to scrutinise the entire scope of a supply chain and offers a methodology for addressing the myopic control in supply chains that has been reinforcing wastage problems in construction (Vrijhoef and Koskela, 1999).

The first signs of SCM were perceptible in the JIT delivery system as part of the Toyota Production Process which aimed to regulate supplies to the Toyota motor factory (noted previously under JIT). The main objective of Toyota was to drastically decrease inventory and to regulate the suppliers' interaction with the production line more efficiently and effectively (Vrijhoef and Koskela, 1999).

After its emergence in the Japanese automotive industry as part of a production system, the conceptual evolution of SCM has become a distinct subject of scientific research in its own right. Along with original SCM approaches, other management concepts (e.g. value chains and extended enterprise) have been influencing the conceptual evolution of SCM (Vrijhoef and Koskela, 1999). SCM combines particular features from various concepts including Total Quality Management (TQM), JIT, Value Management, Risk Management and Partnering.

Definition of Supply Chains

SCM has been defined as *"the network of organisations that are involved, through upstream and downstream linkages, in the different processes and activities that produce value in the form of products and services in the hands of the ultimate customer"* (Vrijhoef and Koskela, 1999).

Handfield and Nichols (1999) have also expanded the supply chain definition to include upstream supplier networks and downstream distribution channels. They have argued that SCM is not as simple as materials demand and supply networks; it also includes the management of information systems, sourcing and procurement, production scheduling, order processing, inventory management, warehousing, customer service and after-market disposition of packaging and materials. The supplier network consists of all organisations that provide inputs, either directly or indirectly, to the host firm. Thus a supply chain includes internal functions, upstream suppliers and downstream customers, defined as:

- a firm's *internal functions* include the different processes used in transforming the inputs provided by the supplier network
- the *upstream supplier network* consists of all organisations that provide inputs, either directly or indirectly, to the host company
- the *external downstream supply chain* encompasses all the downstream distribution channels, processes and functions that the product passes through on its way to the customer.

All organisations are part of one or more supply chain(s). Whether a company sells directly to the end customer, provides a service or manufactures a product, it can be characterised within the context of its supply chain. Supply chains are essentially a series of linked suppliers and customers; every customer is in turn a supplier to the next downstream organisation until a finished product reaches the ultimate end user (Handfield and Nichols, 1999).

Vrijhoef (1998) has defined SCM as: *"the establishment, co-ordination and maintenance of an optimised supply chain that operates effectively and efficiently, fulfilling all its preconditions and goals optimally and involving its stakeholders."*

Supply Chain Management Concepts

SCM forms linkages across the entire supply chain rather than just a linkage to the next process level. It aims to increase transparency and alignment of the supply chain's co-ordination and configuration, regardless of functional or corporate boundaries (Vrijhoef and Koskela, 1999).

Table 7.2 *Characteristic differences between traditional ways of managing the supply chain and SCM (source: Vrijhoef and Koskela, 1999)*

Elements	Traditional management	Supply chain management
Inventory management approach	Independent efforts	Joint reduction of channel inventories
Total cost approach	Minimise firm costs	Channel-wide cost efficiencies
Time horizon	Short-term	Long-term
Amount of information sharing and monitoring	Limited to needs of current transaction	As required for planning and monitoring processes
Amount of co-ordination of multiple levels in the channel	Single contact for the transaction between channel pairs	Multiple contacts between levels in firms and levels of channel
Joint planning	Transaction-based	On-going
Compatibility of corporate philosophies	Not relevant	Compatibility at least for key relationships
Breadth of supplier base	Large, to increase competition and spread risks	Small, to increase co-ordination
Channel leadership	Not needed	Needed for co-ordination focus
Amount of sharing risks and rewards	Each on its own	Risks and rewards shared over the long term
Speed of operations, information and inventory levels	'Warehousing' orientation (storage, safety stock) interrupted by barriers to flows; localised to channel pairs	'Distribution centre' orientation (inventory velocity) interconnecting flows; JIT, quick response across the channel

The shift from traditional methods of managing the supply chain towards SCM includes various elements (Table 7.2). The traditional way of managing is essentially based on a 'conversion (or transformation) view' of production, whereas SCM is based on a 'flow view' of production. The conversion view suggests that each stage of production is controlled independently, whereas the flow view focuses on the control of the total flow of production (Vrijhoef and Koskela, 1999).

Advocated Advantages of Deploying SCM

The theoretical advocated advantages of deploying SCM within a construction operational environment may be summarised as follows:

- cost savings for all members of the supply chain
- increased effectiveness of information transfer
- a drive for continued quality improvements
- enhanced client satisfaction
- a reduction in the number and severity of disputes
- the company embarks upon a learning curve, leading to its continued existence based on stakeholder satisfaction
- increased technical and process innovation
- a reduction in overall project duration
- increased intrinsic motivation related to enhanced job satisfaction.

However, in order to test these advocated advantages, fifty construction organisations were asked to rank them via a postal survey questionnaire. The results of the survey are presented in Figure 7.1.

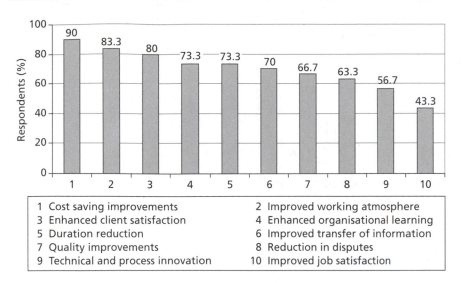

1 Cost saving improvements   2 Improved working atmosphere
3 Enhanced client satisfaction   4 Enhanced organisational learning
5 Duration reduction   6 Improved transfer of information
7 Quality improvements   8 Reduction in disputes
9 Technical and process innovation   10 Improved job satisfaction

Figure 7.1 *Responses from 50 construction firms on advocated advantages of SCM*

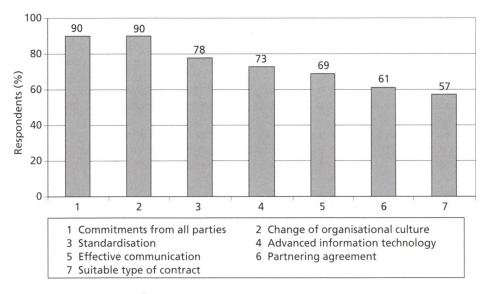

1 Commitments from all parties   2 Change of organisational culture
3 Standardisation   4 Advanced information technology
5 Effective communication   6 Partnering agreement
7 Suitable type of contract

Figure 7.2 *Issues to be considered for SCM deployment*

It is evident from Figure 7.1 that most of the theoretical advocated advantages are attained in practice. However, job satisfaction is not improved to any great extent by the deployment of SCM. The respondents were also asked to identify what they considered to be the most important issues for a successful implementational process. These results are presented in Figure 7.2.

It has been established by the survey that the most vital issue is commitment. Therefore senior management must ensure that the host company has true commitment and that this is effectively communicated to all concerned.

Deployment of Supply Chain Management

SCM is a very flexible concept that can be deployed in conjunction with other techniques. TQM is one such technique and Vrijhoef and Koskela (1999) have argued that the SCM methodology bears resemblance to the

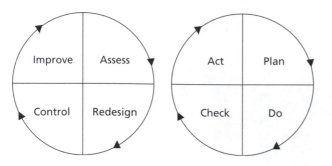

Figure 7.3 *Generic SCM methodology compared to the Deming Cycle (source: Vrijhoef and Koskela, 1999)*

Deming Control Cycle (Figure 7.3). Generically, the methodology of SCM consists of four main elements: (1) supply chain assessment, (2) supply chain redesign, (3) supply chain control and (4) continuous supply chain improvement.

Supply Chain Measurement

Supply chain measurement covers all tools and mechanisms that provide information on the performance of the supply chain process, or series of processes, across functional and organisational areas within a supply chain. It seeks to analyse the current situation in a supply chain and establish how supply chain performance may be improved.

After defining the relationships within a supply chain, the next stage is to assess the current process across the supply chain in order to detect wastage and associated problematic issues and develop solutions.

The critical last phase of supply chain measurement is to install a measurement system which enables analysis and effective control of the supply chain process to be achieved consistently. The system should deliver adequate data to facilitate the detection and removal of the root cause(s) of wastage.

Supply Chain Redesign and Re-engineering

Once causality is understood and having established the root cause, the next stage is to redesign the supply chain in order to address the noted problematic issues. This may include a redistribution of roles, tasks and responsibilities among partners in the supply chain and a review of procedures.

The main focus at this stage is the optimal reallocation of activities, responsibilities and authority among supply chain participants in order to focus attention on optimisation of material and information flows. The objective of a host firm is to enhance the efficiency and effectiveness of the supply chain structurally as a total system in a holistic manner.

Supply Chain Control and Continuous Improvement

The next stage is to control the supply chain according to its new configuration. An important part of the control function is the installation of a monitoring mechanism to continuously assess how the supply chain is functioning. This includes systems to measure and estimate total wastage across the entire supply chain. A company should continuously identify new opportunities and find new initiatives to develop the supply chain. In fact, this continuous improvement implies the on-going evaluation of the supply chain process and the recurring deployment of the previous three stages – assessment, redesign and control.

To maintain and improve the SCM process it is necessary to continually monitor and evaluate the whole process in a holistic manner. This is best achieved by employing the Deming Plan, Do, Check,

Action Dynamic Control Cycle. However, this needs to be adapted to the four main activities of process evaluation and improvement. Hence we have:

Assess = Plan
Redesign = Do
Control = Check
Improve = Action
(see Figure 7.3)

It should be noted that for continuous improvement to be attained each time the cycle is deployed, it should lead to the development of an implementational strategy. Thus, the original Figure 5.16 of the Plan, Do, Check, Action cycle applied to construction activities has been adapted to better relate to the SCM's four key functions. This is shown in Figure 7.4.

Because of the interdependent nature of a supply chain, supply chain partnerships must exhibit attitudes and behaviour indicative of common interests and collaboration. Long-term commitment to the supply chain is critical, and without it a supply chain management relationship will lose any synergistic advantages.

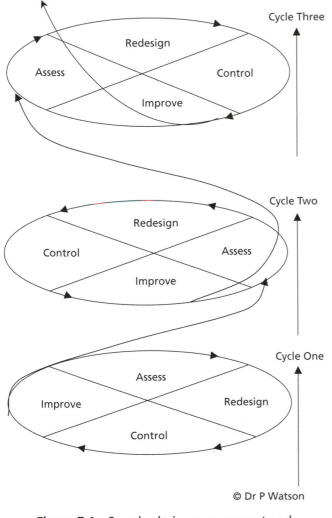

© Dr P Watson

Figure 7.4 *Supply chain management and continuous improvement*

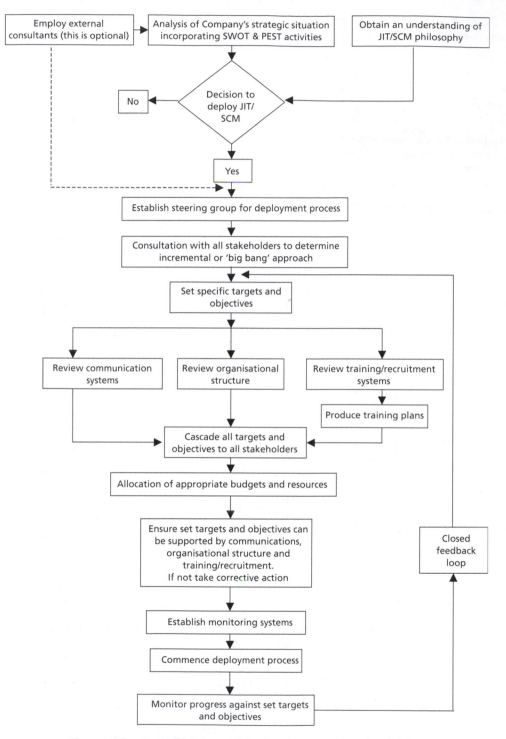

Figure 7.5 *Generic JIT and SCM implementational model*

Conclusions

The concept of SCM represents a logical continuation of previous management developments. It has been influenced by other management concepts such as JIT, TQM, Value Management and Partnering. The most critical issue in SCM is the achievement of a co-operative relationship between all supply chain partners. SCM would enable the efficient and effective utilisation of both plant and materials.

Summary

A critical issue of both JIT and SCM is the achievement of co-operation between supply partners. This requires considerable effort on behalf of all participants and puts pressure on the ability of firms to develop inter-organisational relationships. This can only be attained when there is evidence of understanding, commitment, trust and collaboration. However, there are critical preconditions that have to be addressed before all participants would be willing to reduce their business to just core competencies and contribute complementary assets to a coalition with other construction organisations (Vrijhoef, 1998).

The above noted, there are real and obtainable advantages to be gained for construction companies in pursuing JIT and SCM. These advantages, when realised, such as overall cost savings and the establishment of more effective communication systems, are well worth the efforts required for deploying JIT and SCM. Figure 7.5 provides a generic implementational model based on feedback from the respondents of the postal questionnaires. The model can assist in empowering construction firms to obtain the advantages of JIT and SCM deployment, and thus address the key issues associated with plant and materials management.

References

Champy J. (1993) In Re-engineering, Organisational Change Must Start from Day One, *CSC Insights*, No. 4.

Davenport T.H. (1993) *Process Innovation – Re-engineering Work through Information Technology*. Harvard Business Press, Cambridge, MA, ISBN 0–87584–366–2.

Hammer M. and Champy J. (1993) *Re-engineering the Corporation – A Manifesto for Business Revolution*. Nicholas Brealey Publishing, ISBN 1–85788–029–3.

Handfield R.B. and Nichols E.L. (1999) *Introduction to Supply Chain Management*. Prentice-Hall, Englewood Cliffs, New Jersey.

Harrington H.J. (1991) *Business Process Improvement – The Breakthrough Strategy for Total Quality, Productivity and Competitivensss*. McGraw-Hill Inc., New York, ISBN 0–07–026768–5.

Hellard R.B. (1995) *Project Partnering: Principle and Practice*. Thomas Telford, London, ISBN 0–7277–2043–0.

Ibbs C.W., Kartam S.A. and Ballard G. (1995) Re-engineering Construction Planning, *Project Management Journal*, June.

Koskela L. (1992) Application of the New Production Philosophy to Construction, *CIFE, Technical Report*, No. 72. Stanford University, USA.

Latham M., Sir (1994) *Constructing the Team – Final Report of the Government/Industry Review of Procurement and Contractual Arrangements in the UK Construction Industry*. HMSO, London, ISBN 0–11–752994–X.

Miozzo M., Betts M., Clark A. and Grillo A. (1997) *An IT-Enabled Process Innovation Strategy for Construction*. Department of Surveying, University of Salford, UK.

Mohamed S. and Tucker S. (1996) Options for Applying BPR in the Australian Construction Industry, *International Journal of Project Management*, **14**(6), 279–385. Elsevier Science Ltd and IPMA, Great Britain, ISSN 0263–7863.

Schimming B.B. (1993) The Comming Competition Forum, *Civil Engineering*, **53**(6), 18.

Soles S. (1995) *Work Re-engineering and Workflow – Comparative Methods*. Future Strategies Inc., USA, ISBN 0–9640233–2–6.

Vrijhoef R. (1998) *Co-makership in Construction: Towards Construction Supply Chain Management*, MSc Thesis, Delft University of Technology, The Netherlands [on-line]. Available at URL:http://www.cscm.net. Accessed 11 March 2001.

Vrijhoef R. and Koskela L. (1999) *Roles of Supply Chain Management in Construction* [on-line]. Available at URL:http://www.ce.berkeley.edu/~tommelein/IGLC-7/PDF/Vrijhoef&Koskela.pdf. Accessed 11 March 2001.

Wafa M.A. and Yasin M.M. (1998) *A Concept Framework for Effective Implementation of JIT – An Empirical Investigation*. [Cited 19 July 1999.] Available from internet URL:http://www.emerald-library.com/brev/02418kcl.htm.

8 Progress and Control

Introduction

This chapter examines the management functions of progressing and controlling the siteworks. It links directly to associated Chapter 4, *Planning and Programmes*. Maintaining progress monitoring throughout the works and taking control decisions leading to positive action are fundamental in keeping the project on track to the planned programme. Key to progress and control is the concept of short-term planning which focuses on planning sections or stages of the works contained within the master programme and feeds progress information back from short-term programmes into the master programme. Also key are the practices involved in maintaining a progress chart which keeps the process 'live' and dynamic, and having procedures which record progress efficiently and feed information expeditiously into the monthly project progress meetings.

Basic Concepts of Construction Progress and Control

It was identified and discussed in an earlier chapter how progress control follows the planning and programming of the site works and compares the work undertaken against the programme – allowing for the redistribution of resources, if necessary, to speed up the work if it is falling behind the programme. The fundamental requirement for any progress control approach is therefore the implementation of a set of effective arrangements which guide management and operatives to work to the programme. These arrangements are the management procedures and working instructions developed within the framework of the corporate and project management systems. These were discussed in Chapter 2.

An example of a problem encountered by a contractor working for a major petrochemical-industry client serves to highlight the practicalities of the basic concepts of progress control. A working programme for the construction of a reinforced concrete blast wall indicated that a small turbine could be installed after the blast wall had been built. However, a change in the specification and size of the turbine meant it was necessary for the installation of the turbine to precede the construction of the blast wall. For reasons of manufacture and delivery, installation of the turbine could not be brought forward in the programme and hence a sequencing problem, so common within construction processes, arose. The construction of the blast wall was now delayed, and moreover the delay had knock-on effects to later sequenced activities. Fortunately for the client, the contractor had effective arrangements in place and sufficient flexibility built into the programme so that alternative options, which would have been costly to the client, such as a redesign of the wall or further change in specification to the turbine, were avoided. The contractor reprogrammed later works with shorter durations by deploying greater resources to the operations involved, a common solution to rescheduling construction activities. While it was not possible to quantify accurately the effect on cost of using any of the alternative options, it was determined that the reprogramming of works cost the client an additional 6% on that particular section of the works, which the client deemed acceptable under the circumstances. This example shows the importance of having a flexible programme incorporating effective management procedures to accommodate unforeseen events or circumstances.

In any progress control approach there must be a warning mechanism of those problems which are likely to arise. The early warning of problems which affect progress is essential if progress is to be maintained and the programme met. Comprehensive and frequent reassessment of the programme for the remaining works, in particular through *short-term* planning, will assist in identifying relevant early warnings. An effective control mechanism will maintain the momentum of working to programme and act as a prompt to management and supervisors alike. Management will, therefore, tend to be proactive to challenges rather than be reactive to events.

A further example, drawn from a contractor serving a major process engineering client under a measured and dayworks term contract, illustrates the importance of the early identification of potential progress problems. The site engineer, when setting out for the formwork to a number of machine bases in a particularly restrictive part of the process plant, realised that if the machinery was installed as sequenced on the programme it would obstruct the subsequent casting of machine plinths to the rear of those bases. Consideration of the anticipated problem with the client's works superintendent resulted in a reprogramming of the short-term plans relating to the machine bases and the machine plinths. This resulted in a reorganisation of those sections of the works with a more practical sequence and without any disruption to the priority or sequence of subsequent works. Had the bases been cast first it was unlikely that the plinths could have been cast without undertaking remedial works to the bases and disassembling parts of the machinery fixed to the machine bases to facilitate working access.

The most important lesson that can be learnt from the two examples described is that if a progress control mechanism gives an early warning of problems then it allows management to be tolerant of imperfect construction planning.

Benefits of Effective Progress Control

It was seen earlier, where effective progress control mechanisms are in place the contractor has the ability to:

- rearrange project priorities in the light of changed circumstances
- prevent work overload or underload
- highlight potential under-resource to meet programme needs
- determine the material stock available for use
- check materials on order
- show when further orders should be confirmed
- show when subcontractors are needed on site
- show deployment and labour strengths of subcontractors
- provide information to head office on the project's progress.

Relationship of Programming to Progressing

It was identified earlier that effective construction planning and progress control requires: (1) a planned programme of the method and sequence of the works, and (2) arrangements to compare the works as executed against the programme, i.e. progress control. As programming is a *planning* function and progressing a *control* function, they have been separated while concepts and principles are developed. Planning and programmes were addressed in Chapter 4 and progress and control are dealt with in this chapter.

It is vital that an accurate and realistic sequence of events is formulated at the outset. If it is not, an *ad-hoc* sequence of events is likely to emerge leading to major problems which are likely to result in lost time and money. Effective programming avoids *ad-hoc* management and haphazard methods and sequence of working while good progress control ensures that resources are deployed effectively and that the works are conducted in line with the programme. Both programming and progressing are therefore essential to the efficient and effective management of any construction project.

Progress Control

Progress control is the mechanism which ensures that the work being carried out is in accordance with the programme. Progress control encompasses a number of management tasks such as ensuring: that operations start when intended; that resources are deployed as required; that materials are delivered when needed; that plant is as fully utilised as possible; and that authoritative information has been received as required from the client's representative.

Ensuring that appropriate checks on the on-going works are made and acted upon is essential to effective construction management. Any problems which impede progress on the project will result in delay and possible overrun, and consequently have an effect on contract cost. Effective progress control relies on comparison of achievement (progress) with forecast (programme) such that, if achievement is not meeting the forecast, action can be taken to bring the two back into balance. It is therefore essential, but also very practical, that comparison of progress and programme is depicted in a simple and clear visual manner. This is important because any imbalance between progress and programme must be readily discernible and not masked by a complex mechanism. The implications of a problem to the remaining works should be apparent and the method should facilitate comparison at frequent intervals, sufficient to allow action to be taken quickly before the programme and the works become disorganised.

To develop an effective method of monitoring progress the contractor must consider the following:

  (i) the frequency of monitoring to check progress
 (ii) the method of measuring the work as executed
(iii) the mode of recording progress to facilitate comparison with the programme
(iv) the method of analysing and reviewing progress.

Short-Term Planning and Progress Control

For planning and progressing to be effective, their activities must be continuous throughout the life of a construction project. These activities range from a strategic overview of the work provided by the contract, or master, programme to planning implemented during the daily activities at the production workplace – this is the mechanism termed *short-term planning*.

Short-term planning serves a number of important functions in construction management. Apart from its function as a management tool to convey instructions to works supervisors and site personnel, a short-term plan is an effective technique for controlling the progress of operations, in that it provides a performance yardstick which is simple and easy to understand and apply in the work situation. In addition, a short-term plan is essential to effective scheduling and resource deployment, and in maintaining feedback of information to both the principal contractor and client organisations.

Objectives of Short-Term Planning

Short-term planning should fulfil the following objectives:

- to subdivide the contract, or master, programme into section or stage programmes that will be meaningful to controlling the siteworks in the immediate or short term
- to enable site management and works supervisors to deploy resources most effectively by considering in detail those particular sections of the works that are current
- to provide a current accurate assessment of the short-term materials' requirements
- to allow site management to look ahead, in a micro context, to see likely disruption and delays to progress and manage problems effectively in the short term before there is any adverse effect to the overall contract programme
- to assess progress in the short term and manage the works accordingly, thereby maintaining the proactive function of the contract programme
- to present the opportunity to establish better working methods through the use of method study, applied to particular works sections or operations

- to provide detailed feedback on specific parts of the works to assist in the execution of subsequent activities, and to provide information to both the contractor and client organisations for use in short-term planning on future projects.

Benefits of Short-Term Planning

The principal benefit of short-term planning is that it makes site management more aware of those aspects of the work which are critical to production progress, and does so on a continuing basis. Whereas it is possible to lose sight of project objectives within the overall contract programme, a short-term plan identifies immediate objectives which need to be addressed. For example, short-term planning stimulates management into:

- making clear decisions on methods and sequencing of the works, essential to maintaining progress of the contract programme
- determining the position with regard to labour, plant, materials and project information such as drawings, specification and instructions
- appraising the standards of performance, i.e. workmanship and quality, given the influences and conditions on the project at that time.

Some of the objectives of short-term planning and its challenges can be seen in an example. A site inspection, conducted by the site agent, in conjunction with the foreman joiner, foreman steelfixer and works section ganger, at 1.00 pm on Wednesday, determined that 30 m^3 of concrete should have been cast by noon that day. Only 24 m^3 had, in fact, been poured. Discussion at the workplace revealed that there had been a delivery problem with one of the ready-mix loads which had hindered progress. In addition, a compressor powering a vibrating poker had malfunctioned and there had been a problem with a steel reinforcement cage being positioned out of alignment within a section of formwork.

The site manager was confronted with a number of problems, each requiring a solution if the short-term plan, in this case a one-week concrete pouring schedule, was to be kept on line. Essentially, it was determined that the work would have to be made up within that day's schedule because to leave the outstanding concrete pour would disrupt the work schedule on the following day. To alleviate the problem, work would continue with another section of the pour while the steelfixing gang rectified the bar reinforcement. To make up the concrete pouring schedule an extra delivery of ready-mix was hurriedly ordered and confirmed to arrive on site in the late afternoon, and additional operatives redeployed from other work to place the concrete. The matter of the compressor was solved by borrowing a machine from a neighbouring construction project.

Despite the site manager's laudable attempts, a number of minor tasks were not completed within the daily time frame. However, flexibility within the short-term plan did mean that all the outstanding tasks could be rescheduled for Friday afternoon and Saturday morning, thereby completing the concreting operations within the one-week programme.

Non-Achievement of Short-Term Plans

Where there is any deviation from the short-term programme, it is essential for site management to check for the following circumstances:

(i) any change in activity from the construction method used and work sequence determined when compiling the programme
(ii) any difficulties concerning the supply or deployment of resources
(iii) any problem resulting from a lack of project information or instruction
(iv) any change to climatic or site conditions as against those anticipated.

The identification of any of these circumstances will highlight the reason for any delay in progress and will indicate the kind of remedial action that site management must take to bring the work back on programme. When implementing short-term programming it should be borne in mind that the two most crucial areas to be addressed are labour and construction plant, as these can be manipulated to directly affect the duration of a contract. However, management must also direct due attention to materials as these, if not available on site at the right time in the programme, can be the cause of severe delay. It is important to note that because of the tight schedules used in short-term planning generally, if a major problem does arise in the provision of resources, such as non-delivery of materials, there are likely to be knock-on effects to progress.

Management must also consider supervision when developing a short-term plan. Without adequate supervision of the workforce on a daily basis it may not be possible to keep to the programme. The contracting organisation must also support management to assist in the resolution of any issues concerning short-term plans. By definition, short-term plans are short, and this means that in order to avoid non-achievement management must be constantly vigilant and able to react quickly and effectively to deviations from the programme.

Developing Short-Term Programmes of Work

Construction managers are constantly faced with questions concerning the deployment of resources. Such questions form an everyday part of site management routine and most are answered based on years of site experience and knowledge of work outputs and optimum gang sizes. An important aspect of short-term planning is the ability to estimate the duration of operations and those tasks that go to make up operations, so that the deployment of resources is made effectively. It is essential both not to under-resource nor to over-resource any operation. The former situation would likely lead to an overrun on programme, while the latter would mean that resources were being wasted.

The following practical examples show how management calculate accurately the resources needed to carry out operations in the development of short-term programmes of work.

Suppose a programme shows that the duration for erecting formwork to a floor is 5 days, the labour output is 1 m² per operative hour and the area of the floor is 200 m². How many trade operatives must be deployed to complete the work to programme?

200 m² to be achieved in 5 days = daily output of 40 m².

At a work output of 1 m² per man hour, the total work content is 40 operative hours.

Therefore, based on an 8-hour day: $\dfrac{40 \text{ operative hours}}{8 \text{ hrs/day}} = 5$ operatives

The above determines the trade requirement, i.e. 5 joiners. In addition, trade gangs will be supplemented with trade 'general operatives' and these may be deployed at a given tradesperson/general operative ratio, usually one general operative for every four tradesperson, although this will be determined by the particular requirements of the tasks involved.

Suppose a programme shows concreting operations to be carried out to a floor slab, machine bases and drainage trenches, totalling 19 m³, and these must be completed within the daily schedule. How many cubic metres of concrete must be placed per hour to complete the work to programme?

Based on a working day of 6.5 hours (this allows 1.5 hours for setting-up the mixer, standing/idle time, cleaning the mixer at the end of the day etc.):

$$\text{output} = \frac{19 \text{ m}^3}{6.5 \text{ hrs/day}} = 2.92, \text{ or } 3 \text{ m}^3 \text{ per hour}$$

Drawing up the Short-Term Programme

As a subdivision of the contract programme, a short-term programme may plan for an individual operation or for a group of operations. It may be a programme of tasks extending up to a week or more, or conversely it may be a one-day schedule of activities. Figure 8.1 shows a one-week short-term programme for the construction sequence of twelve inspection chambers. The precise nature of a short-term programme will be determined by the particular characteristics of the work item, and where and how it fits into the schedule of activities in the contract programme.

Irrespective of its specific orientation, a short-term programme should consider the following points when being drawn up:

- that each short-term programme is clearly identifiable with operations or sections of work shown on the contract programme
- that items of work are clearly defined and their start and finish dates determined so that there can be no overrun in the short-term programme that could affect the progress of the contract programme
- that performance outputs used in determining resource deployment are accurate, realistic and moreover fair to the workforce
- that resources are deployed at the right time and in the correct location on site to enable maximum performance to be achieved.

Communicating Work Schedules from the Short-Term Programme

Whereas the contract programme is developed by building production management for use by management, the short-term programme is developed for implementation by works supervisors in deploying their operational gangs in the workplace. Communication is, therefore, an important aspect of short-term programming. Short-term plans must be effectively translated by first-line and second-line supervisors into instructions which are easily and clearly understood by craft and general operatives. A commonly used method is to issue a written instruction – *a work or job sheet*, which conveys

SHORT–TERM PROGRAMME: INSPECTION CHAMBERS (NE SECTION) 12 NO.						
CONTRACT WEEK NO.:	7					
DAY	MON	TUES	WED	THURS	FRI	SAT
DATE:	7 (JULY)	8	9TH	10TH	11TH	12
IC NO.: 1	✕					
2	✕					
3		✕				
4	✕					
5		✕				
6		✕				
7			✕			
8		✕				
9				✕		
10				✕		
11				✕		
12					✕	

Figure 8.1 *An example of a short-term programme for the construction sequence of twelve inspection chambers*

	JOB SHEET NO: 17
	CONTRACT NO: 4295
	DATE OF ISSUE:
	ISSUED BY:

OPERATIVES DEPLOYED:	DESCRIPTION OF WORK:
A Brown – (Chargehand) B Green – Bricklayer C Blue – " W White – " A Black – " J Red – General operative	Brickwork to inspection chambers

WORK LOCATION:	VOLUME OF WORK:
NE section of site	12 IC in section 4m^2 in each (approx.)

DURATION ON CONTRACT PROGRAMME:	COMMENCEMENT:	COMPLETION:	ACTUAL DURATION:
5 Days	Mon 7 July	Fri 11 July	4.5 days

PROGRESS DETAILS :

– All bricklaying works complete

– IC no. 6 in section, still to bed & set cover, wrong size in store, new size ordered 11/7/03, to be delivered on 14/7/03 and fixed – work will be carried out 08.00 am on 14/7/03, no effect on programme.

COMPILED BY:	APPROVED BY:

Figure 8.2 *An example of a work, or job, report detailing progress of the works*

the necessary information from the short-term programme and also serves as a feedback form – *a work or job report*, detailing the progress of the works (see Figure 8.2).

Impact of Subcontractors and Suppliers on Short-Term Programming

Subcontractors and suppliers play an important part in the planning processes, in particular in short-term programming, where they can form a critical link in the preparations for undertaking specific construction operations. On many projects there may be a multiplicity of subcontractors and this can make co-ordination with the main contractor and other subcontractors a daunting task. It is, in practice, essential that subcontractors are brought into the planning processes as early as possible when developing the contract programme and consulted continuously throughout the contract to facilitate short-term planning. The main contractor should determine *programmed stage objectives* with subcontractors so that subcontract work integrates with the main contractor's series of short-term programmes. Suppliers should be managed broadly in the same way as subcontractors when considering short-term planning although, in practice, it is often not possible for the contractor to ensure the same compliance with the programme as that which applies to subcontractors.

A reliable organisational procedure for material procurement is essential to effective contract and short-term planning, as without the timely delivery of materials to site the benefits of good programming will be lost and progress will not be maintained. Often, subcontractors and suppliers can be oblivious to the effects of their work on the principal contractor's programme, unless there has been sufficient discussion to integrate respective work schedules and objectives. It is essential therefore that the principal contractor impresses on all subcontractors and suppliers, at an early stage in the contract, exactly what their work or service entails and the schedule to which it must be undertaken to avoid disruption to the work of the main contractor or others employed on the contract.

Progress Control

It has been seen that programming is one of two essential parts of time management, i.e. the *planning* aspect. The second aspect of time management, i.e. *control*, is concerned with the *progress* of the works. Progress control as a concept and procedure was explained earlier in this chapter. Essentially, progress control, to recap, is concerned with comparing the construction work undertaken against the programme, such that if there is any deviation from the programme then action can be taken to bring the progress of the works back into line with the programme. This means that whatever technique is used to programme the work, a bar chart for example, should be used as the method to record progress – *the progress chart*.

Progress Charts

Progress can be recorded on the planned programme, shown in Figure 8.3, by drawing a second bar, representing the actual time taken for an operation, in the lower half of each operational duration row. Comparison between the bar depicting programmed duration and the bar recording progress gives a simple check on the current position of the work, i.e. it is being carried out to programme, it is ahead of programme, or it is behind programme. On site, it is often an alternative custom to mark progress with coloured pins to reflect the start and finish of operations which are compared with the position of the planned duration shown by the duration bar. Alternatively, some site managers prefer to mark up the plan as an activity progress 'S' curve. In such an application, activities are added for each week and plotted as a line resembling the conventional 'S' curve seen in the progress and financial profiles of many construction projects.

The resulting progress chart, derived from the planned programme which was explained in Chapter 4, *Planning and Programmes*, is again shown in Figure 8.4. To obtain a greater level of detail on progress, one could look at the short-term programme for a particular group and sequence of operations, to establish how work is progressing within the parameters of, say, a weekly programme of site activity. By monitoring the two levels of programme – *contract* and *short-term* simultaneously, site management is able to build up a picture of works progress at both a micro and macro level and take action rapidly should progress deviate from that which was planned.

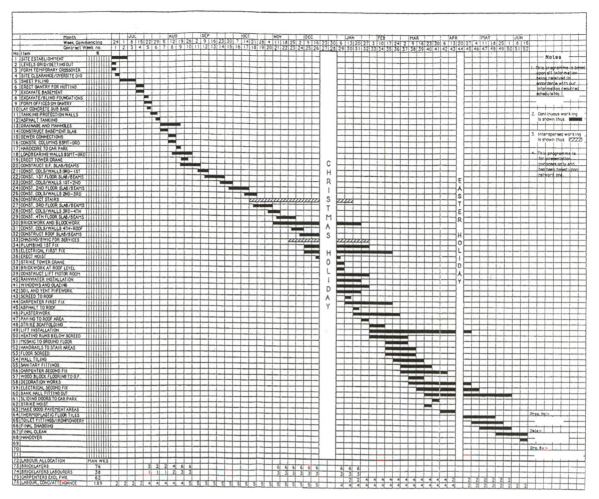

Figure 8.3 *Planned programme*

Making the Progress Chart 'Live'

It is essential that the progress chart becomes and remains a live statement of what is happening on a construction project. In practice, it is all too easy for momentum in planning and control to be lost, and once lost it is difficult to regain. The answer to this is to create the chart in a very simple format so that progress in each operation or section appears directly next to, or below, the programmed duration. It is important that any anomalies between programme and progress are instantly apparent. In addition, it is essential that the 'same' programme is used throughout the contract such that it remains the focal point for work objectives and activities and acts as a centrepiece around which the various short-term programmes are formulated. The programme/progress chart should be the key time management document used at progress meetings and the various contract meetings that will occur throughout the duration of the project. In this way, the progress chart becomes a 'live' statement of the evolution of the works.

The duration bars shown on conventional bar charts, as shown in Figure 8.3, can be substituted by cost data to form a cash-flow spreadsheet, a simple example being shown in Figure 8.5. Although

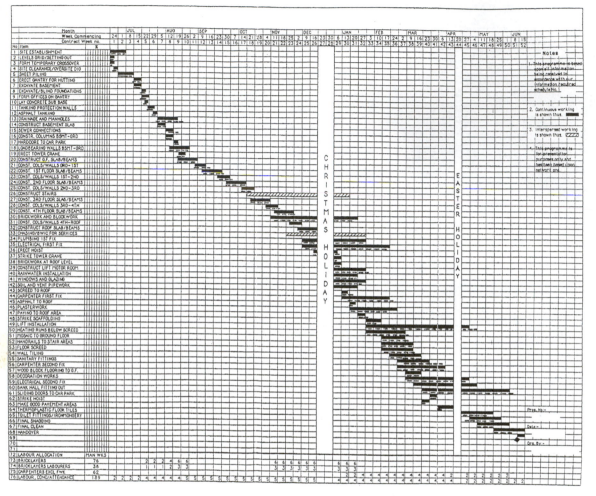

Figure 8.4 *Recording progress on planned programme*

such analysis spreadsheets are used for planning and monitoring project financial information, the relationship between cost and time is clear to see and such spreadsheets can be used to augment time programmes where time and cost matters need to be communicated within the principal contractor's organisation.

Measuring Work Completed

The measurement of progress by time can be misleading as the workforce may accomplish tasks quickly or slowly. Therefore, the most accurate and reliable method of assessing performance is to measure the quantity or volume of work completed and compare that against the programme. When assigning work from the contract programme and short-term programmes, the volume of work encompassed in the operations will be known. This can then be monitored to ensure that a sufficient proportion of work has been achieved at any given point in time, for example at the end of the day's schedule.

ANALYSIS OF CASH FLOW

REF	MONTH NUMBER	1	2	3	4	5	6	7	8	9	10	11	12	TOTALS
1	SITE CLEARANCE/OVERSITE DIG	750												750.00
2	SHEET PILING	16704												16,704.00
3	EXCAVATE BASEMENT/FOUNDS	7972												7,972.00
4	BLIND FOUNDATIONS	900												900.00
5	LAY CONCRETE SUB BASE	3892												3,892.00
6	TANKING PROTECTION WALLS		4916											4,916.00
7	ASPHALT TANKING		2709											2,709.00
8	DRAINAGE AND MANHOLES		2800											2,800.00
9	CONSTRUCT BASEMENT SLAB		3560											3,560.00
10	SEWER CONNECTIONS		2000											2,000.00
11	CONSTR. COLUMNS BSMT-GRD		263											263.00
12	HARDCORE TO CAR PARK		342											342.00
13	LOADBEARING WALLS BSMT-GRD		9534											9,534.00
14	CONSTRUCT G.F. SLAB/BEAMS		4281	4281										8,562.00
15	CONST. COLS/WALLS GRD-1ST			1366										1,366.00
16	CONST. 1ST FLOOR SLAB/BEAMS			13901										13,901.00
17	CONST. COLS/WALLS 1ST-2ND			683	673									1,356.00
18	CONST. 2ND FLOOR SLAB/BEAMS				13901									13,901.00
19	CONST. COLS/WALLS 2ND-3RD				1366									1,366.00
20	CONSTRUCT STAIRS				3333	6666	6666	3333						19,998.00
21	CONST. 3RD FLOOR SLAB/BEAMS					6950	6950							13,900.00
22	CONST. COLS/WALLS 3RD-4TH					1366								1,366.00
23	CONST. 4TH FLOOR SLAB/BEAMS					13901								13,901.00
24	BRICKWORK AND BLOCKWORK					20683	20663	20663						62,049.00
25	CONST. COLS/WALLS 4TH-ROOF					673	673							1,346.00
26	CONSTRUCT ROOF SLAB/BEAMS					13901								13,901.00
27	CHASING/BWIC FOR SERVICES					500	2000	2000						4,500.00
28	PLUMBING 1ST FIX						20000	40000						60,000.00
29	ELECTRICAL FIRST FIX						40000	40000						80,000.00
30	CONSTRUCT LIFT MOTOR ROOM							4000						4,000.00
31	WINDOWS AND GLAZING							60000						60,000.00
32	SOIL AND VENT PIPEWORK													0.00
33	SCREED TO ROOF							3390						3,390.00
34	CARPENTER FIRST FIX							5000	5000					10,000.00
35	ASPHALT TO ROOF							3390						3,390.00
36	PLASTERWORK							5000	10000					15,000.00
37	PAVING TO ROOF AREA							1000						1,000.00
38	LIFT INSTALLATION								30000	30000	30000			90,000.00
39	HEATING RUNS BELOW SCREED								40000	40000				80,000.00
40	MOSAIC TO GROUND FLOOR								10000	10000				20,000.00
41	HANDRAILS TO STAIR AREAS								7500	7500				15,000.00
42	FLOOR SCREED								3000	3000				6,000.00
43	WALL TILING								1250	3750				5,000.00
44	SANITARY FITTINGS								5000					5,000.00
45	CARPENTER SECOND FIX								16666					16,666.00
46	WOOD BLOCK FLOORING TO G.F.								5000					5,000.00
47	DECORATION WORKS								5000					5,000.00
48	ELECTRICAL SECOND FIX										40000			40,000.00
49	BANK HALL FITTING OUT									25000				25,000.00
50	SLIDING DOORS TO CAR PARK										4000			4,000.00
51	MAKE GOOD PAVEMENT AREAS										2000			2,000.00
52	THERMOPLASTIC FLOOR TILES										3000			3,000.00
53	TOILET FITTINGS/IRONMONGERY											5000		5,000.00
54	FINAL SNAGGING											5000	5000	10,000.00
55	FINAL CLEAN											5000	5000	10,000.00
56	STATUTORY SERVICE INTAKES									30000				30,000.00
57	NETT VALUATION AT END OF MONTH	30,218.00	30,413.00	20,231.00	26,223.00	50,739.00	103,923.00	167,796.00	136,759.00	150,916.00	79,000.00	15,000.00	10,000.00	841,209.00
58	PRELIMINARIES 12% OF NETT	3,777.25	3,801.63	2,528.88	3,277.88	6,342.38	12,990.38	23,474.50	17,093.75	18,864.50	9,875.00	1,875.00	1,250.00	105,151.13
59	TOTAL OF NETT PLUS PRELIMS =	33,995.25	34,214.63	22,759.88	29,500.88	57,081.38	116,913.38	211,270.50	153,843.75	169,780.50	88,875.00	16,875.00	11,250.00	946,360.13
60	PROFIT AND OVERHEADS 7.5%	2,549.64	2,566.10	1,706.99	2,212.57	4,281.10	8,768.50	15,845.29	11,538.28	12,733.54	6,665.63	1,265.63	843.75	70,977.01
61	MONTHLY VALUATION	36,544.89	36,780.72	24,466.87	31,713.44	61,362.48	125,681.88	227,115.79	165,382.03	182,514.04	95,540.63	18,140.63	12,093.75	1,017,337.13
62	CASH FLOW	36,544.89	73,325.62	97,792.48	129,505.92	190,868.40	316,550.28	543,666.07	709,048.10	891,562.13	987,102.76	1,005,243.38	1,017,337.13	
63	AVERAGE STRAIGHT LINE GRAPH	84,778.09	169,480.60	254,183.11	338,885.62	423,588.13	508,290.64	592,993.15	677,695.66	762,398.17	847,100.68	931,803.19	1,016,505.70	

Figure 8.5 *Cash-flow spreadsheet derived from bar-chart programme*

The following short-term planning example of constructing twelve inspection chambers illustrates the flexibility needed.

From Figure 8.1 it can be seen that the volume of work, i.e. the brickwork to twelve inspection chambers, with approximately four square metres of brickwork in each, was specified to be accomplished in five days and carried out to the particular sequence shown in the short-term programme. On day one of the short-term programme, three manholes, nrs 1, 2 and 4 were to be constructed. At the end of the first day, the works can be inspected to ensure progress has been sufficient and if remedial action is necessary, say to speed up work because progress fell behind on the first day, this could be implemented on day two. It is clear that flexibility has been consciously incorporated in the short-term programme as the works have been scheduled over five days with leeway in work output incorporated on days three and five. Should there be any problems encountered during the five-day schedule, there is sufficient time available to effect remedial action and ensure progress meets with the requirements of the short-term programme.

Reviewing Progress

If a progressing mechanism is established effectively it can become a live record of the works. It can determine: the volume of work in each operation; the anticipated duration of each operation; and the commencement and completion date of each operation. Moreover: it records the actual work completed; the date when each operation was actually started; how much progress was made during each contract week; and when action was taken to speed up progress to maintain the works to programme. However, through reviewing past experiences on the project, site management can glean much information to assist in progressing future works on the project. For example, by reviewing the actual ordering and delivery of materials for completed stages of the works, reliable logistics for future material procurement can be obtained. Similarly, accurate estimates of demand for labour and plant resources can be made.

Weekly Recording

For progress monitoring and recording to be efficient and effective, monitoring must be almost continuous and recording frequent. The most appropriate schedule for monitoring and recording is to check the on-going works daily and record progress formally at least once per week. *Weekly recording* is commonly adopted throughout the construction process because it provides reliable and therefore usable information to access progress. The principal reason for favouring weekly measurement is that progress during one week of the works cannot diverge too greatly from the intended programme. Therefore, action can be taken quickly should any problem develop. Moreover, action can be taken before the problem escalates and gets out of hand. In contrast, if one implemented monthly recording, irretrievable disorganisation might occur before it was identified by the system and any action taken may be too late.

Charting Progress Throughout the Contract

In undertaking the weekly recording task, the following items should be noted and recorded, where appropriate, on the progress chart (these are in addition to the progress of the operations previously shown):

- the dates by which contract drawings and associated documentation are required such that progress is maintained
- an indicator that orders for resources (labour, plant and materials) should be actioned
- a warning highlight that critical operations or groups of operations are about to commence
- the agreed date for completion of the works and sections of the works, where appropriate.

Further detail could be included on a programme and progress chart. However, it should be remembered that if a chart is too complex it loses its value, because it will not readily be understood. Simplicity is the practical goal of any progressing mechanism – it must convey easily and coherently the state of the works at any point in time, giving indications of how progress will proceed and recording what has happened before. The maintenance of programming and progressing is an important rigour for any building production project, and can add immeasurably not only to the procedural management of a contract but also to the spirit of co-operation and teamwork between all parties.

Stages in Updating the Progress Chart

There are six broad stages in updating a progress chart:

- Collect information concerning progress since the last update
- Identify changes which will affect the future works
- Determine the consequences of taking no action to alter the programme for the remaining works

- Consider if there are any alternative courses of action which will produce a better approach to the remaining works
- Decide which approach to adopt by choosing between alternatives identified
- Ensure that everyone affected by the new programme understands and works to it.

Compiling progress information on a major building project can easily become time-consuming and problematic. Therefore, a simple yet effective method of determining and recording the following needs to be found in practice:

- whether any operations have been recently started and therefore become 'critical' and, if so, their actual start date
- whether any operations have been recently completed and, if so, their actual finish date
- the remaining duration of on-going critical operations
- any change whatsoever in the status of operations, or group of operations, since the last report which will affect the future works.

Problems in Progress Charting

Perhaps the most debilitating problem in progressing is the phenomenon referred to as *the fading bar chart*. A building project which experiences the fading bar chart phenomenon is clearly suffering from a distinct absence of time management, but this may not result from a lack of programming. The phenomenon is recognisable from bar chart programmes that are pinned up in site offices. Experience usually shows that marking-up may not continue throughout the contract and that after a time, work will not proceed in accordance with the programme. If enquiries are made, it is likely to be explained that there were good reasons why things were done differently. Sometimes the reasons are changes of programme introduced because they improved the flow or sequence of the works, but more often they will be changes enforced by other problems, such as changes to the brief during design, changes to design during construction, bad weather or resource unavailability.

Such reasons for the programme represented in the bar chart falling into disuse are common and understandable. The problem is that the time management mechanism on construction projects facing real uncertainties must be flexible. An inflexible programme will fall into disuse inevitably on all but the few most predictable projects. In practice, rigid programmes only work when time management is, in essence, easy to perform. It follows that time management on the majority of construction projects must provide for flexibility – changes of programme are required to represent improved plans for remaining work based on new and better information. The only way to alleviate the fading bar chart problem is to keep drawing up new ones. Each new version should show the project being finished on time or the next milestone being reached on time. To use this approach from project start-up recognises that planning must be continuous if it is to remain a useful component in the management of time. The practice of constantly updating the construction plan is assisted considerably by computer software, as mentioned in an earlier chapter, which can achieve the reconfiguration of plans and progress charts relatively easily.

Altering a Programme Following Progress Monitoring

Any decision to amend a programme should not be taken in isolation from the corporate organisation and also other parties which may be affected. Preferably, they should have the opportunity to contribute to the decision to amend the programme, otherwise they may not fully understand the implications of the change and, in consequence, be unwilling to implement it.

Progress Updating Cycle in Preparation for the Monthly Progress Meeting

The updating cycle for a contract programme should be married to the project's regular progress review meetings, usually the monthly site meeting. This should not prevent the plan being updated more frequently if appropriate. It does mean that the most recent possible updated programme must be considered at such meetings. At progress meetings the current contract programme should be reviewed and decisions made on any changes needed to carry out the remaining works. To make these decisions with the best possible advice a full and thorough updating cycle should have been carried out in the period immediately prior to the meetings, i.e. through the weekly recording mechanisms.

The updating cycle must, in reality, be short. If the latest updated plans tabled at a progress meeting are out of date, decisions about the future works will, at best, be deferred or, at worst, made without the benefit of the best possible forecast.

A practical point in evaluating progress and programme amendment is that the determination of the consequences of taking no corrective action is often ignored or misinterpreted. Investigating alternative courses of action in programming is an activity which is frequently omitted altogether because of lack of time, knowledge or commitment. The reason that programmes fall behind is usually that the production site management team is busy solving current and pressing problems. This should be alleviated by maintaining a realistic contract programme throughout the project. Planning should never stop if *ad-hoc*, or even crisis, management is to be avoided.

Monthly Progress Meetings

The most effective instrument of control for contract management is the client's monthly site meeting usually denoted as the *monthly progress meeting*. The principal purpose of the meeting is for the client's site representative, or principal consultant, to ascertain the progress status of the project. The meeting is the operational focus for all parties associated with the project to consider their position and report on it, e.g. the contractor will report on construction progress, the quantity surveyor on financial status while project participants may also raise such matters as quality, environment, safety, materials and project information. The evaluation of progress on the project will also feed into the development of the interim valuation submitted by the principal contractor to the client for payment of work completed that month.

Participants

The following persons will attend progress meetings:

- client's representative (architect/engineer, depending on project type)
- specialist sub-consultants (where applicable)
- client's quantity surveyor (often referred to as principal quantity surveyor, or PQS)
- contractor's site (project) manager
- contractor's quantity surveyor
- subcontractors (if requested)
- suppliers (if requested)
- secretary (to take written minutes of the meeting discussions).

The monthly progress meetings are *formal* in nature and have recognition within the contract for the project. Minutes should be taken, recorded and circulated in written form to all participants. They should be approved and matters raised that require action should be so actioned by the various parties.

The principal matter of the meeting's agenda is for the contractor to report on and parties to discuss the contractor's progress with the works. This should therefore cover:

- progress
- information availability
- resource position
- requested variations or instructions to the works as originally programmed
- other matters as presented by the participants for collective discussion and agreement, for example quality, health and safety, and environment (these aspects are reviewed elsewhere in this book).

Copies of the master, or contract, programme together with networks and short-term programme/ progress charts should be available at the meeting for the purposes of discussing time management issues while cashflow spreadsheets or 'S' curve diagrams should be available for evaluating the project's financial status. These sources of information were described earlier in this book.

It should be remembered that the monthly progress meeting between the parties is formal, binding and will have actions attached to the matters discussed. Progress monitoring, recording, evaluation and reporting are a total project task and must be maintained throughout the duration of the site works. Only in this way will effective planning, monitoring and control be achieved.

Summary

Progress control is the principal contractor's key mechanism for ensuring that the siteworks carried out are delivered as planned. It ensures that operations commence when they should, that they are given the resources necessary for their undertaking and that they are completed to plan. In this way subsequent siteworks are not disrupted. An important element of progress control is that information is fed back within the principal contracting organisation and outside it to other project participants, ensuring that the best project information at any stage of the project is available for considering subsequent project matters. While the principal contractor will, probably, seek to implement uncomplicated methods of progress monitoring and charting, perhaps the most important facet of progress monitoring and control is to make and keep the progress chart 'live' so that the function remains an on-going and dynamic aspect of managing throughout the project.

9 Quality Assurance

Introduction

Neither the operational environment nor the size of construction firms are rational barriers to the implementation of an appropriate management system. However, an appropriate management system must be built around the 'Quality of Service Provision' provided to the client. A truly service focused quality system will have an in-built mechanism for the attainment of continued organisational improvement.

A further reason for implementing a recognised systematic approach to quality in a construction operational environment is the attainment of a sustainable competitive advantage. This should result in improved quality products and services, delivery and administration, which ultimately satisfy the clients' functional and aesthetic requirements within defined cost and completion parameters.

This chapter concentrates upon three key approaches to achieving a quality service provision. They are:

- Investors in People Standard (IiP)
- BS EN ISO 9001: 2000
- European Foundation for Quality Management Business Excellence Model (EFQM).

Competition in the construction industry is no longer just between firms from the same sector but also those from different sectors. Contractors are exposed to competition from a greater proliferation of companies because of the convergence of construction and non-construction industries. Thus it is imperative that construction-related enterprises make full use of their limited resources in the pursuit of a sustainable competitive advantage. They must implement strategies that exploit their internal strengths by responding to environmental opportunities, while at the same time neutralising external threats and avoiding internal weaknesses. The challenge in obtaining a sustainable competitive advantage is being able to holistically define the nature of a quality provision and then to utilise systematic processes for its attainment and maintenance.

This systematic approach should provide a reduction in operational costs through the elimination of corrective actions in the overall construction process. The above three noted methods provide the necessary systematic approach in varying degrees of complexity.

The IiP Standard concentrates on the linkage between corporate strategy, training, setting objectives and their completion. It is very much people focused but within the context of organisational mission and vision.

BS EN ISO 9001: 2000 is very much process focused when considering the provision of clients' requirements. The demands of BS EN ISO 9001: 2000 are more stringent than the IiP Standard. The people component is still evident within BS EN ISO 9001: 2000. Both recognise that tasks are completed by people and that there must be a link between setting objectives/tasks, training and performance. There exists an emphasis within BS EN ISO 9001: 2000 for continuous improvement informed by client feedback.

The EFQM Business Excellence Model incorporates the elements and philosophies of both IiP and BS EN ISO 9001: 2000. The EFQM Business Excellence Model focuses on a truly holistic approach to quality of the service provision. Specific criteria are allocated for benchmarking organisations and an overall score can be obtained. Again the model empowers continuous improvement but with the rigour of trend analysis based on a comprehensive range of performance criteria.

Investors in People Standard

Investors in People (IiP) was launched in October 1991 as part of the strategy outlined in the 1989 Government White Paper *Employment in the 1990's*. The Standard has four stages with several assessments that have to be attained before accreditation is awarded. These criteria indicators are the measure against which the organisation is judged.

The Investors in People Standard provides a framework which is based on the best practices of many different companies. Its emphasis is on directing employees towards the aims of the organisation so that they may be the best at what they do and that they perform the right tasks. It is essential, therefore, that the company informs its employees of its focus, what it does better than its competitors, and where they are most cost effective. When the company innovates it should be able to draw on the strengths of its employees and know that they will meet the challenge.

Benefits of IiP Application

Construction organisations that set their standards against the IiP criteria for excellence in training and development stand to gain major benefits. Their employees will also benefit from the emphasis on a planned approach to their training and development. The IiP programme has been utilised by construction firms to ensure that their investment in training and development is focused on the achievement of business objectives.

Essentially, the IiP strategy is one which aims to create a learning organisation in which people can understand the barriers that impede progress, and acquire the necessary knowledge, skills and motivation to overcome them.

"IiP will assist in making effective use of all resources, by developing a culture of continuous improvement. This drives the organisation to higher levels of performance and enables it to be managed competitively in the face of continuous operational environmental change" (Investors in People UK, 2000).

Changes

Since its introduction in 1991 IiP has been very successful in improving corporate performance. However, as part of a reflective consultation process commenced in October 1998 and involving 1000 firms, changes to the Standard have been developed. The key drivers for change were that the Standard was not accessible to small and medium-sized enterprises (SMEs). This issue was linked to the assumption that the Standard was too bureaucratic and paper-orientated and in consequence it was not easy to attain the original objective of improved corporate performance. In the revised version the Standard's core values remain the same but now have a clearer focus on outcomes.

The definitive, written version came into operation in April 2000 (Table 9.1). Firms pursuing IiP will now automatically be assessed against the new Standard. The revised Standard has a less complex structure and is expressed in clearer language, drafted in conjunction with the Plain English Campaign. In addition, recent changes to the assessment process have seen a move away from the pass or fail approach to the introduction of a more cumulative form of assessment with greater feedback from the assessor.

In the revised version, when construction firms are assessed to see if they meet the Standard, the assessor will consider the results of the firm and people processes. Businesses must also demonstrate

Table 9.1 IiP standard requirements (source: Investors in People UK, 2000)

Indicator (What the organisation needs to achieve)	Evidence
Commitment *An Investor in People is fully committed to developing its people in order to achieve its aims and objectives*	
1. The organisation is committed to supporting the development of its people	• Top management can describe strategies that they have put in place to support the development of people in order to improve the organisation's performance • Managers can describe specific actions that they have taken and are currently taking to support the development of people • People confirm that the specific strategies and actions described by top management and managers take place • People believe the organisation is genuinely committed to supporting their development
2. People are encouraged to improve their own and other people's performance	• People can give examples of how they have been encouraged to improve their own performance • People can give examples of how they have been encouraged to improve other people's performance
3. People believe their contribution to the organisation is recognised	• People can describe how their contribution to the organisation is recognised • People believe that their contribution to the organisation is recognised • People receive appropriate and constructive feedback on a timely and regular basis
4. The organisation is committed to ensuring equality of opportunity in the development of its people	• Top management can describe strategies that they have put in place to ensure equality of opportunity in the development of people • Managers can provide specific actions that they have taken and are currently taking to ensure equality of opportunity in the development of people • People confirm that the specific strategies and actions described by top management and managers take place and recognise the needs of different groups • People believe the organisation is genuinely committed to ensuring equality of opportunity in the development of people
Planning *An Investor in People is clear about its aims and objectives and what its people need to achieve them*	
5. The organisation has a plan with clear aims and objectives which are understood by everyone	• The organisation has a plan with clear aims and objectives • People can consistently explain the aims and objectives of the organisation at a level appropriate to their role • Representative groups are consulted about the organisation's aims and objectives
6. The development of people is in line with the organisation's aims and objectives	• The organisation has clear priorities which link the development of people to its aims and objectives at organisation, team and individual level • People clearly understand what their development activities should achieve, both for them and the organisation
7. People understand how they contribute to achieving the organisation's aims and objectives	• People can explain how they contribute to the organisation's aims and objectives

Action
An Investor in People develops its people effectively in order to improve its performance

8. Managers are effective in supporting the development of people
   - The organisation makes sure managers have the knowledge and skills they need to develop their people
   - Managers at all levels understand what they need to do to support the development of people
   - People understand what their manager should be doing to support their development
   - Managers at all levels can give examples of actions that they have taken and are currently taking to support the development of people
   - People can describe how their managers are effective in supporting their development

9. People learn and develop effectively
   - People who are new to the organisation, and those new to the job, can confirm that they have received an effective induction
   - The organisation can show that people learn and develop effectively
   - People understand why they have undertaken development activities and what they are expected to do as a result
   - People can give examples of what they are expected to do as a result
   - People can give examples of what they have learnt (knowledge, skills and attitude) from development activities
   - Development is linked to relevant external qualifications or standards (or both), where appropriate

Evaluation
An Investor in People understands the impact of its investment in people on its performance

10. The development of people improves the performance of the organisation, teams and individuals
    - The organisation can show that the development of people has improved the performance of the organisation, teams and individuals

11. People understand the impact of the development of people on the performance of the organisation, teams and individuals
    - Top management understands the overall costs and benefits of the development of people and its impact on performance
    - People can explain the impact of their development on their performance, and the performance of their team and the organisation as a whole

12. The organisation gets better at developing its people
    - People can give examples of relevant and timely improvements that have been made to development activities

that all staff have equal opportunity of access to training and development appropriate to their organisational role.

Implementational Issues

As with most changes to organisations, any implementational process generates a series of problematic issues. The deployment of IiP is no exception and construction companies have encountered the following critical success factors that need to be addressed.

- People may be unclear about what they are supposed to do
- The training policy may need communicating to employees in order that they fully understand their tasks/duties
- A company may find that its job descriptions lack focus
- Existing management may feel threatened by empowering employees
- It may not be easy to convince employees that the standard will be advantageous to the company
- Communication between departments, site and operatives must be effective and efficient
- It must be ensured that the training provided is correct for the employee and the company
- A clear and understandable mission statement must be provided, together with the linked objectives
- There may be a reluctance from site operatives to show initiative and engage in the process
- There may be problems with the directors or senior management staff; senior managers must lead from the front with regard to the implementational process.

Summary

The IiP Standard is based on four key principles:

(1) Commitment to invest in people in order to achieve set business objectives.
(2) Planning how skills, individuals and teams are to be developed in order to achieve the stated objectives.
(3) Action to develop and use necessary skills in a clearly defined and continuing programme directly linked to business objectives.
(4) Evaluating outcomes of training and development for the progress of individuals towards set goals, to value attained and future needs.

The application of these key principles can bring to a construction organisation such advantages as higher profits, cost savings, better practice/processes, improved management skills and enhanced corporate image. The emphasis is clearly on outcomes of organisational activities rather than a concentration on processes. The Standard has no requirement for written processes but organisations must demonstrate via outcomes what benchmarks are used to determine if the Standard is being achieved.

The IiP Standard accreditation can produce the organisational focus required by construction enterprises. Most construction companies are already committed to training and developing their employees: *"IiP allows companies to become more focused on ensuring that what is done is relevant to business needs"* (Druker and White, 1996).

The application process can be a most problematic activity for construction organisations. Thus a generic implementational model, as indicated in Figure 9.1, has been developed to provide assistance during the application of the Standard.

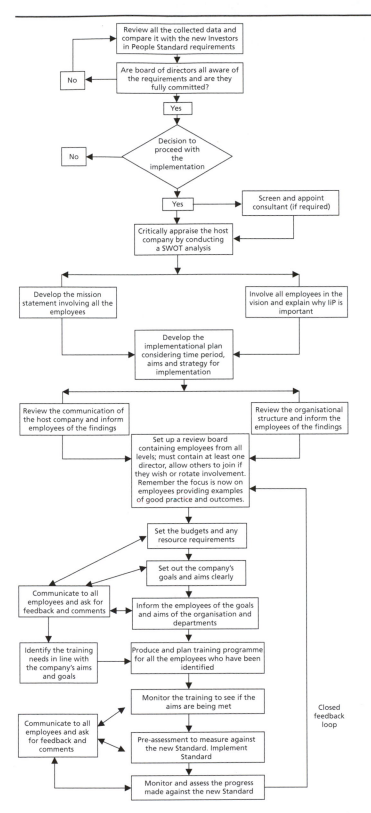

Figure 9.1 *Generic model for the implementation of Investors in People Standard (adapted from Watson and Heath, 1998)*

BS EN ISO 9000: 2000 Quality Management Systems (QMS)

BS EN ISO 9000: 2000 standards are a formal set of QMS. They are international in scope, have a set of guidelines and can be applied virtually to any industry. A formal set of standards which are internationally recognised is always more preferable to customers and suppliers/contractors than setting up one's own set of standards: "*BS EN ISO 9000: 2000 standards are universal and are not simply the decree of one quality guru, but have been devised by an international committee*" (Love and Li, 2000).

"*A QMS embraces all areas of the organisation: marketing, contract acceptance, product design, production, delivery, service, finance and administration*" (Munro-Faure *et al.*, 1995). The objective of a QMS is to ensure that only conforming products and services reach the clients. Love and Li (2000) believe that by being registered to the QMS, "*an optimal level of quality is achieved throughout all stages of the product's quality cycle.*"

Clause 2.1 of BS EN ISO 9000: 2000 clearly explains the rationale for QMS, with the sole purpose of assisting organisations in enhancing customer satisfaction (BSI, 2000a). However, in the introduction to BS EN ISO 9001: 2000 it states that it should be a strategic decision of an organisation to adopt a QMS (BSI, 2000b).

Today's business environment is such that managers must strive for competitive advantage to hold on to their market share, let alone increase it. Price is no longer the major determining factor in customer choice; price has been replaced by quality. In a highly competitive world, customers are becoming more quality conscious. They know that inferior quality has implications long after the joy of low cost has been forgotten. To hold on to customers, producers of products or services require a quality plan. This will directly define its efforts towards customer satisfaction.

The BS EN ISO 9000: 2000 series is unique as a set of standards because it does not deal with a particular product but rather assesses the system as a whole. It is focused on a company's effectiveness in delivering a quality product or service to the customer from the contractual and design phase to installation, storage and maintenance. The standard recognises that a company cannot function smoothly and prove that it has been doing so unless there are clearly kept records of what has been done and what procedures should be followed. Employees cannot be expected to perform at the optimum level if they are never told precisely what they should be doing and there are no guidelines on their specific duties.

BS EN ISO 9000: 2000 specifications may on the face of it seem demanding. However, many construction companies are now certificated. Therefore the effort expended in obtaining the award must be offset by the benefits of certification.

Third Party Certification

The BS EN ISO 9000: 2000 family of standards is not certified by the ISO itself but by a local standards body; in this case, the BSI in the UK exercises accreditation through certification bodies which in turn certify individual organisations. In terms of self-assessment or customer assessment, third party assessment is more objective.

Benefits

A number of researchers around the world affirm the benefits of implementing the BS EN ISO 9001: 2000 standard. They incorporate:

"*1. Meeting customer requirements*
 2. Communicating customer requirements
 3. Staying on tender lists and getting new business
 4. Doing things right the first time"
  (McCabe, 1998)

"*5. Gaining customer loyalty*
 6. Reduced service calls
 7. Higher prices

8. *Greater productivity*
9. *Cost reduction*"
    (McGeorge and Palmer, 1997).

The following quotation illustrates that the new set of standards aims to be a stepping stone towards Total Quality Management, with more emphasis on customer satisfaction, involvement of top management, internal communication and continual improvement: "*It has been long recognised that investment in quality management systems, in addition to responding to customer expectations, has resulted in benefits to the efficiency of the organisation, its operations and economics performance, as well as to the quality of its products and services. Specifically, the revised standards will be of great help for organisations wishing to go beyond simple compliance with Quality Management System requirements for certification purposes. They can be readily applied to small, medium and large organisations in the public and private sectors, and will be equally applicable to users in industrial, service, software and other areas*" (BSI, 2000a).

ISO/TC 176/SC 2 has surveyed existing users on which part of the standards they would like to see improved in the new BS EN ISO 9000: 2000. The results are as follows:

1. *Use simple language and terminology*
2. *Facilitate integration into one management system*
3. *Address continuous improvement*
4. *Use a process model approach to quality management*
5. *Improve compatibility with other management system standards*
6. *Address customer satisfaction more strongly*
7. *Make the standards more business oriented.*

The BS EN ISO 9000: 2000 family of standards has been developed to assist organisations of all types and sizes to implement and operate effective quality management systems. The format consists of:

- BS EN ISO 9000: 2000 providing an introduction to quality management systems and the vocabulary of quality management
- BS EN ISO 9001: 2000 specifying requirements for quality management systems for use where an organisation's capability to provide conforming products needs to be demonstrated
- BS EN ISO 9004: 2000 providing guidance on the implementation of a broadly based quality management system to achieve continual improvement of business performance
- BS EN ISO 10011: 2000 providing guidance on managing and conducting internal and external quality management system audits.

Together they form a coherent set of quality management system standards.

Construction organisations which are currently certificated to BS EN ISO 9000: 1994 were given three years (from December 2000) to address the new standard for continuance of certification. Therefore a comparison of the old standard with the new is provided. This should prove invaluable for certificated construction organisations when engaging in the transitional process. A transitional model is also provided for ease of implementation by construction organisations in Figure 9.3. Both the model and the description of the standards requirements are of value to construction organisations seeking certification for the first time.

The standard has been revised into five main sections as follows:

- Quality Management System
- Management Responsibility
- Resource Management
- Product Realisation
- Measurement, Analysis and Improvement.

Each of these sections is broken down into sub-sections which are addressed in some detail. It is worth noting that none of the requirements of BS EN ISO 9001: 1994 have been removed but are in most cases reorganised into the new standards structure.

Exclusions from the Management System

Currently a construction organisation is able to seek certification against either BS EN ISO 9001, 9002 or 9003: 1994 dependent on the scope of its business. BS EN ISO 9001: 2000 will result in BS EN ISO 9001, 9002 and 9003: 1994 becoming obsolete and will require all organisations to be certified/registered to BS EN ISO 9001: 2000.

However, it is recognised that owing to the nature of an organisation's product or service, customer requirements or applicable regulatory requirements, not all the requirements of BS EN ISO 9001: 2000 will be appropriate for every organisation. BS EN ISO 9001: 2000 will still allow those organisations to claim conformity to the standard but not absolve them from their responsibility to provide a product that meets stated requirements.

Permissible exclusions will be restricted to clause 7, Product Realisation, and will not apply to any other clause of the standard.

This section identifies the changes between BS EN ISO 9000: 1994 and BS EN ISO 9001: 2000. The checklist questions provide a quick check that most of the key aspects have been addressed within a construction organisation's quality management system. To enable cross-referencing to the 1994 version, 1994 clause number(s) are quoted in brackets.

The Main Clauses of the New Standard BS EN ISO 9001: 2000

4 Quality Management System
  4.1 General Requirements
  4.2 Documentation Requirements
      Quality Manual
      Control of Documents
      Control of Quality Records

5 Management Responsibility
  5.1 Management Commitment
  5.2 Customer Focus
  5.3 Quality Policy
  5.4 Planning
  5.5 Responsibility, Authority and Communication
      Responsibility and Authority
      Management Representative
      Internal Communications
  5.6 Management Review
       Review Input
       Review Output

6 Resource Management
  6.1 Provision of Resources
  6.2 Human Resources
  6.3 Facilities
  6.4 Work Environment

7 Product Realisation
  7.1 Planning of Product Realisation
  7.2 Customer Related Processes
  7.3 Design and/or Development
  7.4 Purchasing
  7.5 Product and Service Provision
  7.6 Control of Measuring and Monitoring Devices

8 Measurement Analysis and Improvement
8.1 General
8.2 Measurement and Monitoring
8.3 Control of Non-Conformity
8.4 Analysis of Data
8.5 Improvement

Section 4 Quality Management System

4.1 General Requirements

The requirements of this clause are largely the same as those found in clause 4.2 of the 1994 Standard. However, the emphasis is on identifying and describing the processes involved rather than having formally documented procedures.

- Has the organisation established, documented, implemented, maintained and continually improved the quality management system in accordance with this international Standard? (4.2.1 and 4.5.1).
  (a) Has the organisation identified the processes needed for the quality management system and its application throughout the organisation? (4.2.1)
  (b) Has the organisation determined the sequence and interaction of these processes? (4.2.2 and 3)
  (c) Has the organisation identified the criteria and method for effective operation and control of these processes? (4.2.2/3 and 4.5)
  (d) Are the resources and information necessary to support the operation and monitoring of the process available? (4.1.3 and 4.2.3)
  (e) Does the organisation measure, monitor and analyse the processes to achieve the planned results? (4.1.3)
  (f) Does the organisation implement actions necessary to achieve planned results and the continual improvement of the process?
- Does the organisation manage these processes in accordance with the requirements of the Standard?

Where the organisation chooses to outsource any process that affects product conformity with requirements, the organisation must ensure it has control over such processes. This control will be identified within the quality management system.

4.2 Documentation Requirements

4.2.1 General

The requirements of this clause are the same as those in clause 4.2.2 of the 1994 Standard. However, a distinction is made between the documentation required by the Standard and documentation required by the organisation to control its processes.
Does the quality management system documentation include:
  (a) A quality policy and quality objectives?
  (b) A quality manual?
  (c) The documented procedures required in the International Standard? (4.2.2.a)
  (d) Documents required by the organisation to ensure the effective planning, operation and control of its processes? (4.2.2.b)
  (e) The quality records required by the International Standard?

- Is the extent of the quality management system documentation dependent on:
  (a) size and type of organisation?
  (b) complexity and interaction of the processes?
  (c) competence of personnel?

4.2.2 Quality Manual

A rewording of clause 4.2.1 of the 1994 Standard brings in three new elements covering the scope of the Quality Management System, permitting exclusions and a description of the processes involved.

- Is there a quality manual containing the following:
  (a) the scope of the QMS including details of and justification for exclusions? (4.2.1)
  (b) documented procedures or references to them? (4.2.1)
  (c) a description of the interaction of the processes of the quality management systems? (4.2.1)
- Is the quality manual controlled? (4.5)
- Note – is the quality manual part of the overall documentation of the organisation?

4.2.3 Control of Documents

Although written more succinctly, the requirements of the 1994 Standard remain intact. However, the requirement for changes to be approved by the original authority has been omitted, as has the need to identify the nature of the change (4.5.3 – 1994). A requirement to re-approve documents (b) is new and the wording in (d) has been changed from 'all locations' to 'at points of use'.

- Are relevant documents controlled? (4.5.1)
- Is there a documented procedure that ensures:

  (a) documents are approved for accuracy? (4.5.3)
  (b) they are reviewed and updated as necessary including re-approval? (4.5.3)
  (c) the current revision status and changes to documents are identified? (4.5.2)
  (d) relevant versions of the documents are readily available at points of use? (4.5.2a)
  (e) documents are legible, readily identifiable and retrievable? (4.5.1/4.16)
  (f) documents of external origin are identified and their distribution controlled? (4.5.1)
  (g) there is a system to prevent the unintended use of obsolete documents and they are identified where they are retained for any purpose? (4.5.2.b/c)

4.2.4 Control of Quality Records

The requirements for Quality Records are largely unchanged from the 1994 Standard, with the exception of the need to protect records.

- Are records required for the QMS established and maintained? (4.5.2)
- Are records maintained to provide evidence of conformity to requirements and effective operation of the QMS?
- Is there a documented procedure available for this? (4.16)
- Does it cover identification, storage, retrieval, protection, retention time and disposition of quality records? (4.16)

Section 5 Management Responsibility

5.1 Management Commitment

This clause places a greater emphasis on the commitment of senior management to the quality management system and the meeting of customer needs and regulatory and legal requirements, and how they demonstrate that commitment to the organisation and its stakeholders.

- Is there evidence of senior management commitment to development, implementation and improvement of the quality management system? Have they:
  (a) communicated to everyone the need to meet customer, statutory and legal requirements? (4.1.1)
  (b) established the quality policy?
  (c) established quality objectives? (4.1.1)
  (d) conducted management reviews? (4.1.3)
  (e) ensured the availability of necessary resources? (4.2.2, 4.1.2.2 and 4.9.b)

5.2 Customer Focus

The emphasis of this requirement is on customer needs and expectations. The 1994 Standard required customer needs and expectations to be met but said nothing about senior management responsibility to turn them into specific requirements.

- Have senior management ensured that customer requirements are determined and fulfilled with the aim of achieving customer satisfaction? (4.1.1)

5.3 Quality Policy

These requirements are largely identical to the 1994 Standard. A commitment to continual improvement has been added and the policy need only be communicated and understood at appropriate rather than all levels of the organisation. The policy statement continues to be reviewed and becomes a controlled document.

- Do senior management ensure that the quality policy:
  (a) is appropriate to the purpose of the organisation? (4.1.1)
  (b) includes commitment to complying with requirements and to continually improve the effectiveness of the QMS? (4.1.1, part)
  (c) provides a framework for establishing and reviewing quality objectives? (4.1.3)
  (d) is communicated and understood within the organisation? (4.1.3)
  (e) is reviewed for continual suitability? (4.1.3)
- Is the quality policy controlled? (4.5.1)

5.4 Planning

5.4.1 Quality Objectives

There is a greater emphasis on quality objectives and the responsibility for establishing them now clearly lies with senior management. Quality objectives must be established across the whole spectrum of the organisation's activities including, of course, those needed to meet the requirements of the product. Quality objectives must be measurable.

- Are quality objectives established at relevant functions and levels within the organisation? (4.1.1)

5.4.2 Quality Planning

The text of this clause has been largely rewritten compared to clause 4.2.3 – 1994. The requirements, however, remain the same with the addition of continual improvement and ensuring the effective management of change. Senior management is now responsible for making plans ensuring the objectives of the organisation are met.

- Has senior management ensured that:
  (a) the QMS is planned in order to meet the quality objectives of the organisation? (4.1.2.2 and 4.9)
  (b) the QMS is planned to meet the requirements outlined in clause 4.1?
  (c) the integrity of the QMS will be maintained when changes to the QMA are planned and performed?

5.5 Responsibility, Authority and Communication

5.5.1 Responsibility and Authority

Senior management shall ensure that responsibilities and authorities are defined and communicated within the host company.

5.5.2 Management Representative

This clause is less detailed than 4.1.2.3 – 1994 in that sub-paragraphs (a) to (e) are replaced by (a) to (d). These contain two new responsibilities: identifying the need for improvement (b) and promoting awareness of customer requirements (c).

- Has senior management appointed a member of management with the required authority to:
  (a) ensure the processes of the QMS are established and maintained? (4.1.2. 3.a)
  (b) report to senior management on the performance of the QMS including the need for improvement? (4.1.2.3.b)
  (c) promote awareness of customer requirements throughout the organisation?

5.5.3 Internal Communications

This is largely a new requirement although elements were covered in 4.3 of the 1994 Standard.

- Has senior management ensured that:
  (a) communication processes have been established within the organisation?
  (b) communication takes place regarding the effectiveness of the QMS?

5.6 Management Review

5.6.1 General

The 1994 Standard required management reviews to be conducted at 'defined intervals'. The new term 'planned intervals' appears less onerous. On the other hand, there is now an explicit requirement for senior management to be involved.

- Does senior management review the continuing suitability, adequacy and effectiveness of the quality management system?
  (a) Are the frequencies of reviews planned? (4.1.3)
  (b) Does it assess opportunities for improvement?
  (c) Does it evaluate the need for changes to the QMS? (4.1.3)
  (d) Does it include the quality policy and objectives? (4.1.3)
- Are records of management reviews maintained?

5.6.2 Review Input

This is additional text that provides a useful list of matters to be considered at management reviews.

- Do inputs to the management review include current performance and improvement opportunities and:
  (a) results of audits? (4.14.3.a)
  (b) customer feedback? (4.14.3.a)
  (c) process performance and product conformity? (4.14.3.a)
  (d) status of preventative and corrective action? (4.14.2/3)
  (e) follow up actions from previous management reviews? (4.14.3.c)
  (f) planned changes that could effect the quality management system? (4.14.2.c/d)
  (g) recommendations for improvement?

5.6.3 Review Output

This is additional text that provides a useful list of the results of management review.

- Do outputs from management review include:
  (a) improvement of the quality management system and its processes? (4.1.3)
  (b) improvement of product related to customer requirements?
  (c) resource needs?

Section 6 Resource Management

6.1 Provision of Resources

The requirement to allocate the appropriate resources is now expressed in much broader terms and covers the full range of resources.

- Has the organisation determined the resources it needs and are they provided in order to:
  (a) implement and maintain the quality management system?
  (b) continually improve the effectiveness of the quality management system?
  (c) address customer satisfaction?

6.2 Human Resources

6.2.1 General

While very similar to the 1994 requirements in that 'Personnel shall be qualified', the word 'qualified' is substituted by 'competent', which carries a very specific meaning.

- Are personnel performing work affecting quality competent on the basis of education, training, skills and experience? (4.1.2.2)

6.2.2 Competence, Awareness and Training

While similar to the requirements of 4.18 – 1994, this clause goes further by seeking competence, an evaluation of training and an awareness of the contribution employees make in achieving quality objectives. The emphasis on competency to do specific jobs is completely new.

  (a) Have competency needs for personnel performing activities affecting quality been established?
  (b) Have training or other alternatives been provided, where necessary, to satisfy these needs? (4.18)
  (c) Has the effectiveness of the actions taken been evaluated?
  (d) Has management ensured employees are aware of the relevance and importance of their activities and how they contribute to achieving the quality objectives?
  (e) Has management maintained records of education, training, skills and experience? (4.18)

6.3 Facilities

This clause expands the requirements previously found in 4.9.b (Suitable Equipment and Working Environment).

- Has management identified, provided and maintained the facilities it needs to achieve the conformity of product, including where necessary: (4.9.b)
  (a) buildings, workspace and associated utilities? (4.9.b)
  (b) process equipment, hardware and software? (4.9.e/g)
  (c) supporting services (e.g. transport, communication)? (4.9)

6.4 Work Environment

This is a new requirement that continues to expand on the 4.9.b – 1994 requirement for a suitable working environment.

- Does the organisation identify and manage the factors of the work environment needed to achieve conformity of product?

Section 7   Product Realisation

7.1 Planning of Product Realisation

Product realisation is that sequence of processes and sub-processes required to achieve the product.
   This clause brings together a number of requirements from the 1994 Standard where reference was made to Quality plans or a Contract requirement. The new clause is much more explicit in terms of what should be taken into consideration when planning the control of 'production processes' and planning to meet contractual requirements.

This is the only section of the Standard where exclusions are permissible.

  (a) Are there quality objectives and requirements for the product? (4.3.2)
  (b) Has management established processes, documents and resources specific to the product? (4.3.2 and 4.9.e/f)

(c) How does management verify, validate, monitor, inspect and test activities specific to the product and the criteria for acceptability? (4.3.1.d and 4.9.f)

(d) Are there adequate records that provide evidence that realisation processes and resulting product meets requirements? (4.3.4 and 4.16)

7.2 Customer Related Processes

7.2.1 Determination of Requirements Relating to the Product

This requirement is the initial stage of Contract Review in the old 1994 Standard. The aim is to establish the needs and requirements of the customer and may be prior to receiving an order or contract. Emphasis is given to the implied needs of the customer and end users.

- Does the organisation determine:
  (a) the requirements specified by the customer, including the requirements for delivery and post-delivery activities, the requirements for the availability and support of the product? (4.3)
  (b) requirements that are not specified by the customer but are necessary for the intended use? (4.3.2)
  (c) obligations related to product, including statutory and regulatory requirements?
  (d) what other requirements it may have in respect of the product?

7.2.2 Review of Requirements Relating to the Product

These requirements are identical to those of Contract Review in the 1994 Standard.

- Does the organisation review the requirements relating to the product?
- Has the review to cover this been carried out before the decision or commitment to supply the product to the customer? (4.3.1)
  (a) Have product requirements been clearly defined? (4.3.2.a)
  (b) Have, where the customer provides no documented requirement, the offered requirements been confirmed with the client? (4.3.2.b, implied)
  (c) Have contract or other requirements differing from those previously expressed been resolved? (4.3.3)
  (d) How has management confirmed that the organisation has the ability to meet the requirements? (4.3.2.c)
  (e) Have the results of the review and subsequent follow-up actions been recorded? (4.3.4)
  (f) When a product requirement has been changed, has the organisation ensured all relevant documentation is amended? (4.3.3 and 4.5.3)
  (g) Have all relevant personnel been advised of changed requirements? (4.3.3 and 4.5.2.a)

7.2.3 Customer Communication

This is a new requirement to establish arrangements for communication with customers. It brings together elements found in 4.3 Contract Review and 4.14 Customer Complaints in the old 1994 Standard.

- Has the company identified and implemented effective arrangements for communicating with the customer? (4.3.3)
- Does it cover:
  (a) product information?
  (b) enquiry, contract or order processing, including amendments?
  (c) customer feedback, including customer complaints? (4.14.2.a)

7.3 Design and/or Development

7.3.1 Design and/or Development Planning

These requirements encompass all the elements of Design Control, Design and Development Planning and Organisational Interfaces found in the 1994 Standard.

(a) Have the stages of design and/or development processes been determined? (4.4.1)
(b) Have the review, verification and validation activities appropriate to each design and/or development stage been developed? (4.4.2)
(c) Has management clearly defined the responsibilities and authorities for design and/or development activities? (4.4.2)
(d) Have the interfaces between the different groups of design and/or development functions been managed to ensure communication is effective and responsibilities are clear? (4.4.3)
(e) Is planning output updated, as appropriate, during the progress of the design and/or development process? (4.4.2)

7.3.2 Design and/or Development Inputs

This requirement is almost identical to the 1994 clause with the exception that resolving incomplete or ambiguous requirements directly with those responsible for imposing them has been omitted.

(a) Have requirements for product function and performance been defined and documented?
(b) Have the applicable statutory and legal requirements been defined and documented? (4.4.4)
(c) Is there applicable information derived from previous similar designs?
(d) Have any other requirements essential for design and/or development been defined and documented? (4.4.4)
(e) Have these documented inputs been reviewed for adequacy? (Incomplete, ambiguous or conflicting requirements shall be resolved.) (4.4.4)

7.3.3 Design and/or Development Outputs

Two changes appear in this requirement: the addition of information for production and service operation and the omission of disposal requirements.

- Are the outputs of the design and/or development process documented to ensure verification of the design and/or development inputs? (4.4.5)
  (a) Design and/or development outputs shall meet the design and/or development input requirements. (4.4.5.a)
  (b) Design and/or development output shall provide appropriate information for production and service provision.
  (c) Do design and/or development outputs contain or reference product acceptance criteria? (4.4.5.b)
  (d) Does design and/or development define the characteristics of the product that are essential to its safe and proper use? (4.4.5.c)
  (e) Are design and/or development output documents approved prior to release? (4.4.5)

7.3.4 Design and/or Development Review

Several minor changes appear in this clause: 'formal documented reviews' has been replaced by 'systematic reviews'. The requirement to include 'other specialist personnel as required' has been replaced by 'shall include representatives of functions concerned with the design or development stages being reviewed'. A specific requirement to identify problems and propose follow-up action has been added.

- Are systematic reviews of the design and/or development conducted at suitable intervals to:
  (a) evaluate the ability to fulfil requirements? (4.4.6)

(b) identify problems and propose follow-up? (4.4.6)
(c) do the participants of the review include representatives of functions concerned with the design and/or development stage(s) being reviewed? (4.4.6)
(d) are the results of the reviews and the subsequent follow-up actions recorded? (4.4.6/4.16)

7.3.5 Design and/or Development Verification

These requirements are identical to the 1994 Standard with the addition that 'subsequent follow-up action must be recorded and held as a quality record'.

7.3.6 Design and/or Development Validation

This clause has been rewritten, and now requires validation to confirm that the resulting product meets the requirements for the intended use, as opposed to 'defined user needs or requirements'. The basic intent remains the same.

(a) Is validation undertaken to verify the resulting product is capable of meeting the requirements of the intended use? (4.4.8)
(b) Is validation completed prior to delivery or implementation of the product? (4.4.8)
(c) Where it is impractical to perform validation, is it performed to the extent applicable? (4.4.8)
(d) Are the results of validation and any follow-up actions recorded? (4.16)

7.3.7 Control of Design and/or Development Changes

The requirements of this clause have been expanded to include the evaluation of changes on constituent parts and delivered product. The terms verified and validated have been added to approval, significantly enhancing the activities involved in controlling changes in design.

(a) Are design and/or development changes identified and recorded? (4.4.9)
(b) Does it cover the evaluation of the effects of change on constituent parts and delivered products?
(c) Are the changes verified and validated, as appropriate, and approved before implementation?
(d) Are the results of the review and subsequent follow-up actions recorded? (4.4.9)

7.4 Purchasing

7.4.1 Purchasing Control

This clause pulls together the requirements previously found in 4.6.1 and 4.6.2 of the 1994 Standard. A change of emphasis results from the new wording 'the type and extent of control being dependent upon the effect on subsequent processes' now being applied to all purchased items, not just subcontractor activity as in the 1994 Standard. A criterion for periodic evaluation has been added to the evaluation of suppliers.

- Does the company control its purchasing to ensure purchased products conform to requirements? (4.6.1)
- Is the extent of control appropriate to the effect on the subsequent realisation processes and outputs? (4.6.2.b)
- Has the organisation evaluated and selected its suppliers based on their ability to supply products in accordance with requirements? (4.6.2.a)
- Are the criteria for the selection of and periodic evaluation of suppliers defined? (4.6.2.b)
- Are results of evaluation and follow-up actions recorded? (4.6.2.c and 4.1.6)

7.4.2 Purchasing Information

This clause is much less prescriptive than 4.6.3 in the previous 1994 Standard in that the detail in sub-clauses a) and b) is replaced by a list of five headings. The requirement to review and approve purchasing documents prior to release is replaced by a requirement to ensure their adequacy. This seems a less onerous requirement.

- Does the purchasing document contain information describing the product to be purchased, including where appropriate:
  (a) requirements for the approval or qualification of product, procedures, processes, facilities and equipment?
  (b) requirements for qualification of personnel?
  (c) quality management system requirements?
- Does the organisation ensure the adequacy of specified requirements contained in the purchasing documents prior to their release? (4.6.3)

7.4.3 Verification of Purchased Product

The requirements of this clause are identical to those in the 1994 Standard but are written much more succinctly. This clause incorporates the requirement for Goods Receiving Inspection (previously found in 4.10.2).

- Has the organisation established and implemented the inspection or other activities necessary for ensuring that purchased product meets specified requirements? (4.6.4/4.10.2)
- Has the organisation specified the intended verification arrangements for the product where the verification takes place at the supplier's premises? (4.6.4.1 and 2, also 4.10)
- Are the intended verification arrangements and method of product release included in the purchasing information? (4.6.4.1)

7.5 Product and Service Provision

7.5.1 Product and Service Provision Control

This clause encompasses a wide range of existing requirements and expresses them in a far less prescriptive manner. The Standard now defines six key elements of controlling production:

- information on product characteristics
- work instructions
- equipment
- measuring and monitoring devices
- measuring and monitoring activities
- processes for release, delivery and after sales service.

- Does the organisation carry out production and service provision under planned and controlled conditions, including, as applicable:
  (a) the availability of information that specifies the characteristics of the product? (4.9.c)
  (b) the availability of work instructions (where necessary)? (4.9.a/4.9.c)
  (c) the use of suitable equipment for production and service provision? (4.9.b/4.9.g)
  (d) the availability and use of measuring and monitoring devices? (4.11.2.a)
  (e) the implementation of monitoring activities? (4.9.d/4.11.2.b)
  (f) the implementation of defined processes for release, delivery and applicable post delivery activities? (4.12)

7.5.2 Validation of Special Processes

These requirements were previously referred to in clause 4.9 Process Control. The text has been rewritten but the requirements remain the same. The term 'validation' is used rather than 'verification'. This implies that the processes must conform to their stated intended use.

- Has the organisation validated any production and service processes where the resulting output cannot be verified by subsequent measurement or monitoring? (4.9)

- Is there provision for processes where deficiencies may become apparent only after the product is in use or the service has been delivered? (4.9)
- Does validation demonstrate the ability of the process to achieve planned results?
- Does the organisation define arrangements for validation that includes, where applicable:
  (a) approval/review of processes? (4.9)
  (b) approval of equipment and qualification of personnel? (4.9.f/g)
  (c) use of defined methods and procedures? (4.9.c)
  (d) requirements for records? (4.9)
  (e) re-validation?

7.5.3   Identification and Traceability

This clause now addresses both product identification (4.8) and inspection status (4.12). The onus is placed on the organisation to decide whether it is necessary to identify product and at what times (e.g. at receipt, during production and delivery and installation or servicing). The identification should not only define the product but also indicate its status in terms of inspected, tested, acceptable or not.

- Has the organisation identified, where appropriate, the product throughout the product realisation? (4.8.a)
- Has the status of the product been, with respect to measurement and monitoring requirements, identified throughout the process? (4.12)
- Has the organisation controlled and recorded the unique identification of the product, where traceability is required? (4.8)

7.5.4   Customer Property (4.7)

These requirements are identical to those of the 1994 Standard. However, the paragraph stating that the customer is responsible for providing acceptable product has been dropped. A note extends customer property to cover Intellectual Property such as confidential information.

Customer property is an item or material that the customer wants the organisation to incorporate into the product or service being provided by the host organisation. Management may decide that a procedure is required for the control of customer property, particularly if intellectual property is involved and specific security measures are required.

- Does the organisation deal with customer supplied products? (4.7)
- Does the organisation exercise proper care over customer property?
- Is it identified, verified, protected and maintained properly prior to incorporation into the product? (4.7)
- Is there a process available for dealing with damaged or otherwise unsuitable customer supplied products? (4.7)
- Is there a method of recording and reporting customer supplied products found to be unsuitable? (4.7)
- Does the customer supply any intellectual property and if so is it properly controlled?

7.5.5   Preservation of Product

These requirements are identical to those found in 4.15 in the 1994 Standard but are far less prescriptive. The requirements have been extended to cover 'in process activities' and include 'constituent parts' of the product. The term 'protection' is used in place of 'preservation'.

- Does the organisation preserve conformity of product with customer requirement during internal processing and delivery to the intended destination? (4.15)
- Does preservation extend to constituent parts?

7.6   Control of Measuring and Monitoring Devices (4.11)

This clause has been significantly rewritten and as a result is much more succinct. The title has changed and monitoring devices replace inspection and test equipment. The clause now contains a requirement to identify the measurements to be made in order to establish product conformity.

- Has the organisation established processes to ensure that measuring and monitoring activities are capable of and are carried out in a manner that is consistent with the monitoring and measurement requirements? (4.11.2.a)
- Has the organisation identified suitable measuring devices to assure conformity of product to specified requirements? (4.11.2.b)
- Where required to maintain valid results, are measuring and monitoring devices:
  (a) calibrated or verified at specified intervals or prior to use, against measurement devices traceable to National/International Standards and adjusted where and as necessary? (4.11.2.b/c)
  (b) safeguarded from adjustments that would invalidate the measurement result? (4.11.2.i)
  (c) protected from damage and deterioration during handling, maintenance and storage? (4.11.2.h)
  (d) equipped so that the results of calibration are recorded? (4.11.2.e)
  (e) checked for validity of previous results and where necessary corrective action taken? (4.11.2.b/f)

is and Improvement

1 4.10.1 and 4.20 – 1994 and extends them to cover contin-

mented monitoring, measurement, analysis and continual d to:
ıct?
agement system?
eness of the quality management system? (4.10)

use of applicable methodologies (including statistical

ring

perceptions and determining the methods used to obtain eferred to customer satisfaction only in its scope (clause 1) to customers' needs and expectations. Customer complaints (4.14.2) corrective action.

rmation on customer perceptions of whether or not the ements? (4.14.2.a/b)
and using information determined? (4.14.2.c)
customer satisfaction data it needs to collect?

vith two small but significant changes. Personnel other than eing audited can now conduct audits. This is particularly useful to small organisations where finding an independent auditor was difficult. Internal Audits must now also confirm conformance to the Standard. Internal Audits previously focused on compliance with internal procedures.

- Do the audits check that the QMS complies with the planned arrangements, quality management system requirements and the requirements of this Standard? (4.17)
- Do the audits check that the QMS has been effectively implemented and maintained? (4.17)

- Are the audits planned taking into account the status and importance of the area being audited? (4.17)
- Do audits have a clear scope and do they take into account the previous audit findings when setting the frequency and methodology?
- Are audits conducted by personnel whose choice will ensure the objectivity and impartiality of the audit process? (4.17)
- Does the process ensure that auditors do not audit their own work?
- Is there a procedure covering audit requirements and does it define the responsibilities and requirements for conducting audits and ensuring their independence? (4.17)
- Are results recorded and reported to management? (4.17/4.1.2.3)
- Do management take timely corrective action on deficiencies found during the audit (4.17) and do follow-up actions verify the implementation of corrective actions? Is this reported? (4.17)

8.2.3 Measurement and Monitoring Processes

This requirement is very similar to that in Process Control (4.9.e) and also draws on Statistical Techniques (4.20) of the 1994 Standard.

- Are there suitable methods for measuring and monitoring the processes necessary to meet customer requirements? (4.10.3.a)
- Do these methods confirm the continuing ability of each process to satisfy its intended purpose? (4.10.3.b)

8.2.4 Measurement and Monitoring of Product

The clause encompasses the requirements previously found in clause 4.10 Inspection and Testing. Written far more succinctly, the requirements retain the same intent.

- Are the characteristics of the product measured and monitored to verify that the requirements for the product are met? (4.10.4)
- Are these checks carried out at appropriate stages of the process? (4.10.4)
- Is there documented evidence of conformity with the acceptance criteria? (4.10.5)
- Do records indicate the authority for release of the product? (4.10.5)
- Is it clear that product release cannot proceed until all specified activities have been completed satisfactorily? (4.10.4/5)

Is there a process by which concession approval by the customer may be obtained?

8.3 Control of Non-Conformity

The requirements of this clause are virtually identical to clause (4.13) of the 1994 Standard. It is however less explicit by omitting the sub-clauses regarding rework etc. A documented procedure is required for controlling non-conformity in accordance with the above requirements. However, 'notification to the functions concerned' has been expanded to include customers, end users and regulatory bodies, with sub-clause (4.13.2) (Review and Disposition) being withdrawn.

- Are products that do not conform identified as such? (4.13.1)
- Are they prevented from unintended use or delivery? (4.12/4.13.1)
- Are these processes defined and documented in procedure? (4.13.1)
- Are non-conforming products corrected and subject to re-verifications to demonstrate conformity? (4.13.2)
- When non-conformity of product is detected after delivery or use does the organisation take appropriate action regarding the consequences?
- Is it clear where rectification of a non-conforming product needs to be reported to the customer, end user, regulatory body or other? (4.13.2)

8.4   Analysis of Data

This is a new requirement although various clauses, (such as 4.14 Corrective Action and 4.20 Statistical Techniques) of the 1994 Standard refer to 'analysing potential causes of non-conformance and verifying process capability'.

- Does the organisation determine, collect and analyse appropriate data to establish the suitability and effectiveness of the QMS? (4.14.1/4.14.2.d)
- Does this identify where continual improvements can be made? (4.14.2.c)
- Does this data include measuring and monitoring activities etc.? (4.20.1)
- Does the organisation analyse the data to provide information on:
  (a) customer satisfaction? (4.14.2)
  (b) conformance to product requirements?
  (c) characteristics of processes, product and their trends?
  (d) suppliers? (4.6.2.c)

8.5   Improvement

8.5.1   Continual Improvement

This is a new and explicit requirement which permeates the BS EN ISO 9001: 2000 Standard.

- Does the organisation facilitate the continual improvement of the QMS by use of its:
  (a) quality policy? (4.1.3)
  (b) objectives? (4.1.3)
  (c) audit results? (4.17)
  (d) analysis of data? (4.9)
  (e) corrective and preventative actions? (4.14)
  (f) management review? (4.3.1)

8.5.2   Corrective Action

This requirement is virtually identical to clause 4.14.2 of the 1994 Standard; a documented procedure remains a requirement.

- Is there evidence that the organisation has taken corrective action to eliminate the cause of non-conformities? (4.12.2.c)
- Does this action prevent recurrence? (4.14.2.d)
- Is it appropriate to the impact of the problems encountered?
- Does the documented procedure for corrective action define requirements for identifying non-conformities (including customer complaints) (4.14.2.a) and
  (a) determining the causes of non-conformity? (4.14.2.b)
  (b) evaluating the need for actions to ensure that non-conformities do not recur? (4.14.2.c)
  (c) determining and implementing the corrective action needed? (4.14.4.d)
  (d) recording results of action taken?
  (e) reviewing corrective action taken? (4.14.3.c)

8.5.3   Preventative Action

This is identical to the same clause in the 1994 Standard. A documented procedure is required to prescribe the activities related to taking preventative action.

- Has the organisation determined preventative action to eliminate causes of potential non-conformities to prevent occurrence? (4.14.3.a)
- Is preventative action appropriate to the impact of the problem? (4.14.3.b)
- Is there a documented procedure? (4.14.3)
- Does the documented procedure for preventative action define requirements for:
  (a) determining potential non-conformities and their causes? (4.14.3.a)
  (b) evaluating the need for action to prevent occurrence of non-conformities? (4.14.3.b)
  (c) determining and implementing action needed?
  (d) recording results of action taken? (4.14.3.d)
  (e) reviewing preventative action taken? (4.14.3.d)

New Philosophy

The new Standard is based on the Deming model of 'Plan', 'Do', 'Check' and 'Act', shown in Figure 9.2:

- **Plan**: Identify customer needs and expectations. Set strategic objectives
- **Do**: Implement and operate processes
- **Check**: Collect business results. Monitor and measure the processes, review and analyse
- **Act**: Continually improve process performance.

The strategic planning process should be built into the 'Quality System' (see Figure 5.16).

Bearing the above in mind, the model shown in Figure 9.3 provides a format for implementing the changeover process from BS EN ISO 9001: 1994 to BS EN ISO 9001: 2000. This model is also of value for initial implementation and indicates the linkage with the Deming Cycle.

Management

To meet this requirement the people who create policy, steer and direct the construction organisation will have to show how they participate in the development and direction of the system. This may be by direct involvement in the process, attendance at meetings, delivery of presentations and communication sessions, or by any other activity that shows leadership with respect to the quality system.

Quality Policy and Objectives

The quality policy is the driving force of the system and commits the organisation to both meeting requirements and improvement. This will become one of the key documents against which the performance of the quality system will be judged. The translation of the quality policy into practice is then facilitated by the definition of supporting objectives. Quality objectives are now a clear requirement in their own right as opposed to a part of quality policy. They must be established widely within the organisation, support the policy, be measurable and focus on both meeting product requirements and achieving continual improvement.

Quality Planning

Quality planning now functions at two levels. The responsibility of senior management is to ensure the planning of the QMS, achievement of continual improvement and the planning for achievement of quality objectives. It is a much clearer and unavoidable requirement than that of the old 1994 Standard.

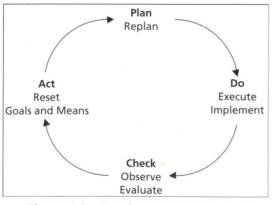

Figure 9.2 *Deming Dynamic Control Loop Cycle (cited by McCabe, 1998) [adapted by author]*

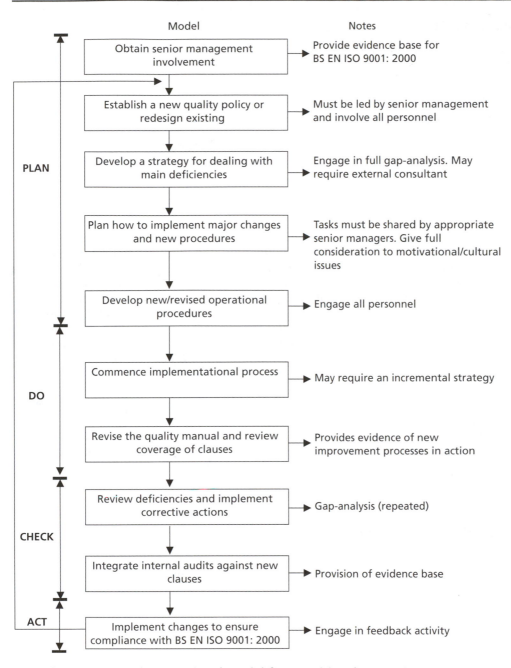

Figure 9.3 *Implementational model for transitional process/new implementation of BS EN ISO 9001: 2000*

At a lower level an organisation's documented quality planning for realisation processes is mandatory, although the format is optional. There is little change in the content at this level apart from the reference to objectives for product, project or contract and less prescriptive content; the overall effect is to make the requirement more auditable.

Legal Requirements

The Standard now makes it clearer that in determining customer needs and expectations this must include applicable regulatory and legal requirements. Compliance with such requirements is then invoked throughout contract review, design, process control etc.

Training and Competence

The emphasis is clearly on competence rather than just training. The evaluation of the effectiveness of training and the need for employee awareness are new requirements and would be items reviewed during the transition check.

Process Approach

The Standard still requires organisations to have a quality manual, which includes the documented procedures or references to them. It also has to include a description of the sequence and interaction of the processes that make up the quality management system. The scope of the system has to be defined including the basis for the use made of the 'permissible exclusions' clause. Most manuals will require some changes or additions but hopefully will not need to be extensively rewritten. The fact that a manual is laid out to match the twenty system elements of the previous 1994 Standard does not matter and need not be considered when deciding what, if any, changes are required.

Overall, the effect of the requirements of the new Standard is to reduce the instances where documented procedures are mandatory and to allow a construction organisation the freedom to determine the type and extent of documentation needed to support the operation of the processes that make up the quality management system.

Process Thinking

The reverse of '*unitary thinking*' is '*process thinking*' which, rather than concentrating on outputs and numerical measures, concentrates on processes. Behind '*process thinking*' is the idea that the cause of success and failure lies in the systems that produce the results and ultimately in the processes that make up the systems. This is reflected in the BS EN ISO 9001: 2000 process model (Figure 9.4).

Information and Communication

The Standard now specifically requires construction organisations to ensure effective internal communication between functions regarding system processes and external communication with customers, not only at the contract stage but also with respect to the provision of product information and obtaining feedback.

Customer Perception

Customer perception is now addressed by requiring sufficient information on satisfaction or dissatisfaction to be gathered to enable the organisation to monitor customer perception on whether or not customer requirements are being satisfied. Having no complaints may mean that the organisation has no information, not that customer satisfaction has been achieved.

Analysis of Data for Improvement

This has been separated from the body of corrective and preventative action and made a much more specific issue. The requirement to plan and operate the system to facilitate the achievement of improvement makes specific a requirement that was only previously implied.

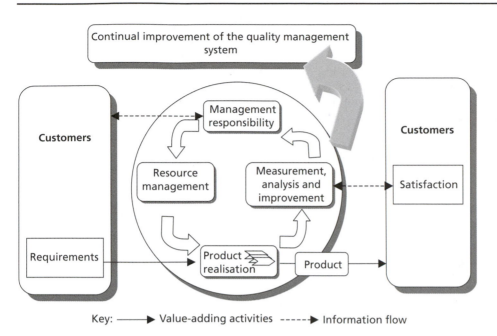

Figure 9.4 *Model of a process-based quality management system (BSI, 2000a)*

There may well be some resistance to changes in the organisation. Coalitions of resistance can develop and if they are linked to a power base they may impede the implementation process. The quality facilitator should try and overcome resistance by allaying employees' fear of change.

Senior management must take an active role in both designing and implementing the quality system, with support coming from the very start of the deployment process. Senior management can show its commitment through the development of organisational policies which involve it in designing and implementing the quality system. Managers within the organisation must 'manage'. They should not abdicate the responsibility to the quality facilitator (team) without providing adequate authority and resources.

Even before the deployment process begins staff members need to be made aware of the benefits of certification. They need to be convinced that the introduction of a quality system is worthwhile and can provide advantages for them and the organisation. It is, therefore, senior management's duty to echo the rationale for the advantages of certification. This is an important issue since people tend to have an in-built resistance to change.

The co-operation of staff is vital for a successful deployment, and in order for them to co-operate two issues require consideration:

- staff must want to be co-operative
- staff have to be allowed to co-operate.

If staff are not coerced into being co-operative they will make a more vital contribution to the deployment process.

It can be concluded that senior management support is a vital component at all stages of the design and implementational process. If this support is not provided a successful outcome to the project will be problematic.

Inappropriate Corporate Culture

Quality systems require a corporate culture based on trust and a desire to identify problems in order to eliminate them. The concept of empowerment is a key component of an effective corporate culture.

If a climate of distrust exists between senior management and the rest of the firm the implementation of BS EN ISO 9001: 2000 is doomed to fail. Organisational culture dictates the way a business operates, and how employees respond and are treated. Organisational culture contains such elements as a guiding philosophy, core values, purpose and operational beliefs. It must be understood that just following documented procedures complying with the BS EN ISO 9001: 2000 series will not guarantee success. Only if the correct culture exists will the true benefits be attained.

Summary

Whoever is charged with the task of designing and implementing a quality system must have the total support of the organisation. This total support involves not only senior management but also the employees (the people who perform the documented tasks). If those responsible for application can obtain this total support for the system then implementation is possible. An important part of obtaining total involvement is to inform people of what the system is about and to keep them informed throughout the design and deployment process. Successful implementation of any quality model depends upon the co-operation of the people who are involved.

The following is an overall generic strategy for quality assurance system implementation:

- obtain the total support of the organisation
- set realistic objectives, i.e. the timescale for the implementational process
- plan programme activities ahead of time
- maintain internal and external contacts with key personnel
- establish a clear review process
- be flexible and willing to sacrifice time and other resources to obtain improvements
- do not expect a great improvement in the saving of resources immediately
- use expert opinions and advice when necessary
- do not expect too great an immediate return on investment. Some improvement projects may have key benefits because they provide customer satisfaction and assist the long-term survival of the organisation.

BS EN ISO 9001: 2000 provides a holistic framework encompassing both the operational and strategic activities of construction firms. This facilitates the effective and efficient utilisation of organisational resources in the attainment of a sustainable competitive advantage based on continued client satisfaction.

Companies have also developed a greater understanding of the importance and real meaning of quality over the last few years. They realise that the monitoring of quality is not a dreary routine but rather a proactive way to achieve greater customer satisfaction and higher profits. As a result, companies have accepted the necessity for an effective quality system.

A summary of the advantages of applying BS EN ISO 9001:2000 is tabulated below:

   (i) provides a marketing focus
  (ii) provides a means of achieving a top-quality performance in all areas of the organisation
 (iii) provides operational procedures for all staff
 (iv) performs critical audits, allowing for the removal of non-productive activities and the elimination of waste
  (v) provides a quality advantage as a competitive weapon
 (vi) develops group/team spirit within the company
 (vii) improves the communication systems within an organisation
(viii) reduces inspection costs
 (ix) makes more efficient use of scarce resources
  (x) recognises certification from countries outside the UK
 (xi) recognises the need for customer satisfaction, i.e. provides the required customer quality every time.

There are some limitations in implementing a quality standard within the construction industry itself:

"The requirements of BS EN ISO 9000 standards (not principles and philosophy behind) will do little or nothing to clarify these project specific requirements except where they relate to the manufacturer of products such as bricks, tiles and sanitary fitments used in the building" (Hellard, 1995).

After implementation, there are also problems in maintaining the standard within the construction industry. These are summarised below (Low and Omar, 1997):

"1. Organisation structure
2. Employer's attitude
3. Employees' attitude
4. Lack of resources
5. Education and training
6. Inadequate supervision
7. Performance of suppliers and subcontractors
8. Engineering and construction problems
9. Co-ordination and communication"

The above noted, the benefits of application do outweigh the disadvantages. However, a firm's organisational structure and culture need to support and not undermine the system.

Application of BS EN ISO 9001: 2000 to Construction Related Organisations

Construction companies can relate to the various clauses of the new Standard by following the framework indicated in Table 9.2. Table 9.2 provides a good starting point for addressing the Standard and ensuring compliance in readiness for engaging in the certification process.

Project Quality Plans

Quality has to be made relevant to personnel engaged in the on-site construction process, thus managing quality on site is a vital activity. The Corporate Quality System, therefore, has to be translated into a 'Project Quality Plan'. Project Quality Plans (PQPs) are prepared for each construction project; they detail the procedures to be adopted for a specific construction project.

Project Quality Plans are prepared by both main and subcontractors. The various subcontractor plans are added to the contractor's plan to form the total PQP for the project. The PQP is a document often presented in tabular form and communicates to all involved in the project a summary of the proposed actions to be taken to ensure that the specified quality for the project is attained. The main aim of the PQP is to make sure that everyone examines their specified tasks objectively. It should also ensure that the requirements of the contract documents are achieved first time, every time. The PQP should not be prepared in isolation but rather be developed in conjunction with the contract programmes, method statements, materials and plant schedules and cost budgets.

The Project Quality Plan brings together the project information and the company's policies, procedures and inspection routines. This process will help to identify clearly:

- the scope of the work
- responsibilities of one to another
- the procedures used to achieve the specified quality
- the records necessary to prove that the required quality has been achieved.

Because it is documented, the project quality plan can be monitored and regulated to adapt to changing conditions or circumstances. In outline a PQP for a particular project should include:

- quality objectives
- allocation of responsibilities and authority

Table 9.2 *Construction activities related to BS EN ISO 9001: 2000*

CLAUSES OF BS EN ISO 9001: 2000	CONSTRUCTION ACTIVITIES
4 **Quality Management System**	
4.1 General Requirements	
4.2 Document Requirements	• Quality Plans
4.2.1 General	
4.2.2 Quality Manual	
4.2.3 Control of Documents	
4.2.4 Control of Records	• Document Control Procedures
5 **Management Responsibility**	
5.1 Management Commitment	• Policy Statement
5.2 Customer Focus	• Customer Care
5.3 Quality Policy	
5.4 **Planning**	• Planning and Construction Procedures
5.4.1 Quality Objectives	• Quality Objectives
5.4.2 Quality Management System Planning	• Policy Statements and Organisational
5.5 Responsibility, Authority and Communication	Charts
5.5.1 Responsibility and Authority	
5.5.2 Management Representative	
5.5.3 Internal Communications	
5.6 Management Review	• Personnel and Training Procedures
5.6.1 General	
5.6.2 Review Input	
5.6.3 Review Output	
6 **Resource Management**	
6.1 Provision of Resources	• Management Review
6.2 Human Resources	• Operational Procedures
6.2.1 General	• Planning, Estimating and Construction
6.2.2 Competency, Awareness and Training	Procedures
6.3 Infrastructure	
6.4 Work Environment	• Health and Safety Processes
7 **Product Realisation**	
7.1 Planning of Product Realisation	
7.2 Customer-Related Processes	
7.2.1 Determination of Reqts Related to the Product	
7.2.2 Review of Reqts Related to the Product	
7.2.3 Customer Communication	
7.3 Design and Development	
7.3.1 Design and Development Planning	
7.3.2 Design and Development Inputs	• All Operational Procedures and
7.3.3 Design and Development Outputs	Processes
7.3.4 Design and Development Review	
7.3.5 Design and Development Verification	• Design Build Activities
7.3.6 Design and Development Validation	
7.3.7 Control of Design and Development Changes	• Purchasing and Commercial
7.4 Purchasing	Management Activities
7.4.1 Purchasing Process	
7.4.2 Purchasing Information	• Quality Plans and Records
7.4.3 Verification of Purchased Product	
7.5 Production and Service Provision	
7.5.1 Control of Production and Service Provision	
7.5.2 Validation of Processes for Construction	
7.5.3 Identification and Traceability	
7.5.4 Customer Property	
7.5.5 Preservation of Product	
7.6 Control of Monitoring and Measuring Devices	
8 **Measurement, Analysis and Improvement**	
8.1 General	• System Management Procedure
8.2 Monitoring and Measurement	• Performance Measures and
8.2.1 Customer Satisfaction	Management Review Processes
8.2.2 Internal Audit	• Management Review
8.2.3 Monitoring and Measurement of Processes	
8.2.4 Monitoring and Measurement of Product	
8.3 Control of Non-Conforming Product	
8.4 Analysis of Data	
8.5 Improvement	
8.5.1 Continual Improvement	
8.5.2 Corrective Action	
8.5.3 Preventative Action	

- procedures, method statements and work instructions for key activities
- inspection and testing arrangements
- project review and update procedures
- project records to be maintained during the course of the work to verify that quality achieved is acceptable.

The Project Quality Plan is prepared before work commences on site. It involves managing quality by:

- identifying what needs to be done
- specifying how it should be done
- defining by whom it should be done
- planning when it will be done
- executing it with skill and integrity
- proving that it has been done correctly.

The aim is to improve communication by ensuring that:

- specifications are easily understood
- drawings are of such quality that they can be read clearly and adhered to without the contract staff having to guess the designer's intentions
- written instructions are clear and will achieve the required purpose
- verbal instructions are clearly understood and acted upon correctly.

The first step in preparing a Project Quality Plan is to collect the necessary information. This would generally include, from external sources: contract documents, specifications and drawings; and from internal sources: company Quality Manual including a Policy Statement and procedures, company organisation charts and/or job descriptions, project programme and Method Statements.

Contents of a Project Quality Plan would generally contain a brief outline of the project and the identification of any special requirements, which might include particular problems such as boundaries, access, liaison with adjoining owners etc. By defining these issues at the beginning of the Plan, project participants are made aware of the requirements which have to be addressed if the project is to progress efficiently and effectively.

Quality objectives ensure that the construction management team construct the works in accordance with the contract documents, in line with the client's requirements. The inclusion of, or referral to, a schedule of all applicable drawings and specifications should also be included within the quality objectives. In addition it is good practice to include a statement which defines the company's quality objectives, covering such issues as:

- achieving the requirements of specifications, drawings and other instructions within cost and time boundaries
- creating the right conditions for successful project completion
- maintaining a good team spirit incorporating main and subcontractors
- procuring the materials on time and to specification
- completing the work and achieving zero defects
- maintaining effective supervision
- maintaining and improving customer confidence
- improving profitability.

Organisation

Whether the project is large or small, it is important for the person in charge to produce a project organisational structure. The following questions should be considered and answers provided:

- What is the function, responsibility and authority of each member of the site team?
- Who is responsible for ensuring that the company's quality management system is being implemented throughout the Project Quality Plan?

- Who interfaces with external bodies such as the statutory authorities etc.?
- Who is to be the site contact with the client's representative?
- What is the procedure for covering absences in the management team?
- What are the responsibilities of head office and who is their main contact?

It is usual to provide an organisational chart, including a brief description of functions, responsibilities and authority of each post, rather than a named person. A separate list of names and telephone numbers can be used and kept up to date.

Procedures

One should include a schedule of the standard company procedures which apply to the project and any amendments which have been agreed should be identified. Occasionally it may be necessary to produce a special procedure for an operation unique to a particular project. Care must be taken to ensure that such a procedure does not become transferred to another project in error.

Inspection and Testing

This relates to a point beyond which an activity or operation should not proceed unless there is an inspection or verification record issued which confirms that the required quality at that stage has been achieved. Inspections can be of two kinds. Standard tests are those specified or expected to be carried out. Examples include water tests on drains, air tests on ventilation ducting and inspection of foundations. Inspections are carried out to minimise risk. However it is also well known that things can and do go wrong even when care has been taken. Such inspections should be kept to the minimum that will ensure requirements are met. They may cover such things as checks on formwork, or inspection of drains under a factory floor before concrete is laid.

Records

It is important to specify in the PQP who will be responsible for identifying the schedule of records, what records are required, who is responsible for maintaining the records, where records should be kept and how long they are to be retained. The project manager should gain agreement at each contract review on the records required, in addition to the normal records that the company keeps for commercial purposes. A schedule of records should be maintained and regularly updated as the project proceeds.

Document Control

Accurate and up-to-date information is essential for meeting specified requirements. The PQP should incorporate the company's document control procedures. Drawing registers should be maintained and an issue and receipt system implemented. The withdrawal of obsolete drawings and information has to be recorded and notified to all relevant personnel.

Auditing

The PQP extends the company's QMS to a particular project. Auditing is, therefore, an inherent part of the process of maintaining relevant operations. Auditing on a project is concerned with monitoring and recording detailed activities carried out according to the PQP.

Special Processes

There are some situations in construction where the process of creating an element is the critical issue in controlling quality. Once completed, such elements may become incorporated into others or covered up and are not capable of being inspected. Many of the site-formed products such as concrete, mortar, asphalt etc. could fall into this category. Once they are made there is little that can be done to change their quality. Management of these special processes needs very careful attention. Detailed planning of the process and provision of effective control and verification during production are essential to their success. Because the results of such processes may not be known for a long period afterwards, it is necessary to keep full records of the methods used to monitor the processes. The PQP should identify those processes to be treated as 'special'.

Product Identification and Traceability

There may be occasions when it is not possible to assess the quality of a building element until all the work on it has been completed. Records should be maintained to show where each batch of materials or components has been used in the structure, what process was involved, the suppliers and the source of basic materials.

Non-Conformance

The PQP should define how non-conforming products and work are to be dealt with and where non-conformance arises, and a variety of practical responses are possible. Line managers should be briefed to act in the light of the circumstances they face at the time.

Corrective Action

This is more than just remedial work. Corrective action is having a procedure which not only puts right what has been done incorrectly but also ensures that recurrence is prevented. Thus, where possible, the cause of the problem should be established. This is necessary for the effective removal of the problematic issue.

Packaging and Handling

These issues must be discussed and agreed with the supplier prior to the placing of orders. If they are not, it is very likely that an additional charge could be made on the contractors. A strategy for handling has to be developed as part of the PQP and this is particularly important for large and expensive items liable to damage. Protection and handling on site are a management problem. Responsibility for it should be delegated and the person selected held accountable, and the company's procedures should make this clear. It is good practice to use checklists for offloading and handling to assist the person responsible. For example, a checklist may be issued to a storekeeper for inspecting second fix timber at the time of delivery. The PQP should make reference to any special packaging or handling requirements.

Storage

Appropriate methods must be planned and controlled by the person given the responsibility for storage. The type of storage will depend on the material or components but may be classified as follows:

* in the open air with no special requirements
* in the open air requiring a special/level surface
* in the open air but protected from moisture
* in the open air but protected from full sunlight
* within a building protected from moisture
* within a building insulated against freezing temperatures.

Some materials may require one or more types of storage. Protection also has to be provided against wind, traffic, fire, chemicals, dirt, mud, vermin and collapse of stacks. The degree of security required will reflect the quality requirements of materials.

Inspection of the Completed Building

If all the foregoing activities have been established and implemented then the completed building should conform to the specified requirements. The PQP needs to define who is responsible for carrying out the completion and handover procedures. Some contracts may require this to be done in stages. In some cases test records will have been requested (e.g. for services) and these should be collated and filed. Such records will enable the company to assure the client that the building conforms to requirements.

It is evident that the Project Quality Plan is an essential document for maintaining the critical link between corporate quality assurance and on-site level activities. Thus the Project Quality Plan is an integral part of any operations strategy.

European Foundation for Quality Management Business Excellence Model (EFQM)

This section is aimed at establishing how the European Foundation for Quality Management Model (EFQM) can provide a means of implementing Total Quality Management (TQM) within the UK construction industry.

Many companies in the UK are increasingly having to face competition from those in other countries. They also have to accept that more choices are available to construction clients and that they must be assiduous in seeking goods and services with better quality and at a more competitive price in their search for value for money. As Bounds *et al.* (1994) remarked: *"traditional approaches to management are inadequate for keeping up with change."* Increased global competition and improved communications have lead to greater customer expectations.

Total Quality Management (TQM) is a powerful tool supporting the attainment of a sustainable competitive advantage through meeting client expectations. The Latham Report (1994) cited the definition of TQM adopted by the Henderson Committee (1992), which led to the formation of the British Quality Foundation:

"Total quality management is a way of managing an organisation to ensure the satisfaction at every stage of the needs and expectation of both internal and external customers, that is shareholders, consumers of its goods and services, employees and the community in which it operates, by means of every job, every process being carried out right, first time and every time."

(Henderson Committee, 1992. Cited by Latham Report, 1994)

van der Wiele *et al.* (1997) noted that *"TQM is dynamic in nature, based on continuous improvement and change and aims to achieve complete customer satisfaction by identifying and building on best practice in processes, products and services."* Moullin (1994) stated that *"TQM is developing towards the concept of value and is therefore meeting customer requirements at an acceptable price."*

The EFQM, founded in 1988, is committed to promoting quality as the fundamental process for continuous improvement within a business. It is dedicated to stimulating and assisting management in applying innovative principles of TQM appropriate to the European environment. Its aim is to improve the competitiveness of European private and public sector organisations, and over 10,000 firms now incorporate the EFQM Business Excellence Model in their overall corporate management process. In 1999, 60% of the top 25 companies in Europe (and 30% of the top 100) were members of the foundation.

Companies apply the EFQM Business Excellence Model, since the pursuit of business excellence through TQM is a decisive factor in allowing them to compete in today's global market. Although the Model is a relatively simple operational tool, very few UK construction companies have applied it. Thus this section examines the problems associated with applying the Model and offers some practical solutions.

The implementation of any quality system will incur significant costs and use valuable organisational energy. Therefore, it is important to ensure that the theoretical advantages actually exist in practice and this has been investigated.

The Historical Development of EFQM

EFQM, a non-profit-making organisation, provides various networking, benchmarking and training events to help members keep up to date with the latest trends in business management and research in TQM. It launched the European Quality Award in 1991 to stimulate interest, and this is awarded to those who have given 'exceptional attention' to TQM.

EFQM's mission (European Foundation for Quality Management, 2000a) is:

- to stimulate and assist organisations throughout Europe to participate in improvement activities, leading ultimately to excellence in customer satisfaction, employee satisfaction, knowledge management, impact on society and business results
- to support the managers of European organisations in accelerating the process of making TQM a decisive factor for achieving global competitive advantage.

According to the EFQM, therefore, the main reason for companies to apply the EFQM Business Excellence Model is to pursue business excellence through TQM, thereby allowing them to compete successfully in European and global markets.

EFQM Excellence Model Criteria

RADAR

A new key concept of the Business Excellence Model is RADAR, which is the essential business logic underlying the model and determining the success of the search for performance improvements. The fundamental elements of the concept are Results, Approach, Deployment, Assessment and Review (see Figure 9.5).

RADAR Logic

The RADAR logic states that an organisation needs to:

- determine the **R**esults the firm is aiming for as part of its policy and strategy-making processes – these include the performance of the organisation, both financially and operationally, and the perception of its stakeholders
- plan and develop an integrated set of sound **A**pproaches to deliver the required results
- **D**eploy the approaches in a systematic way to ensure full implementation
- **A**ssess and **R**eview the approaches followed based on monitoring and analysis of the results achieved and on on-going learning activities. Based on this assessment, companies should identify, prioritise, plan and implement improvements where required.

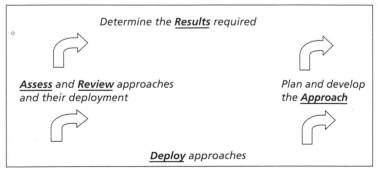

Figure 9.5 *The criteria underpinning the RADAR concept [© EFQM].*

The EFQM Business Excellence Model

Having recognised that corporate excellence is measured by an organisation's ability to both achieve and sustain outstanding results for its stakeholders, the enhanced version of the EFQM Business Excellence Model was developed. The fundamental advantages of the new Business Excellence Model included:

- *increased cost effectiveness*
- *a results orientation*
- *a customer focus*
- *partnership developments*
- *knowledge management*
- *performance and learning.*

The new Business Excellence Model was designed to be:

- simple (easy to understand and use)
- holistic (in covering all aspects of an organisation's activities and results, yet not being unduly prescriptive)
- dynamic (in providing a live management tool which supports improvement and looks to the future)
- flexible (being readily applicable to different types of organisation and to units within those organisations)
- innovative.

The EFQM Business Excellence Model consists of 9 criteria and 32 sub-criteria. The five criteria on the left-hand side of Figure 9.6 are called 'enablers' and are concerned with how the organisation performs various activities. According to Hillman (1994): "*The enablers are those processes and systems that need to be in place and managed to deliver total quality.*" The four criteria on the right of Figure 9.6 are concerned with the 'results' that the organisation is achieving with respect to different stakeholders. Hillman (1994) added that "*results provide the measure of actual achievement of improvement.*"

Watson (2000) stated that "*the EFQM Model provided a truly service focused quality system which had an inbuilt mechanism for the attainment of continued organisational improvement*" and advocated further that "*the model presented a route for construction firms to follow in their pursuit of a sustainable quality of service for their clientele and hence gain a sustainable competitive edge.*"

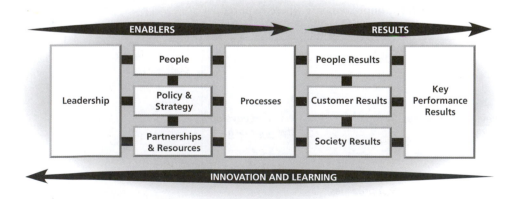

Figure 9.6   *Enhanced EFQM Business Excellence Model*
(European Foundation for Quality Management, 1999) [© EFQM]

van der Wiele *et al.* (1997) identified that "*the criteria of the model helped managers to understand what TQM means in relation to managing a company. However, managers also have to work at making the general descriptions of the criteria more specific to fit their particular situation and give them meaning within the context of their business activities.*"

The improved EFQM Model is principally focused on (European Foundation for Quality Management, 1999):

- **greater emphasis on the customer** and on other stakeholder groups whose importance has been increasing in the late 1990s, such as partners
- **increased visibility of the value chain**, including the increasingly important role of partnerships within the chain
- the emerging **importance of the management of knowledge** within organisations, the learning organisation culture, and innovation, as providing a key competitive advantage
- the **alignment of all corporate activity to the results** being sought and consequently to the organisation's policy and strategy.

The Need for a Model

To be successful, regardless of their size or operation, construction organisations need to establish an appropriate management system. The EFQM Business Excellence Model can provide a practical tool for managing and ensuring that they take the right path, leading to corporate excellence.

Function of the EFQM Business Excellence Model

Hillman (1994) suggested that "*the EFQM Model provided a tried and tested framework, an accepted basis for evaluation and a means to facilitate comparisons both internally and externally.*"

European Foundation for Quality Management (2000b) states that the functions of the model may be split in four ways:

- as a framework which organisations can use to help them develop their vision and goals for the future in a tangible, measurable way
- as a framework which organisations can use to help them identify and understand the systematic nature of their business, the key linkages, and cause and effect relationships
- as the basis for the European Quality Award, a process which allows Europe to recognise its most successful organisations and promote them as role models of excellence from which others can learn
- as a diagnostic tool for assessing the current health of the organisation.

Through this process an organisation is better able to balance its priorities, allocate resources and generate realistic business plans.

All these functions allow the model to be used for a number of activities, for example self-assessment, third party assessment, benchmarking and as the basis for applying for the European Quality award. However, van der Wiele *et al.* (1997) noted that the model could be used as a measurement method but not as a 'how-to' model. The application of the model would depend on the nature of the host organisation.

The Constituent Parts of the EFQM Business Excellence Model

The model incorporates business criteria, which are crucial to the company pursuing business excellence: "*More than ever, the EFQM Excellence Model offers an operational tool for the pursuit of excellence in performance and results*" (European Foundation for Quality Management, 1999).

Oakland (1993) stated that the European Quality Award "*recognises that processes are the means by which a company or organisation harnesses and relates the talents of its people to produce results. Moreover, the processes and the people are the enablers which produce results.*"

The Excellence Model is based on the concept that both customer/people satisfaction and positive impact on society are achieved through leadership driving policy and strategy, people management, partnership and resources, and processes leading ultimately to excellence in business results.

'Enablers' and 'Results'

Enablers of the Excellence Model are leadership, people, policy and strategy, partnering and resources, and processes. The four 'results' are people results, customer results, society results and the key performance results. EFQM has provided definitions for the 'enablers' and 'results', and has also developed several sub-criteria to support each criterion. Watson (2000) considered that the "*sub-criteria posed a number of questions that should be considered in the course of an organisation's assessment of its activities.*"

Enablers

(a) Leadership

Definition   How leaders develop and facilitate the achievement of the mission and vision, develop values required for long-term success and implement these via appropriate actions and behaviours, and are personally involved in ensuring that the organisation's management system is developed and applied.

Sub-criteria   Leadership covers the following four sub-criteria:

(1) leaders develop the mission, vision and values and are role models of a culture of excellence
(2) leaders are personally involved in ensuring the organisation's management system is developed, implemented and continuously improved
(3) leaders are involved with customers, partners and representatives of society
(4) leaders motivate, support and recognise the organisation's people.

Involvement of senior management is vitally important to the success of the implementation of the Excellence Model.

(b) Policy and Strategy

Definition   How the organisation implements its mission and vision via a clear stakeholder focused strategy, supported by relevant policies, plans, objectives, targets and processes.

Sub-criteria   Policy and strategy cover the following five sub-criteria:

(1) based on the present and future needs and expectations of stakeholders
(2) based on information from performance measurement, research, learning and creativity related activities
(3) developed, reviewed and updated
(4) deployed through a framework of key processes
(5) communicated and implemented.

(c) People

Definition   How the organisation manages, develops and releases the knowledge and full potential of its people at an individual, team-based and organisation-wide level, and plans these activities in order to support its policy and strategy and the effective operation of its processes.

Sub-criteria   People cover the following five sub-criteria:

(1) people resources are planned, managed and improved
(2) people's knowledge and competencies are identified, developed and sustained
(3) people are involved and empowered
(4) people and the organisation have a dialogue
(5) people are rewarded, recognised and cared for.

McCabe (1998) suggested that *"any quality effort should be based on the principle that if the people involved are convinced of the importance of improvement, they are more likely to be motivated to achieving it"*.

(d) Partnerships and Resources

Definition   How the organisation plans and manages its external partnerships and internal resources in order to support its policy and strategy and the effective operation of its processes.

Sub-criteria   Partnerships and resources cover the following five sub-criteria that should be addressed and managed:

(1) external partnerships
(2) finances
(3) buildings, equipment and materials
(4) technology
(5) information and knowledge.

(e) Processes

Definition   How the organisation designs, manages and improves its processes to support its policy and strategy and to fully satisfy, and generate increasing value for, its clients and other stakeholders.

Sub-criteria   Processes cover the following five sub-criteria:

(1) processes are systematically designed and managed
(2) processes are improved, as needed, using innovation in order to satisfy fully and generate increasing value for customers and other stakeholders
(3) products and services are designed and developed based on customer needs and expectations
(4) products and services are produced, delivered and serviced
(5) customer relationships are managed and enhanced.

In his study on business processes, Jones (1994) established that *"processes are the essential link between the customer's or client's requirements and the delivery of products or services. Processes are the means whereby the organisation and its employees fulfil their purpose or 'mission'."*

Results

(a) Customer Results

Definition   What the organisation is achieving in relation to its external customers.

Sub-criteria   Customer results cover the following two sub-criteria:

(1) perception measures
(2) performance indicators.

(b) People Results

Definition   What the organisation is achieving in relation to its people.

Sub-criteria   People results cover the following two sub-criteria that should be addressed:

(1) perception measures
(2) performance indicators.

(c) Society Results

Definition   What the organisation is achieving in relation to local, national and international society as appropriate.

Sub-criteria   Society results cover the following two sub-criteria that should be addressed:

(1) perception measures
(2) performance indicators.

(d) Key Performance Results

Definition   What the organisation is achieving in relation to its planned performance.

Sub-criteria   Key performance results cover the following two sub-criteria:

(1) key performance outcomes
(2) key performance indicators.

Depending on the purpose and objectives of the organisation, some of the measures contained in the guidance for key performance outcomes may be applicable to key performance indicators, and vice versa.

The EFQM Business Excellence Model provides a construction organisation with a way of achieving a top-quality performance. The 'enablers' and 'results', together with their sub-criteria, provide a guide for the construction manager on which area to focus in order to succeed in satisfying client needs on a construction project. It is important for a construction company to understand and identify the components included in the Excellence Model. The five enablers are closely related to the daily operational activities of a construction company.

The European Quality Award

Figure 9.7 shows how points are allocated to model criteria. Organisations perform a self-assessment based on this model. Points are awarded to each category and different categories have different maxima. The total points tally available is 1000.

Scoring Against the EFQM Business Excellence Model

The scoring system for allocating the possible 1000 points is based on the RADAR approach.

RADAR Expanded upon

Results

Under the results heading the host organisation, when conducting self-evaluation, considers trends, targets, comparisons and elements.

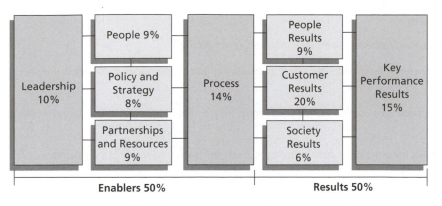

Figure 9.7   *The European Quality Award [© EFQM]*

The above are applied to the four key areas of 'results' within the model. The scoring criteria as defined within the model are as follows:

Trends: trends are positive and/or there is sustained good performance
Targets: targets are achieved; targets are appropriate
Comparisons: comparisons with external organisations take place and results compare well with industry averages or acknowledged 'best in class'
Causes: results are caused by approach
Scope: results address relevant areas.

The following apply to the 'enablers' only:

Approach

The approach must be sound and integrated as defined below.

Sound
- approach has a clear rationale
- there are well-defined and developed processes
- approach focuses on stakeholder needs.

Integrated
- approach supports policy and strategy
- approach is linked to other approaches as appropriate.

Deployment

The deployment element is divided into the two sub-areas of implementing and systematic:

- implementing: approach is implemented
- systematic: approach is deployed in a structured way.

Assessment and Review

This heading is subdivided into measurement, learning and improvement:

Measurement Regular measurement of the effectiveness of the approach and deployment is carried out.

Learning Learning activities are used to identify and share best practice and improvement opportunities.

Improvement Output from measurement and learning is analysed and used to identify, prioritise, plan and implement improvements.

Applying the Model

The following example is based on a fictitious construction organisation and is only applied to the 'leadership' enabler and 'performance' results criteria (to engage in the whole process would require approximately 60 pages – beyond the requirements for this book). Readers are advised to obtain a copy of the full scoring document from EFQM, Brussels, or visit the website address noted in the Reference section of this text.

Leadership

How leaders develop and facilitate the achievement of the mission, vision, develop values required for long-term success and implement these via appropriate actions and behaviours, and are personally involved in ensuring that the organisation's management system is developed and implemented.

Leaders Motivate, Support and Recognise the Organisation's People

This may include:

- personally communicating the organisation's mission, vision, values, policy and strategy, plans, objectives and targets to people
- being accessible, actively listening and responding to people
- helping and supporting people to achieve their plans, objectives and targets
- encouraging and enabling people to participate in improvement activities
- recognising both team and individual efforts, at all levels, within the organisation, in a timely and appropriate manner.

The above have been used to evaluate the host organisation and a score allocated by (trained) senior management. The sub-areas are graded and a final score awarded, for example Approach in this case is 55%, as shown in Table 9.3.

In the example (Table 9.3), Leadership is awarded a score of 55% for Approach, 45% for Deployment, 40% for Assessment and Review and an overall score of 45%. It is the overall score of 45% that is carried forward to the scoring summary sheet. However, before we engage in this process let us apply the scoring system to 'Performance Results'.

Performance

Key Performance Results

What the organisation is achieving in relation to its planned performance.

Table 9.3 *Leadership enabler example (adapted from EFQM, Scoring Matrix)*

Elements	Attributes	Scores 0%	25%	50%	75%	100%
Approach	**Sound:** – approach has a clear rationale – there are well-defined and developed processes – approach focuses on stakeholder needs	No evidence or anecdotal	Some evidence	Evidence	x Clear evidence	Comprehensive evidence
	Integrated: – approach supports policy and strategy – approach is linked to other approaches as appropriate	No evidence or anecdotal	Some evidence	x Evidence	Clear evidence	Comprehensive evidence
		0% 5% 10%	15% 20% 25% 30% 35%	40% 45% 50% 55% 60%	65% 70% 75% 80% 85%	90% 95% 100%

Elements	Attributes	Scores 0%	25%	50%	75%	100%
Deployment	**Implemented:** – approach is implemented	No evidence or anecdotal	x Implemented in about 1/4 of relevant areas		Implemented in about 3/4 of relevant areas	Implemented in all relevant areas
	Systematic: – approach is deployed in a structured way	No evidence or anecdotal	Some evidence	x Evidence	Clear evidence	Comprehensive evidence
		0% 5% 10%	15% 20% 25% 30% 35%	40% 45% 50% 55% 60%	65% 70% 75% 80% 85%	90% 95% 100%

Elements	Attributes	Scores 0%	25%	50%	75%	100%
Assessment and Review	**Measurement:** – regular measurement of the effectiveness of the approach, deployment is carried out	No evidence or anecdotal	x Some evidence	Evidence	Clear evidence	Comprehensive evidence
	Learning: – learning activities are used to identify and share best practice and improvement opportunities	No evidence or anecdotal	Some evidence	x Evidence	Clear evidence	Comprehensive evidence
	Improvement: – output from measurement and learning is analysed and used to identify, prioritise, plan and implement improvements	No evidence or anecdotal	Some evidence	x Evidence	Clear evidence	Comprehensive evidence
		0% 5% 10%	15% 20% 25% 30% 35%	40% 45% 50% 55% 60%	65% 70% 75% 80% 85%	90% 95% 100%

Overall Total		0% 5% 10%	15% 20% 25% 30% 35%	40% 45% 50% 55% 60%	65% 70% 75% 80% 85%	90% 95% 100%

Key Performance Indicators (KPIs)

These measures are the operational ones used in order to monitor, understand, predict and improve the organisation's likely key performance outcomes. Depending on the purpose and objectives of the organisation and its processes, they may include those relating to:

- processes
  - performance
  - deployment
  - assessments
  - innovations
  - improvements
  - cycle times
  - defect rates
  - maturity
  - productivity
  - time to market
- external resources including partnerships
  - supplier performance
  - supplier price
  - number and value-added of partnerships
  - number and value-added of innovative products and service solutions generated by partners
  - number and value-added of joint improvement with partners
  - joint improvement with partners
  - recognition of partners' contribution
- financial
  - cash-flow items
  - balance sheet items
  - depreciation
  - maintenance cost
  - return on equity
  - return on net assets
  - credit ratings
- buildings, equipment and materials
  - defect rates
  - inventory turnover
  - utility consumption
  - utilisation
- technology
  - innovation rate
  - value of intellectual property
  - patents
  - royalties
- information and knowledge
  - accessibility
  - integrity
  - relevance
  - timeliness
  - sharing and using knowledge
  - value of intellectual capital

Again, the above has been used to evaluate the host organisation in order to allocate a score (Table 9.4).

Table 9.4 *EFQM scoring matrix for results (performance) example*

Elements	Attributes	Scores: 0%	25%	50%	75%	100%
Results	**Trends:** • trends are positive and/or there is sustained good performance	No results or anecdotal information	Positive trends and/or satisfactory performance on some results	Positive trends and/or sustained good performance on many results over the last 3 years x	Strongly positive trends and/or sustained excellent performance on most results over at least 3 years	Strongly positive trends and/or sustained excellent performance in all areas over at least 5 years
	Targets: • targets are achieved • targets are appropriate	No results or anecdotal information	Favourable and appropriate in some areas x	Favourable and appropriate in many areas	Favourable and appropriate in most areas	Excellent and appropriate in most areas
	Comparisons: • comparisons with external organisations take place and results compare well with industry averages or acknowledged 'best in class'	No results or anecdotal information	Comparisons in some areas	Favourable in some areas	Favourable in many areas x	Excellent in most areas and 'best in class' in many areas
	Causes: • results are caused by approach	No results or anecdotal information	Some results	Many results X	Most results	All results. Leading position will be maintained
	TOTAL	0% 5% 10%	15% 20% 25% 30% 35%	40% 45% 50% 55% X 60%	65% 70% 75% 80% 85%	90% 95% 100%

Elements	Attributes	Scores: 0%	25%	50%	75%	100%
Results	**Scope:** • results address relevant areas	No results or anecdotal information	Some areas addressed	Many areas address	Most areas addressed	All areas addressed
	TOTAL	0% 5% 10%	15% 20% 25% 30% 35%	40% 45% 50% 55% 60%	65% X 70% 75% 80% 85%	90% 95% 100%

OVERALL TOTAL		0% 5% 10%	15% 20% 25% 30% 35%	40% 45% 50% 55% 60%	65% 70% 75% 80% 85%	90% 95% 100%

Applying the same rationale as before we have obtained an overall score of 60% for 'Performance Results' from Table 9.4. It should be noted that when scoring against the criteria, management will find it most useful to consider areas of:

• strengths: areas of good/best practice that could be disseminated throughout the organisation
• areas for improvement so that corrective actions can be employed
• evidence which supports the awarded percentage points.

Upon completion of the scoring related to the five enablers and four results (with sub-criteria), the scores are carried forward to the scoring summary sheet (Table 9.5).

In the example used, the scores for Leadership and Performance along with a completed analysis have been inserted so a final score can be obtained (Tables 9.5, 9.6 and 9.7):

• enter the score awarded to each criterion (of both sections 1 and 2 above)
• multiply each score by the appropriate factor to give points awarded
• add points awarded to each criterion to give total points awarded for applications.

In order to put the score of 508.45 points in the context of best practice it should be noted that EFQM will conduct a site visit on an organisation obtaining over 500 points. The EFQM award for excellence is usually awarded to organisations obtaining a score between 750 and 850 points. Therefore, a score of 508.45 points is a very respectable score but we must remember that the objective is continuous improvement.

Table 9.5 *Enablers criteria*

EXAMPLE
Scoring summary sheet — Section 1

Criterion number	1	%	2	%	3	%	4	%	5	%
Sub-criterion	1a	45	2a	50	3a	60	4a	50	5a	45
Sub-criterion	1b	40	2b	50	3b	35	4b	50	5b	60
Sub-criterion	1c	45	2c	40	3c	40	4c	55	5c	60
Sub-criterion	1d	50	2d	30	3d	40	4d	40	5d	50
Sub-criterion			2e	45	3e	50	4e	35	5e	50
Sum		180		215		225		230		265
		÷ 4		÷ 5		÷ 5		÷ 5		÷ 5
Score awarded		45		43.2		45.2		46		53

Note: The score awarded is the arithmetic average of the percentage scores for the sub-criterion. If applicants present convincing reasons why one or more parts are not relevant to their organisation it is valid to calculate the average on the number of criteria addressed. To avoid confusion (with a zero score), parts of the criteria accepted as not relevant should be entered 'NR' in the table above.

Table 9.6 *Results criteria*

Criterion number	6	%	7	%	8	%	9	%
				Section 2				
Sub-criterion	6a	$50 \times 0.75 = 37.5$	7a	$60 \times 0.75 = 45$	8a	$50 \times 0.25 = 12.5$	9a	$60 \times 0.50 = 30.0$
Sub-criterion	6b	$50 \times 0.25 = 12.5$	7b	$50 \times 0.25 = 12.5$	8b	$60 \times 0.75 = 45$	9b	$55 \times 0.50 = 27.5$
Score awarded		50		57.5		57.5		57.5

Table 9.7 *Calculation of total points allocated*

Criterion	Score awarded	Factor	Points awarded
1. Leadership	45	× 1.0	45
2. Policy and Strategy	43.2	× 0.8	34.6
3. People	45.2	× 0.9	40.7
4. Partnerships and Resources	46	× 0.9	41.4
5. Processes	53	× 1.4	74.2
6. Customer Results	50	× 2.0	100.0
7. People Results	57.5	× 0.9	51.8
8. Society Results	57.5	× 0.6	34.5
9. Key Performance Results	57.5	× 1.5*	86.25
Total points awarded			508.45

* Note these are the factors from the model.

The scoring summary sheet provides a useful overall picture of the organisation. However, the data has been further developed to provide more detailed information for the host company. It would be very useful for a company to know the profile related to:

- approach
- deployment
- assessment and review
- criteria
- results.

This would allow an organisation to focus its efforts for improvement on highlighted areas.

An example of the above approach follows (Tables 9.8 and 9.9). Note, average scores have been calculated for the noted areas and 'Results' have been divided into 'Results' and 'Scope', thus providing more detail enabling more effective corrective actions. For ease of presentation, this data can be represented on a RADAR Pentagonal Profile. This is presented in the example as Figure 9.8. The pictorial representation of the RADAR Pentagonal Profile enables instantaneous understanding of the current state of the company.

It is also a very quick and accurate method of benchmarking an organisation. Senior managers must remember that the self-evaluation process is designed to develop continuous improvement. Therefore, the benchmarking activity must be conducted on a regular basis so that corrective actions can be evaluated. Readers are advised to see EFQM's RADAR Model incorporated in their *Assessment Scoring Handbook*.

The EFQM Business Excellence Model provides a valuable framework for addressing the key operational activities of construction organisations. It is useful because it enables a link to be made between people, organisational objectives and improvement processes, all encompassed under the umbrella of continued improvement.

The scoring methodology is simple to apply but senior construction managers are advised to obtain some formal training before applying the model.

The model, when implemented, does provide detailed information for employing constant and consistent benchmarking activities.

The EFQM Business Excellence Model is a non-prescriptive framework based on nine criteria – 5 'enablers' and 4 'results'. It can be used to assess an organisation's progress towards excellence. The Model provides a non-prescriptive framework to guide a construction company to achieve a top quality performance via the attainment of a sustainable competitive advantage. Within the non-prescriptive framework, certain fundamental concepts underlie the Model.

Table 9.8 Summary assessment sheet for 'RADAR' company profile

Criteria	Approach score	Deployment score	Assessment and review score
1a	55	45	45
1b	60	55	35
1c	60	55	45
1d	55	55	40
2a	65	60	20
2b	50	55	15
2c	70	80	35
2d	55	70	50
2e	65	75	40
3a	55	65	50
3b	65	70	60
3c	50	35	25
3d	50	65	25
3e	30	40	30
4a	55	55	40
4b	55	65	40
4c	40	35	15
4d	30	40	25
4e	65	60	25
5a	70	85	40
5b	50	60	30
5c	45	50	25
5d	60	60	40
5e	65	75	50
Totals and average (mean score)	$\dfrac{1319}{24}$	$\dfrac{1415}{24}$	$\dfrac{845}{24}$
	= 54.9 score	= 58.9 score	= 35.2 score

Table 9.9 Summary assessment sheet for 'RADAR' company profile

Criteria	Results	Scope
6a	55	65
6b	40	30
7a	60	50
7b	40	55
8a	25	20
8b	45	35
9a	55	45
9b	50	60
Totals and average (mean score)	$\dfrac{370}{8}$	$\dfrac{360}{8}$
	= 46.3 score	= 45 score

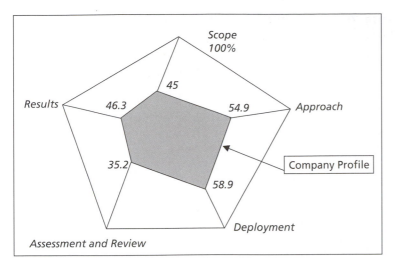

Figure 9.8 *RADAR pentagonal profile [© Dr P Watson]*

The Implementation of the EFQM Business Excellence Model within the Construction Industry – Advocated Advantages and Related Issues

Theoretical advocated advantages associated with the implementation of the EFQM Business Excellence Model have been identified by various authors, including the EFQM. However, it must be considered whether these advantages can be obtained in practice, for a construction firm needs to be confident that it can obtain the advocated benefits if it is to invest resources in the deployment process. There is an opportunity cost associated with this resource investment. Therefore, field research, based on a structured questionnaire, has been carried out to test whether the noted benefits can be obtained in practice. The theoretical advocated advantages of applying the EFQM Business Excellence Model have been compared with the result of the field research and conclusions drawn.

Theoretical Advantages of Applying the EFQM Model

The advantages of the model have been identified earlier. In a study on self-assessment, Hillman (1994) elaborated further on the benefits of the EFQM Model, stating:

- It is not a standard but allows interpretation for all aspects of the business and all forms of organisation.
- Its widening use facilitates comparison between organisations. This provides the potential to learn from others in specific areas by using a common language.
- The inclusion of tangible results ensures that the focus remains on real improvement, rather than preoccupation with the improvement process, i.e. it focuses on achievement, not just activity.
- Training is readily available in the use and scoring of the model.
- It provides a repeatable basis that can be used for comparison over several years.
- The comprehensive nature and results focus, broken down into discrete elements, helps develop a total improvement process specific for each organisation – it is a model for successful business.

Benefits Derived from the Implementation of the EFQM Business Excellence Model

The following provide the underpinning rationale for companies pursuing a competitive strategy through the application of the EFQM Business Excellence Model. The model is recognised as:

- providing a marketing focus
- being a means of achieving a top-quality performance in all areas of the organisation
- providing operating procedures for all staff
- allowing for the review of organisational self-assessment performance
- providing a competitive weapon via a quality approach
- developing group/team spirit within the company
- enabling an improvement in communication systems
- reducing inspection and corrective action costs
- facilitating more efficient utilisation of a company's resources
- creating easier loan arrangement through financial institutions
- providing customer satisfaction, i.e. providing the required customer quality every time.

Empirical Research

Research has been conducted to test, via a questionnaire, the various issues related to the theoretical advantages, and the associated problems of application.

Selecting the sample for the survey was problematic because only a few construction companies in the UK are registered members of the EFQM. However, because the Business Excellence Model is universally applicable, the implementation issues are generic, and data collected from other industries is appropriate. The results are based on a sample size of fifty companies.

Data Analysis

The results relating to the advantages of the model and the testing of the theoretical advantages for its implementation are summarised in Tables 9.10 and 9.11 respectively.

The results of the research show that the majority of sampled companies found that the Model was simple, holistic, dynamic and flexible. They also agreed that the model did enhance the understanding of TQM among senior management and enabled the identification of a company's strengths and weaknesses. The main reason offered by the sampled companies for applying the EFQM approach to quality was self-assessment. This empowered organisations to achieve a top-quality performance in all areas, in other words, to achieve TQM within their organisation. The research results established that most of the theoretical advantages relating to the benefits derived from the application of the Model could be achieved in practice. The research also established some problems that construction firms could face during implementation, and these are addressed below.

The sampled companies were asked to identify any problems related to the application of the EFQM Business Excellence Model (see Table 9.12). Information was also gathered from structured interviews and a case study carried out with a major construction company. The results of this consultation process have been utilised in the production of two generic models designed to assist construction-related organisations in their implementation of the EFQM Model (Figures 9.9 and 9.10).

One-third of the quality managers interviewed experienced considerable resistance from staff with regard to documentation gathering and deployment. Reluctance to participate in the project may have been due to resistance to change. One strategy that could be employed is for firms to commence implementation within a business unit keen to go for the system. When the benefits of implementation are proved, the quality department can then use this achievement to sell the idea to the other business units. This is an incremental strategy based on initial success. There must always be someone available to provide all necessary support to those departments that wish to be involved. Quality departments

Table 9.10 *Summary of responses for theoretical advantages of the EFQM Business Excellence Model Application*

Theoretical advantages of the EFQM Business Excellence Model	Response to question (%)		Advantage proved		
	Yes (x)	No	To a great extent (x > 60)	Partly (60 > x > 40)	Not to any great extent (x < 40)
1. Simple – easy to understand and use?	Q32-1 47	53			
2. Holistic – in covering all aspects of an organisation's activities and results, yet not being unduly prescriptive?	Q32-2 75	25	*	*	
3. Dynamic – in providing a live management tool which supports improvement and looks to the future?	Q32-3 75	25	*		
4. Flexible – being readily applicable to different types of organisation and to units within those organisations?	Q32-5 83	17	*		
5. Innovative?	Q32-5 58	42		*	
6. Helpful in defining TQM in a way which management can more easily understand?	Q27 92	8	*		
7. Provides a definition and description of TQM which gives a better understanding of the concept, improves awareness and generates ownership for TQM among senior managers?	Q28 83	17	*		
8. Enables measurement of the progress towards TQM to be made?	Q29 92	8	*		
9. Provides an objective measurement, gains consensus on the strengths and weaknesses of the current approach and helps to pinpoint improvement opportunities?	Q30 75	25	*		
10. Provides a number of sub-criteria and areas to address which make it easier to construct a questionnaire/assessment list that covers the entire model for quantitative measurements?	Q31 83	17	*		
11. Assists in the implementation process?	Q33 67	33	*		

Q = Question number on questionnaire.

Table 9.11 Summary of responses for theoretical advantages for implementation of the EFQM Business Excellence Model

Theoretical advantages for implementation of EFQM Business Excellence Model	Response to question (%)					Advantage proved		
	Greatly	Hardly	Not at all	Yes (x)	No	To a great extent (x > 60)	Partly (60 > x > 40)	Not to any great extent (x < 40)
1. Provides a marketing focus?	Q3 70	30				*		
2. Provides a means of achieving a top-quality performance in all areas of the organisation?	Q4 50	50					*	
3. Provides operating procedures for all staff? Before implementation				Q5 25	75			
After implementation				Q6 67	33	*		
4. Provides a quality advantage as a competitive weapon?	Q8 81	19				*		
5. Develops group/team spirit within the company?	Q9 50	33	17				*	
6. Improvement of communication systems within organisation?	Q10 83	17				*		
7. More efficient utilisation of company's resources?				Q7 75	25	*		
Q12 a	67	33						
b	16	42	42					*
c	17	66	17					*
d	25	50	25					*
e	58	42						
8. Customer satisfaction?	Q13 84	16		Q14 17	83	*	*	
9. Improvement of business results? Q15 a	42	50	8				*	
b	58	42					*	
c	25	50	25				*	
d	50	50					*	
10. Enhancement of continuous improvement process?				Q16 67	33	*		
11. Benefits derived outweighed implementation cost?	Q17 56	44				*		

Q = Question number on questionnaire.

Table 9.12 *Summary of responses for problems in the implementation of the EFQM Business Excellence Model*

Problematic areas for implementation of EFQM Business Excellence Model	Response to question (%)							
	A lot	Some	None	Yes	No	Great amount	Hardly any	None
Q19 Problems in interpreting the Model?	8	67	25					
Q20 Problems in documentation?	25	42	33					
Q21 Adequate time allocation for implementation?				50	50			
Q22 Sufficient funding for implementation?				58	42			
Q23 Resistance from staff?						(a) 33	67	
						(b) 33	67	
Q24 Sufficient authority empowerment?				83	17			
Q25 Sufficient senior management support?				83	17			

need to monitor the progress of each department in order to keep them on target and continually motivated.

Sufficient training must be provided for all staff in order to overcome the fear of change. Once having acquired the necessary skills to face the challenge, employees will become supportive of the quality system. Training and education of employees are required in order to develop a culture of support for the organisation's continuous improvement process.

Generic Implementation Model

The following generic models (Figures 9.9 and 9.10) are provided for construction companies in order to guide them through the implementation process. The models have been developed based on feedback from the empirical studies.

In order to successfully apply the Business Excellence Model, there are three main issues to be considered:

- Training and development of people – the organisation should not concentrate purely on financial issues during implementation. Emphasis needs to be given to people development. Personnel must be competent to perform their tasks if standards are to continually improve. Ultimately this enhanced competitiveness will lead to increased financial performance.
- Improved communication – efficient communication channels should be created to encourage two-way communication from top to bottom, and vice versa. A 'no blame' culture should be developed so that people within the organisation are encouraged to reveal their errors/problematic issues, as only by adopting this approach will they be removed in such a way that they will not recur. Employees should be encouraged to generate ideas for the continuous improvement of the organisation. This is important because people who actually perform the work are best placed to suggest improvements.
- Getting to know the client – clients are the final judges of an organisation's performance. Therefore, meeting client specifications and needs is the only formula to delight the customer and obtain a sustainable competitive edge.

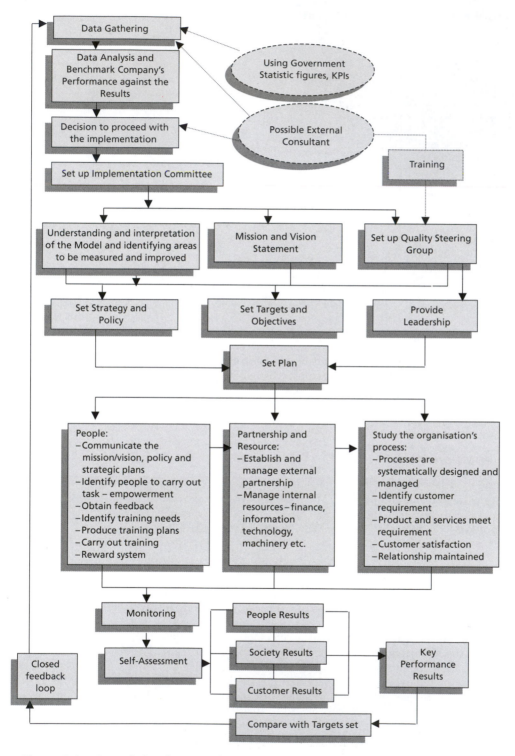

Figure 9.9 *Generic implementation model of EFQM Business Excellence Model as an approach toward TQM (Watson and Seng, 2001)*

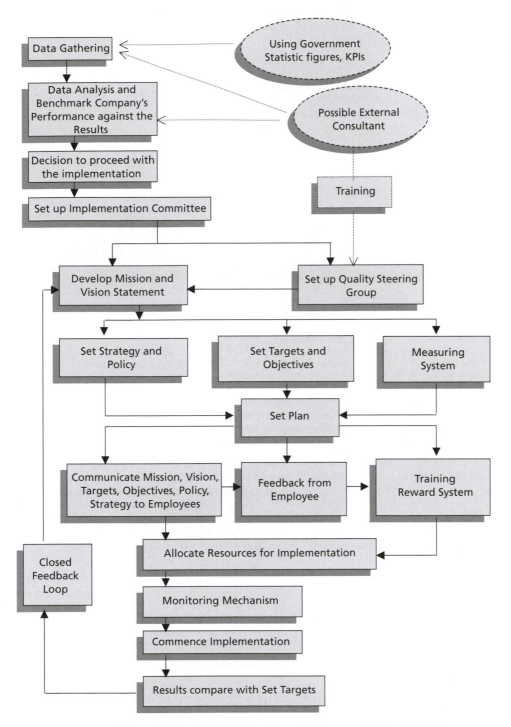

Figure 9.10 *Generic implementation model for EFQM Business Excellence Model (Watson and Seng, 2001)*

Summary

The EFQM Business Excellence Model is a European approach toward TQM. It provides a framework and guidance to assist construction organisations that wish to develop a TQM culture. The fundamental concepts of the Excellence Model are:

- a result orientation
- a customer focus
- leadership and constancy of purpose
- management by processes and facts
- people development and involvement
- continuous learning, innovation and improvement
- partnership development
- public responsibility.

The results of the research established that the majority of the sampled companies found the Model simple, holistic, dynamic and flexible. They agreed that the model enhanced the understanding of TQM among senior management and enabled identification of the company's strengths and weaknesses via the self-assessment approach.

The main problems faced by quality managers during the implementation process were:

- inexperience in the implementation of the Model
- difficulties during the documentation and data-gathering stages
- insufficient time and funds allocated to the project
- resistance from employees within the host company.

Even though the research showed that senior management provided quality managers with full support and sufficient authority during the design and implementation phases, quality managers faced problems of insufficient funds and time allocation for the project. A possible explanation may be that full responsibility had been delegated without adequate authority.

van der Wiele *et al.* (1997) suggest that *"award models are helpful in defining TQM in a way which management can more easily understand."* They also help organisations to develop and manage their continuous improvement activities. For example:

- they provide a definition and description of TQM which gives a better understanding of the concept, improves awareness and generates ownership of TQM among senior managers
- they enable measurement of the progress towards TQM to be made, along with benefits and outcomes
- the scoring criteria provide an objective measurement, assist in gaining a consensus on the strengths and weaknesses of the current approach and help to pinpoint improvement opportunities
- appropriate benchmarking and organisational learning are facilitated
- training in TQM is encouraged.

The research established that a construction firm can obtain sustainable competitive advantages from the application of the EFQM Business Excellence Model. What the construction industry requires is more examples of best practice in the application of new management concepts so that it does not continually lag behind the manufacturing sector. The generic models developed should provide valuable stepping stones for EFQM Model implementation and consequent best practice activities. The case study described later in this chapter provides a good example for construction firms.

The Requirements for Attaining/Maintaining a Competitive Advantage

The requirements may be classified by considering the following proposed elements of competitiveness (Sheldon *et al.*, 1991):

(a) Timeliness
(b) Quality
(c) Affordability.

(a) Timeliness: this has always been an area of concern in the construction industry. Jones (1984) highlights two aspects of performance for which construction has been much criticised as time and cost overruns in contract completion and the innovative performance of the industry. He argues that the cost and time overruns are numerous but are in part concerned with the level of site efficiency in getting men and materials on-site on time and completing work on schedule. A further area of concern often cited by researchers is the lack of trained manpower in the construction industry. All three systems, IiP, BS EN ISO 9000: 2000 and EFQM Business Excellence Model emphasise the importance of training and could therefore resolve this problematic issue.

Hillebrandt (1984) describes the construction process as often being fraught with disruptions and delays.

(b) Quality: will the product satisfy customer requirements in terms of durability, aesthetics and reliability?

(c) Affordability: most construction costs are established before the contract commences. However, cost overruns and claims from the contractor could be passed on to the client.

According to Sheldon *et al.* (1991), the above three dimensions largely determine the value of a product to the customer.

The systems are based on inspiring employees at every level to continuously improve what they do, thus rooting out unnecessary costs. The competitive advantage results from concentrating resources on controlling costs and improving customer service (both internal and external).

All three systems noted in this chapter enable a construction company to fully identify the extent of its operational activities and focus them on customer satisfaction. Part of this service focus is the provision of a significant reduction in costs through the elimination of poor quality in the overall construction process. This empowers the host organisation in the attainment of a competitive advantage.

The high failure rate associated with the implementational process of Quality Management Systems is due to the pursuit of a post-modernist concept within a modernist organisational environment. Therefore to achieve a successful implementation, organisations must fully understand the basic requirement for its deployment.

Research in the field of construction management however, and in particular appertaining to TQM, is not utilised to its full potential. Research conducted within the manufacturing sector has identified the importance of adopting a post-modernistic paradigm when implementing quality systems. This fact seems not to have been embraced within the construction industry, thus the resulting high failure rate in implementing TQM within a construction operational environment. Empirical studies have shown that 80% of quality initiatives fail within one to three years. Construction-related enterprises must fully appreciate that the old-style morphostatic change processes are not capable of sustaining an effective and efficient Quality Management System.

Empirical research has been conducted to test some of the key elements of the deployment process. The first issue to be tested via the sampled companies' (Table 9.13) questionnaire related to an organisation's ability to respond to change. This encompasses criteria such as:

- empowerment
- innovation
- flexibility
- cultural dynamics.

Table 9.13 *Companies' sampled distribution*

Size of firms	Sample size
Small	50
Medium	25
Large	25
Total sampled	100

Since the previous studies on this topic there has been a great improvement for construction firms in their ability to respond to changing environments (Figure 9.11). This has a positive correlation with the number of firms now operating a post-modern philosophy.

The analysis in Table 9.14 indicates that morphostatic processes are common and that the amount of proactive participation has not filtered down through the sampled firms. Therefore there are still changes required in organisational cultural dynamics in order to fully obtain a sustainable competitive advantage.

Construction firms require variety in their approach and hierarchical authoritarian organisations are poorly equipped to provide such variety. Only business enterprises based on a post-modernist model with vastly reduced bureaucratic control, a rich array of horizontal communication channels, and in which personnel are given a substantial share of authority to make choices and to develop new ideas, can survive under current global market conditions.

Both the IiP and BS EN ISO 9001: 2000 Standards can be encapsulated within the EFQM Business Excellence Model (Table 9.15). Therefore an incremental application to the full deployment of the EFQM Business Excellence Model could be via a strategy of implementing IiP, followed by BS EN ISO 9001: 2000 leading ultimately to EFQM model deployment.

This strategic incrementalist approach to deployment would avoid the 'big bang' application with its associated cultural issues which are most noticeable and problematic in construction organisations.

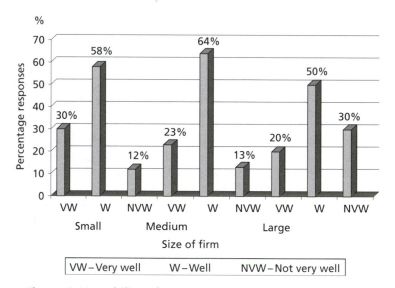

Figure 9.11 *Ability of organisations to respond to change (as a percentage)*

Table 9.14 *Analysis of morphostatic and morphogenic process*

Organisations, as a percentage having morphostatic process	Organisations allowing employees to be proactive at the following levels, as a percentage		
	Senior management	Middle management	Operational level
Small, 33%	60	30	10
Medium, 56%	50	35	15
Large, 11%	50	42	8

Table 9.15 *EFQM Business Excellence Model Deployment: advantages for construction organisations (note: not mutually exclusive)*

Key deployment issues	Resulting benefits
• Process improvement	• A clear understanding of how to deliver value to clients and hence gain a sustainable competitive advantage via operations
• Attaining organisation's objectives	• Enabling the mission and vision statements to be accomplished by building on the strengths of the company
• Benchmarking Key Performance Indicators (KPIs)	• Ability to gauge what the organisation is achieving in relation to its planned performance (Plan, Do, Check, Act)
• Development of clear, concise action plans resulting in a focused policy and strategy	• Clarity and unity of purpose so the organisation's people can excel and continuously improve
• Integration of improvement initiatives into normal operational activities	• Activities interrelated and systematically managed with a holistic approach to decision making
• Development of group/team dynamics	• People development and involvement – shared values and a culture of trust, thus encouraging empowerment in line with a post-modernist company

Case Study – EFQM Business Excellence Model Deployment

Thus far an outline of the EFQM Business Excellence Model (BEM) has been provided. However, what construction firms need is a suitable methodology for its deployment. The following describes such a methodology and is based on an actual case study. Data collected through model deployment is confidential but it is the process of implementation that is useful to construction firms and this is not confidential. Therefore both the process used and examples of the documentation employed are explored.

Commencing Model Deployment

It is imperative when deploying the model that senior management fully supports the process. Having senior management support for the project is vital because TQM requires the total commitment of all staff within the host organisation. Furthermore, staff must be adequately resourced. These two key aspects are dependent on senior management support (Oakland, 1993). Ensure that at least one person has undergone the EFQM Business Excellence Model training course. This is a comprehensive two-day event based on an actual organisation's results. The training will ensure an understanding of the Model and assist in its application. Figure 9.12 is a pictorial representation of the deployment process utilised in the case study documented.

The person who has undertaken the training should be appointed team leader. As depicted in Figure 9.12 (at **7**) a team should be set up and consist of a representative sample of the organisation/ department. This group of staff needs to be introduced to the model and its implementational aspects. This is best done by way of a workshop of approximately five hours' duration. Prior to this first workshop, reading material describing the model should be distributed for staff to study. At the first workshop the constituent parts of the model and its advantages should be explored. Once all the staff has a reasonable understanding of the model, individuals should be allocated into teams and given one or more criteria on which to gather data. There are nine criteria and some staff will, therefore, work on more than one. It is preferable that at least two people are allocated to each criterion. Some mutual support can be obtained by working in teams of two. Overall support can be provided by the team leader.

Figure 9.12 (at **10**) refers to interviews. It is important that a representative sample of staff is selected for a semi-structured interview; great consideration, therefore, should be given to this critical activity. Within this case study sixty interviews were conducted from 180 staff. These provided representation at all levels and functions. In order to assist other organisations in deploying the model,

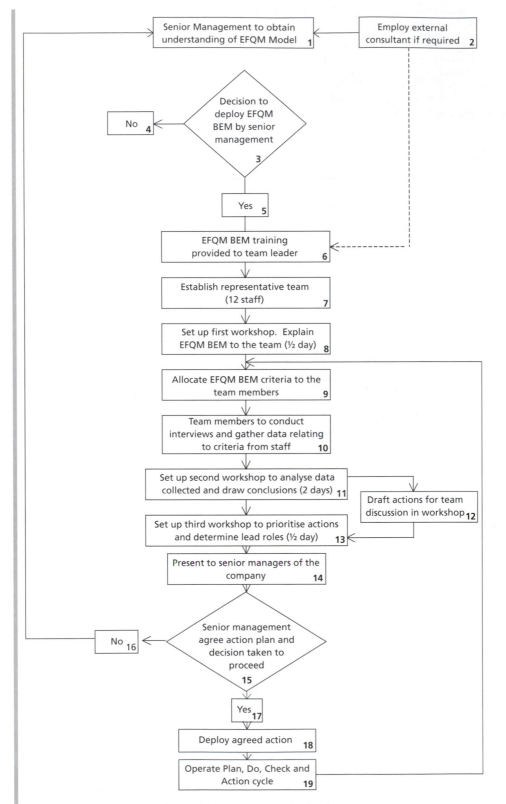

Figure 9.12 *Case study deployment process*

proformas are included. However, because of the space limitations of this book, only samples can be incorporated here. In the first stage of conducting interviews, therefore, the criteria of 'Leadership' (enabler) and 'Key Performance Results' have been presented for the reader.

Table 9.16 (Leadership) shows the format of the proformas utilised during the semi-structured interviews. People were asked to comment on the table headings. As an example, some fictitious data has been incorporated.

Table 9.17 incorporates responses relating to the strength (and weakness) areas for improvement. Again sample data is included.

Table 9.18 provides the opportunity to establish specific actions. It also allows for scoring. However, it is suggested that scoring is not undertaken during the first round of model application. Poor results could act as a demotivator and cause implementational problems. At the end of Figure 9.12 (Deployment Process Model) the final box states 'Operate Plan, Do, Check and Action cycle'. This box (**19**) has a connecting arrow line back to box **9** which is the allocation of criteria. This is the commencement of the second round of deployment, and will start once the actions established during the first round have been addressed. Scoring can commence during round two (see Figure 5.16).

Only leadership sub-criteria (a) have been provided in the example, this again being due to space limitations. However, the process will be the same for the rest of the sub-criteria.

Table 9.19 provides the format for responses related to key performance with some examples included. Table 9.20 again is a partial SWOT analysis with suggested actions for improvement. The above process has to be conducted on all criteria with the sampled staff base. Once this has been done all the

Table 9.16 *Leadership 1a – How leaders develop the mission, vision and values and are role models of a culture of excellence*

What is the top-level approach, philosophy or vision that addresses this issue?	What are the activities that you undertake which turn your overarching approach into action?	How widely is the *approach* actioned across all areas and down through the staff structure?	Is the *approach* (not the activity itself) checked to ensure it is still effective, lessons are learned and improvements made?	Are there any key issues highlighted, external to organisation or implications, or validating comments?
Actively encouraged involvement with stakeholders	• Business planning • Customer surveys	Varied dependent on specific activity but generally well deployed	Not in a holistic manner	Requires a coherent strategy

Table 9.17 *Leadership 1a – How leaders develop the mission, vision and values and are role models of a culture of excellence: strengths and areas for improvement*

Strengths (according to the evidence available)	Areas for improvement (according to the evidence available)
• Leaders knowledge of the external environment plus consultation with views and needs of staff to develop a vision that is not in conflict with staff	• Leaders to engage in personal improvement activities, for example 360° appraisal

Table 9.18 *Leadership 1a – How leaders develop the mission, vision and values and are role models of a culture of excellence: action areas*

Suggested actions for tackling areas for Improvement	Score achieved	
1. Development of a 'Best Practice Guide' covering aspects such as leadership style, communication, motivation and delegation	Approach Deployment Assessment Review Overall	% % % %

Table 9.19 *Key performance results 9a*

Results measured	How result is measured	Results		Trends				Comparison against best in class				Key causes/comments
		Actual (unit)	Agreed target	Difference	Is there evidence to show that results are getting better over a 3 year or more period?			Are our results good or bad compared to those identified as best in class (where appropriate)?				Which approach directly impacts on this result?
What are the results that are actually being measured in this area?					Yr1	Yr2	Yr3	Yr1	Yr2	Yr3		
• Profitability • Liquidity	• Cash-flow analysis • Ratio analysis											

data has to be co-ordinated into a coherent document. A second workshop is necessary to work through all the data. It is suggested that the team be provided with the complete set of data one week before the workshop. Also an electronic version of the data should be used at the workshop to enable alterations to be made during the discussions.

At the second workshop, criteria leaders are asked to talk through collected data and a team agreement be obtained.

It is important to programme the two days. Tables 9.21 and 9.22 were used during the actual case study workshop. (Criteria leader names have been removed.) The key point to note is on Table 9.22, 12.45–2.30 Action Planning session. This is a critical session. However, before this session, during the lunch period 12.00–12.45, the team leader has to group similar actions together. To explain this further, during the interview process, some actions would have been noted under more than one criterion. This grouping activity allows similar items to be brought together. For example, under leadership, issues relating to people would probably be noted, and vice versa.

Table 9.20 *Key performance results 9a: strengths and areas for improvement (source: British Quality Foundation, 1998)*

GUIDANCE NOTES
Review your descriptions of the <u>results</u> as a whole, then record your views on where things have been successful, where improvements could be made and actions that could be taken based on your analysis of the current situation. <u>Your views must be based only on the evidence gathered and recorded in the previous page not on your existing knowledge of what you yourself understand may be the current situation which may not be evidenced</u>

STRENGTHS (according to the evidence available)	AREAS FOR IMPROVEMENT (according to the evidence available)
• Maintaining profitability • Secure workload	• Reduce staff stress levels

SUGGESTED ACTIONS FOR TACKLING AREAS FOR IMPROVEMENT	SCORE ACHIEVED	
1. Engage in Work Life Balance Activities	Results	%
	Scope	%
		%
	Overall	%

Table 9.21 *EFQM Self-Assessment Workshop 2 Agenda (day one)*

Time	Activity	Led by
9.30	**Refreshments**	
9.45	**Introduction**	
	Overview of the workshop format and process	
	Link into the business planning process	
	what is expected of participants	
	Review of the Criteria	
10.00	*Policy and strategy*	
11.00	*Processes*	
12.00	*Key performance results*	
12.30	**Lunch**	
1.00	*Leadership*	
2.00	*People*	
3.00	**Refreshments**	
3.15	*People results*	
3.45	**Review of Action Points from the Day**	
	Review of process for today	
4.00	**Close**	

Table 9.22 *EFQM Self-Assessment Workshop 2 Agenda (day two)*

Time	Activity	Led by
9.30	**Refreshments**	
9.45	**Introduction**	
	Resumé of previous day	
	Plan of events for today	
	Review of the Criteria	
10.00	*Partnership and resources*	
11.00	*Customer results*	
11.30	*Society results*	
12.00	**Lunch**	
	Action planning	
12.45	*Key actions grouped by criteria*	
	Prioritisation of actions	
	Actions allocated and owned	
2.30	**Refreshments**	
3.00	**Close**	

Table 9.23 (one sheet only is presented out of eight produced in the case study) shows actions and related criteria. After this has been done and copies printed off, the EFQM team of twelve staff were split into three groups of four and each group asked to take specific aspects of the actions from Table 9.23. The groupings were:

Group 1
• Key performance indicators
• Management information systems
• Processes
• Customers

Group 2
• Strategy
• Partnerships
• Resources
• Customers

Group 3
• Leadership
• People and Values
• Communication

During the session 12.45–2.30, each group was asked to draft actions by grouping similar data under their allocated headings. At the end of this session each group presented their action points. Figure 9.12, box 11 moving to box 12, establishes where the action points were written up and if necessary (as was the case) grouped. This was done in order to produce a coherent set of actions for discussion and prioritisation at Workshop 3.

On completion of this task the team leader wrote up in a coherent fashion the action points ready for Workshop 3. Table 9.24 is the proforma used for this process. Table 9.24 contains one example for KPIs; the other ten action areas were also dealt with in the same manner.

Once this process had been completed by the team leader the set of actions was disseminated to the whole team one week before Workshop 3. The team was asked to read and identify its scores relating to each action for Impact, Difficulty, Resource and Timescale. For example, for KPIs action Impact rated

Table 9.23 *Summary sheet from data collected under headings*

All Actions		one of...
1a	Development of a 'Best Practice Guide' covering aspects such as leadership style, communication, motivation and delegation	Leadership/People
1a	Dissemination of current best practice that is evident in some areas of the company	Communication
1a	Clear and regular communication and become a people focused organisation	Communication
1a	Move towards the characteristics of a learning organisation	Leadership/People
1a	Develop an environment in which change is accepted and getting staff to realise we are a business	People
1a	Review the organisational structure of the company	Structure

Table 9.24 *Key action points – Workshop 3*

Key action areas	Impact H/M/L*	Difficulty H/M/L*	Resource H/M/L*	Timescale S/M/L†	Ranking
KPIs					
1. Identify appropriate, relevant and balanced set of KPIs to assess current performance, linked to core processes, and disseminate results to inform future actions (feed forward and trend analysis) To include:					
• Customers/clients					
• Internal and external stakeholders					
• Staff					
• Community					

* H = high, M = medium, L = low.
† S = short, M = medium, L = long.

medium; Difficulty medium; Resources low; Timescale short-term; and ranked (out of the eleven action areas) sixth (see Table 9.25).

At Workshop 3 a full and frank discussion was held and the resulting outcome was to agree the team's view regarding the scores for the column headings and rankings. Also a member of senior staff was identified to champion the action point(s) (at **13** on Figure 9.12). As indicated on Figure 9.12 (at **14**) the next stage was to present the action points to the senior management team and the director. This was done and the agreed action points ranked in priority order were then deployed, commencing with a full explanation to all staff (at **18** on Figure 9.12).

The company is currently at the deployment stage and will then move to Figure 9.12 (at **19**) to employ the Plan, Do, Check and Action control cycle. This will require reworking the whole cycle to test the impact of the deployed actions. During this second round the full model should be used, incorporating the scoring system. Only the trained team leader and/or external consultants should engage in scoring out of the 1000 points. It is very important not to forget that the model is based on continuous improvement.

The methodology has worked very well and the team feels that real progress has been made. The model has certainly enabled senior staff to focus on the key issues facing the company.

The organisation feels confident that the advocated advantages noted by the EFQM are attainable in practice. Once the company has completed the first full cycle of deployment it is fully intending to commence the second cycle (see Figure 5.16).

Table 9.25 EFQM case study results matrix

Criteria ranking based on lowest mean scores	Mean	Impact frequency			Difficulty frequency			Resource frequency			Timescale frequency		
		H	M	L	H	M	L	H	M	L	S	M	L
1. Strategy	1.7	18	4		7	11	4	5	9	8	14	7	1
2. Leadership	2.36	13	2		8	7		4	8	3	9	6	
3. People and Values	3.54	13	10	1	6	14	4	5	15	4	15	9	
4. MIS	4	8	1		8	1		2	6	1	8	1	
5. Processes	4.4	9	7		5	9	2	1	10	5	7	8	1
6. Communication	4.9	6	7		5	6	2	3	6	4	4	8	2
7. KPIs	5.7	6	3		3	5	1	2	5	2	2	5	2
8. Customers	6.5	4	5		7	2		7	2		2	7	
9. Resources	7.2	5	4		4	2	3	1	6	2	4	4	1
10. Partnership	7.6	4	14	2	5	9	6	3	7	10	6	8	6
11. Society	9.6	1	6	7	3	5	6	3	3	8	1	5	8

Case Study – EFQM Business Excellence Model Implementation

Introduction

Many companies in recent years have enthusiastically set about implementing the EFQM Business Excellence Model. Sometimes there are fatal flaws in how the implementation is planned, or in how it is deployed. This was the case in a traditional construction company in the UK. Two years after the start of EFQM Business Excellence Model implementation, an investigative group concluded:

- A major effort was needed to improve the fundamentals of the business. The first priority was to establish effective disciplines and controls within each organisational function of the company
- EFQM Model deployment consumes a great deal of management time and resources. Additional time and effort in further development should be deferred until the priority issues have been addressed.

This marked the end of EFQM Model development at the company. However, the background to the investigative group's conclusions are highly informative.

Some Critical Issues

Early initiatives directed at improving product quality had been very successful but the Board realised that progress had plateaued and new ideas were needed to propel the company into the 'world class' league. It was decided that a study team of senior managers would be set up to examine the company's approach to 'putting quality first', and to report in three months' time. The study group concluded that there was no magic solution to improving quality. Although previous one-off initiatives had yielded some benefits, several examples were quoted where an inconsistent and uncertain management approach still prevailed:

- while management preached quality, increased productivity was always the first priority
- while management emphasised 'right first time', corrective actions were still prominent features of the business
- new tools and techniques, like Statistical Process Control (SPC), had been introduced in a disparate way – some areas used SPC but did not know how to use the data collected for improvement purposes.

The study group identified over 20 specific, quality-related issues which needed to be addressed as a matter of urgency. These included the need to:

- review the training policy and procedures
- review the role of the supervisor
- improve discipline and control on working practices
- reduce the number of suppliers, some of whom had a poor quality service record
- improve the control of design changes
- obtain more regular and valid client feedback to inform the decision-making process.

The study group went on to stress the need to commit everyone in the company by means of a single management quality philosophy which encompassed the elimination of waste and an integrated human resource plan. In order to bring about progress on this point, it was recommended that the Board should seek advice. This culminated in the appointment of consultants to help facilitate the initial stages of EFQM Business Excellence Model implementation.

The time taken from presentation of the internal study group report (which the Board had accepted in its entirety) to the decision to go ahead with EFQM Business Model deployment had been less than three months. It is important to note some of the key issues:

- The Managing Director was totally convinced that acceptance of the EFQM Business Excellence Model was the only way forward. He became very enthusiastic. When it was stressed that adoption of some EFQM Model ideas and perceived benefits could take four years to bear fruit, he replied '*We will do it in half the time*'.

- While some Board members shared the Chairman's enthusiasm, others were totally unconvinced about EFQM Model implementation.
- The need to address the list of over 20 quality-related issues referred to above as a matter of urgency was quickly forgotten in the rush to progress with EFQM Model application.

The company, therefore, had already embarked on the road to failure.

The new system mirrored the existing organisational structure so that the transition from process improvement being a 'special' activity to becoming a normal way of life was managed by the same group of people. The following is a brief summary of the structure:

- The steering team was the Board and its task was to lead the transition by supporting the work of other groups
- Local steering committees were set up to manage the improvement activities in each function and each department. They were chaired by a member of the steering team
- The statistical methods office provided technical support in the areas of training, behavioural science and statistics, and worked on the structure and systems of the change process
- Process improvement leaders were statistical facilitators who helped members of project teams and local steering committees
- Culture change teams identified areas where there was conflict with existing management culture. The teams' recommendations were championed by a Board member for implementation.

While this was the theory of the new Quality Management System, the reality was somewhat different. The lukewarm commitment of some Board members meant that some of the steering committees floundered and that some of the culture change teams never got off the ground.

Training and Education

This was carried out as a 'cascade' exercise, with consultants and the Statistical Methods Office training the Board, who in turn trained staff who were below them in the hierarchy. The method was a huge success. But training in philosophy was not followed up with training in tools and techniques for some months. Philosophy training emphasised attitude changes, like driving out fear and treating everyone as if they want to do a good job. But no new systems were in place to support the new philosophy. Also, without tools and techniques, nothing could be done with philosophy training alone.

Quality Planning and Focus

This was intended to cover priorities and measures of quality improvement, integration of quality initiatives like SPC and improvement of management processes like reward systems and appraisal. This whole area proved to be a particular problem. Because existing processes were in many cases not under adequate control, there was little that could be done to improve them. Major changes in systems were required. Appraisal systems were addressed by replacing management by objectives and target numbers with a less specific 'process improvement plan' which managers never properly understood and issues of work measurement were never resolved.

Management Culture and Style

This covered corporate and functional missions, values and goals, the interpretation of Deming's principles and work on the recommendations of the culture change teams. While the corporate missions, values and goals were well handled, functional efforts depended very much on the attitude of the Board members concerned. Interpretation of Deming's principles was never undertaken. Failure of the culture change teams referred to above led to lack of progress in defining operating philosophy and in reviewing company policies.

Communication and Recognition

This was handled by means of conferences, management briefs, company newsletters and handouts to explain the new philosophies. A conscious effort was made not to overplay the publicity, but to ensure that people understood what was happening and why.

Question for the Reader

Comment on the Company's approach to the implementational process of the EFQM Business Excellence Model and on what lessons can be learned from their experience. Key points are summarised below for the reader's reference, after this question has been attempted.

Question – Main Points

- The company failed to fully appreciate that organisational change processes consume a vast quantity of 'organisational energy'. Therefore there is an associated opportunity cost to be paid.
- Cultural change is a very problematic activity and is best managed via an incremental change process. For the Managing Director to state that the company could obtain the advocated benefits 'in half the time' indicated by more knowledgeable consultants demonstrated a lack of understanding on this critical issue.
- The company did not really move towards a 'morphogenic' culture in line with a truly post-modernist approach to organisational development. Therefore what appeared to be fundamental changes taking place in the organisation were, in fact, only superficial. They had not become part of the company's cultural dynamic and accepted work practices.
- Organisational structure should follow strategy, yet the company persisted with the same structure. This was a formidable barrier to EFQM Business Excellence Model deployment.
- The concept of a truly quality focused organisation was not established. This is because the fundamental rationale for EFQM Business Excellence Model implementation had not been fully understood by senior management. Thus, senior management could not articulate this rationale to all organisational employees, clients and suppliers, and this accounts for the 'inconsistent and uncertain management approach'.
- The company still focused on increased productivity and short-term gain. Thus the realisation that EFQM Business Excellence Model implementation was a long-term strategic development and not a short-term tactical approach was not understood or resourced.
- SPC and other tools had been introduced without a coherent strategy. The 'disparate' application of SPC and other analytical tools only served to confuse staff. These tools are valuable in conducting trend analysis and monitoring outputs but staff need to be fully trained in their use. If staff are not fully conversant with their application, incorrect decisions could result.

Lessons to be Learned

Senior managers should develop an implementational strategy. This strategy (plan) needs to be communicated and explained to all stakeholders, both internal and external. They need to remember that consultation should precede implementation. They failed to realise that consultation has two vital components. First, it is an information-gathering process to inform decision making. Secondly, it has a psychological component: people who are consulted and allowed to contribute are more likely to support the initiative as they feel part of the process. Remember, it is vital to avoid 'coalitions of resistance' in any 'change process'.

Senior management should have considered an incremental approach to the application process: taking one part of the company first and proving the advantages of EFQM Model application and then disseminating 'best practice' to other departments/sections. Having communicated the rationale and developed a plan for application, senior managers needed to ensure that all necessary resources and training programmes had been allocated/set up. These need to be monitored to ensure that they are adequate for model implementation. The three key aspects for successful implementation are:

- resources
- systems
- people.

These need to be considered within a holistic framework; they are not mutually exclusive.

Senior managers have to be seen to lead from the front on this issue. They must demonstrate a consistant and supportive attitude throughout model application and operation.

Change Management

One thing that is obvious from reading the above is that a move to becoming a TQM company will involve change at all levels of the organisation. Change, however, will not happen effectively by itself, it needs to be managed. Managing change means taking control of the process and shaping the direction that the change will take. The management of change is a four-stage process, as advocated by Baden (1993):

- Establishing the need for a change
- Gaining and sustaining commitment
- Implementation
- Review.

One important aspect in the effective management of change is that it must involve all members of the organisation. Even when change is managed barriers still exist and include cost, lack of time, employee perceptions, industry culture and lack of ability. Recognition and understanding of these barriers to change are an important step in overcoming them.

Kaizen

One item which needs to be noted in relation to TQM is the Japanese concept of kaizen. It is probably the single most important concept in Japanese management. The message of kaizen is that not a day should go by without some kind of improvement being made somewhere in the company. In Japan, management is perceived as having two functions: maintenance and improvement. Kaizen signified small improvements made in the status quo as a result of on-going efforts. It is different from TQM in that it operates within existing cultures and rarely requires cultural shifts.

The Case Study company would have benefited greatly from the two models presented in this chapter (Figures 9.9 and 9.10) as they provide a generic strategy which addresses the problematic issues they encountered. Thus the models would have helped them (and future construction firms) avoid the pitfalls of applying the EFQM Business Excellence Model.

Summary

Both the Latham Report (1994) and the Egan Report (1998) have identified that the UK construction industry has a problem in supplying/maintaining a quality product/service to its clientele. Latham (1994) advocated the utilisation of TQM as a means of addressing the industry's ills. However, a 'big bang' approach to the deployment of TQM may not be suitable for many construction firms. An incremental implementational strategy with its associated reduction of possible culture shock may be a more realistic approach. Within this chapter a strategy incorporating IiP and BS EN ISO 9001: 2000, leading to the EFQM Business Excellence Model application, has been adopted. Both IiP and BS EN ISO 9001: 2000 can be encapsulated within the EFQM Business Excellence Model. It is for the host construction company to conduct a 'strategic analysis' and then take a decision based on such an analysis relating to which approach should be adopted. This is necessary because each approach places different demands on an organisation's resources. There is no doubt that maximum benefits can only be attained from the full application of the EFQM Business Excellence Model. After all, the Model is customer focused and a crucial issue raised by Latham (1994) and Egan (1998) was a need to fully address clients' requirements in an efficient and effective manner. Readers are recommended to address again the Case Study exercise on the deployment of the EFQM Business Excellence Model. It will allow them to appreciate many of the critical issues incorporated within this chapter.

References

Baden H.R. (1993) Total Quality in Construction Projects. *Achieving Profitability with Customer Satisfaction*. Thomas Telford, London.

Bounds G., Yorks L., Adams M. and Ranney G. (1994) *Beyond Total Quality Management: Towards the Emerging Paradigm*. McGraw-Hill International Editions.

British Quality Foundation (1998) Self Assessment Techniques for Business Excellence. *Identifying Business Opportunities*. BQF, London.

British Standards Institution (1994) *BSEN ISO 9001*. BSI, Milton Keynes.

British Standards Institution (2000a) *BSEN ISO 9000: 2000 Quality Management Systems – Fundamental and Vocabulary*. BSI, Milton Keynes.

British Standards Institution (2000b) *BSEN ISO 9001: 2000 Quality Management Systems – Requirements*. BSI, Milton Keynes.

Druker J. and White G. (1996) *Managing People in Construction*. Cromwell Press, Wiltshire, p 16, ISBN 0–85292–642–1.

Egan J. (1998) Rethinking Construction, *Report of the Construction Task Force to the Deputy Prime Minister, John Prescott, on the Scope for Improving the Quality and Efficiency of UK Construction*. Department of the Environment, Transport and the Regions, London, ISBN 1–85112–094–7.

European Foundation for Quality Management (1999) *Radar and the EFQM Excellence Model*. EFQM Press Releases & Announcements [on line]. Last accessed on 12 June 2000 at URL: www.efqm.org/

European Foundation for Quality Management (2000a) *History of the EFQM* [on line]. Last accessed on 12 June 2000 at URL: www.efqm.org/history.htm

European Foundation for Quality Management (2000b) *EFQM and Self-Assessment* [on line]. Last accessed on 12 June 2000 at URL: www.efqm.org/selfas.htm

Hellard R.B. (1995) *Project Partnering: Principle and Practice*. Thomas Telford, London, ISBN 0–7277–2043–0.

Hillebrandt P.M. (1984) *Analysis of the UK British Construction Industry*. Macmillan (now Palgrave), Basingstoke.

Hillman G.P. (1994) Making Self-assessment Success, *Total Quality Management*, 6(3), 29–31.

Investors in People UK (2000) *Investors in People Standard*. London.

Jones C.R. (1994) Improving Your Key Business Process, *Total Quality Management*, 6(2), 25–9.

Jones T. (1984) *Structure and Performance of Industries*. Philip Allan, ISBN 0860036367.

Latham M., Sir (1994) *Constructing the Team – Final Report of the Government/Industry Review of Procurement and Contractual Arrangements in the UK Construction Industry*. HMSO, London, 0–11–752944–X.

Love P.E.D. and Li H. (2000) Overcoming the Problems Associated with Quality Certification, *Construction Management and Economics*, 18, 139–49.

Low S.P. and Omar H.F. (1997) The Effective Maintenance of Quality Management Systems in the Construction Industry, *International Journal of Quality & Reliability Management*, 14(8), 768–90.

McCabe S. (1998) *Quality Improvement Techniques in Construction*. Addison Wesley Longman, Harlow, Essex and Chartered Institute of Building, Ascot, Berkshire, ISBN 0–582–30776–7.

McGeorge D. and Palmer A. (1997) *Construction Management – New Directions*. Blackwell Science, Oxford, ISBN 0–632–04258–3.

Moullin M. (1994) Redefining Quality, *Proceedings of the 38th EOQ Annual Congress*, Lisbon. European Organisation for Quality.

Munro-Faure L., Munro-Faure M. and Bones E. (1995) *Achieving the New International Quality Standards: A Step-by-Step Guide to BS EN ISO 9000*. Pitman, London, ISBN 0273–61977–2.

Oakland J.S. (1993) *Total Quality Management: The Route to Improving Performance*. Butterworth Heinemann, London.

Sheldon L., Huang G.Q. and Perks R. (1991) Design for Cost – Situation Report, *Proceedings of the International Conference on Engineering Design*, ICED 91, Zurich, 17–29 August, pp 1509–20.

van der Wiele A., Dale B.D. and Williams A.R.T. (1997) ISO 9000 Series Registration to Total Quality Management: the Transformation Journey, *International Journal of Quality Science*, 2(4, June), 236–52.

Watson P. (2000) Applying the European Foundation for Quality Management (EFQM) Model, *Journal of the Association of Building Engineers*, 75(4, April), 18–20.

Watson P. and Heath P. (1998) Implementation of Investors in People Standard in Construction, *Construction Paper No. 97*, Chartered Institute of Building.

Watson P. and Seng L.T. (2001) Implementing the European Foundation for Quality Management Model in Construction, *Construction Information Quarterly*, 3(2), Paper No. 23.

10 Health and Safety

Introduction

The construction industry continues to be one of the most hazardous industries within which to work. Each year, a considerable number of construction workers on site are injured, many seriously and some fatally, as a result of their work and the work of others. Over the last twenty-five years the construction industry has suffered a poor health and safety record. Although the rate of accidents declined during the 1990s, the industry still has a long way to go to ensure that the health, safety and welfare of its workers are constantly safeguarded. Everyone involved with a construction project has a responsibility for health and safety, none more so than site management. It is the duty of managers and supervisors to ensure that working conditions on site are safe and healthy so that operatives, other project participants and persons near to or visiting the site are not placed at risk.

Effective health and safety provision concentrates on ensuring that the project site, its environs and the works themselves are carefully planned, organised, monitored and controlled. The purpose of this chapter is to outline key health and safety law as it affects construction and explain the responsibilities on the principal contractor under health and safety regulations. It then examines the management of health and safety through the implementation of a health and safety management system and plan for the siteworks to comply with occupational health and safety management standard BSI-OHSAS 18001.

European Legislation and UK Law

As a member of the European Union (EU), the UK has to comply with EU legislation. EU legislation becomes UK law through Acts of Parliament and Enabling Acts which in turn give rise to national regulations. European Directives therefore form the basis for UK law within which regulations, including those which influence construction, are created.

EU legislation takes four forms:

(i) Directives – these set minimum standards for legislation at EU level with the modes of implementation at the discretion of the national governments of member countries.
(ii) Regulations – these apply directly to member countries and automatically become part of national legislation.
(iii) Decisions – these are binding deliberations of the European Court of Justice.
(iv) Recommendations – these state the opinions of institutions or bodies and are advisory on matters in question.

Within EU health and safety legislation, Article 118A of the Treaty of Rome (1957) requires that member countries *"pay particular attention to encouraging improvements, especially in the working environment, as regards the health and safety of workers."* As a result of this, EC Directive 89/391/EEC (European Commission, EC, 1989), commonly referred to as The Framework Directive because it provides a framework for other Directives, was born. Within the UK, this Directive was enacted through The Management of Health and Safety at Work Regulations 1992, revised in 1999 (HSE, 1992a).

Health and Safety Law

The Management of Health and Safety at Work Regulations 1992 (HSE, 1992a) [revised 1999] (MHSWR), sometimes denoted as the 'six pack', enacted a set of six EC Directives implementing national health and safety law at the workplace. These Directives are as follows:

- *Directive 89/654/EEC*: the 'workplace Directive' – health and safety requirements for the workplace
- *Directive 89/665/EEC*: the 'use of work equipment Directive' – health and safety requirements for the use of equipment at work
- *Directive 89/656/EEC*: the 'personal protective equipment Directive' – health and safety requirements for use by workers of personal protective equipment at the workplace
- *Directive 90/269/EEC*: the 'manual handling of loads Directive' – health and safety requirements for the manual handling of loads where there is a risk of injury
- *Directive 90/270/EEC*: the 'display screen equipment Directive' – health and safety requirements for work with display screen equipment.

These Directives have been implemented by the following Statutory Instruments within the UK:

- *The Workplace (Health, Safety and Welfare) Regulations 1992* (HSE, 1992b)
- *The Provision and Use of Work Equipment Regulations 1992* (HSE, 1992c)
- *The Personal Protective Equipment at Work Regulations 1992* (HSE, 1992d)
- *The Manual Handling Operations Regulations 1992* (HSE, 1992e)
- *The Health and Safety (Display Screen Equipment) Regulations 1992* (HSE, 1992f).

The sixth Directive within the Framework Directive is *Directive 92/57/EEC*: the 'construction sites Directive'. This addresses the health and safety requirements for temporary or mobile construction sites. This is implemented in the UK by *The Construction (Design and Management) Regulations 1994*, otherwise known as 'The CDM Regulations' (Figure 10.1).

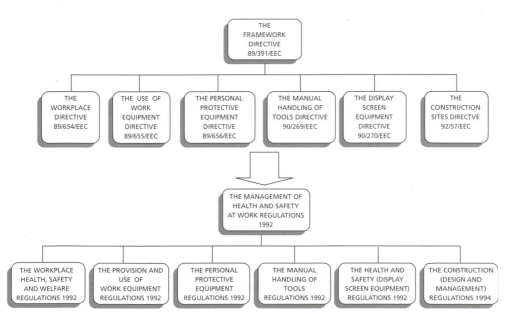

Figure 10.1 *European Directives adopted under Article 118A of the Treaty of Rome leading to UK legislation within The Management of Health and Safety at Work Regulations 1992 (revised 1999) [adapted from Griffith et al., 2000]*

Principal Health and Safety Laws Influencing the Construction Industry

All sectors of industry are regulated by health and safety law. Within construction there are key pieces of legislation – Government Acts of Parliament and Regulations – which are central to the management of 'health and safety' during all construction works, the prominent ones being:

- The Health and Safety at Work, etc. Act 1974
- The Management of Health and Safety at Work Regulations 1992, revised 1999
- The Construction (Design and Management) Regulations 1994.

In addition, there are a number of prominent regulations concerned with the 'welfare' of persons involved with construction. These are:

- The Construction (Health, Safety and Welfare) Regulations 1996
- The Personal Protective Equipment at Work Regulations 1992
- The Reporting of Injuries, Diseases and Dangerous Occurrences Regulations 1995
- The Control of Substances Hazardous to Health (Amendment) Regulations 1999.

The regulations require that health, safety and welfare aspects are effectively planned, managed and controlled. The regulations listed are the most significant pieces of legislation aimed at meeting that endeavour. They specify the duties to be performed by employers, employees, project participants and all who have an involvement with the activities that take place throughout the life of a construction project. There are, in addition, many individual regulations which have jurisdiction over the construction industry and its many specific activities.

The Health and Safety at Work, etc. Act 1974

The Health and Safety at Work, etc. Act 1974, or HSWA (HSE, 1974), applies to all work activities. The Act requires that an employer ensures the health and safety of its employees, as far as reasonably practicable*, together with that of other persons at work and those who might be affected by their work or put at risk by their work.
    The Act imposes duties on the employer and other parties as follows

- on employers towards employees
- on employers and the self-employed towards persons other than their employees
- on people in control of premises
- on people who design, manufacture, supply and install plant, equipment and substances
- on every employee
- on everybody.

All employers must take steps to ensure that their employees and others are aware of and fully understand matters relating to health and safety. Employers should have a health and safety policy and where they employ five or more persons then that policy must be stated in writing. It is essential that employees can see health and safety matters in the context of the organisation for which they work and can appreciate their contribution to the fulfilment of that policy.
    Self-employed persons are required to take steps to protect both themselves and other workers. They must ensure that risks are minimised and that their work does not place others at risk.
    Employees have a duty to contribute towards health and safety practice and to co-operate with their employer. They are required not to do anything on site that places themselves at risk or exposes others to risk. There is a requirement on employers to ensure that employees are aware of site dangers and are trained and instructed in health and safety matters in the course of carrying out their tasks.

---

* Reasonably practicable – a duty carried out having considered the balance of that duty against inconvenience and cost involved.

The HSWA is an important basic piece of legislation affecting all industries. Breaches of the Act can result, depending on circumstance, in civil and criminal liability, punishable by fines or imprisonment. Enforcement of the Regulations is undertaken by HSE inspectors and by local authority inspectors who have wide-ranging powers from simply providing advice to serving prohibition notices and even commencing proceedings to prosecute where health and safety infringements are warranted as serious.

The specific duties on parties are presented in designated sections of the Act. The reader is therefore directed to the Act, together with other guidance presented in the Reference section.

The Management of Health and Safety at Work Regulations 1999

The Management of Health and Safety at Work Regulations 1999, or MHSWR (HSE, 1999a), replaced the original MHSWR 1992. The Regulations apply to everyone at work including those within the construction industry. They require employers to plan, organise, monitor, control and review their work in the context of health and safety provision.

Thirty regulations make up MHSWR. At the heart of the regulations lies the core theme of risk assessment and the development of prevention mechanisms to mitigate the likelihood of accidents or injuries to workers. Underpinning risk prevention is the requirement for the employer to provide effective health and safety training and instruction.

In satisfying the requirements to develop health and safety policy under the HSWA, and plan, organise, monitor, control and review health and safety provision under MHSWR, many contracting organisations within construction implement formal health and safety management systems (H&SMS) within both company, or corporate, management and project site management. Such an approach is described subsequently.

The Construction (Design and Management) Regulations 1994

The Construction (Design and Management) Regulations 1994, or CDM (HSE, 1994), are without doubt the most significant piece of legislation influencing construction works. CDM require that health and safety is considered throughout all stages of a construction project, from conceptual design and planning through to production on site and subsequent maintenance and repair of the completed product.

CDM affect everyone who is involved with the construction process – the client, consultants and contractors. CDM introduced two new roles, that of the planning supervisor – appointed by the client to co-ordinate health and safety during the design and planning phase of the project – and that of the principal contractor – appointed by the client to plan, manage and control health and safety as part of the construction phase of the project.

The Regulations recognise that a comprehensive health and safety plan is developed and implemented for the construction project and that a health and safety file is maintained. CDM require that everyone contributes to improving health and safety on a project and as such they become a shared management commitment.

CDM apply to most building and civil engineering works and include design work for construction purposes including demolition and dismantling works. The Regulations do not apply to construction work where the local authority is the enforcing authority for health and safety nor where works last 30 days or less and involve fewer than five persons on site at one time.

Planning lies at the core of CDM implementation. The Regulations promote a two-phase approach with the objective of developing and implementing a health and safety plan for and throughout the construction project. The first phase – the pre-tender health and safety plan – is the responsibility of the client and a principal duty of the planning supervisor. The second phase, which rolls on from the pre-tender plan – the construction health and safety plan – is the responsibility and duty of the principal contractor. Through its two phases, the health and safety plan provides the basis for considering health and safety aspects during the tendering process and the basis for considering health and safety management provision for the construction siteworks. Information gathered during both phases of planning, together with documented records compiled throughout the total construction process for the project, is collated to form the health and safety file – a project record of health and safety information and experiences retained by the client at the end of the project.

The pre-tender health and safety plan must be available to potential principal contractors at the start of the project tendering or selection process. The plan informs tenderers of the health, safety and welfare aspects that they need to appreciate when planning and pricing for the siteworks. The plan contains information which is project specific and which enables the tendering contractor to look forward and plan for the safe systems of work that it will implement during the works on site.

In meeting this requirement, the pre-tender health and safety plan will probably include:

- general description of the works
- timescales for completion
- known health and safety risks
- information needed by contractors such that they can demonstrate that they are competent and have the necessary resources to carry out the works
- information on which the principal contractor can develop the construction health and safety plan.

Planning for the construction phase involves the principal contractor in developing a health and safety plan which considers those aspects which impinge on the site-management of health, safety and welfare as the project is undertaken. The construction health and safety plan will therefore probably include:

- arrangements for ensuring health and safety of all persons who may be affected by the project
- arrangements for the management of health and safety of the project and for monitoring compliance with health, safety and welfare law
- arrangements for ensuring that the welfare requirements of the project will be met.

CDM place responsibilities and duties on designated parties to a project. The principal contractor is appointed by the client to manage health and safety, that is, to plan, monitor and control all aspects when engaging the siteworks.

Under the CDM Regulations the principal contractor's key duties are to:

- develop and implement the construction health and safety plan
- arrange for competent and adequately resourced contractors to carry out works as required
- ensure co-ordination of and co-operation by contractors
- provide information to contractors on project risks
- ascertain from contractors how they propose to carry out high-risk operations
- ensure that contractors comply with any site rules
- monitor health and safety performance throughout the project
- ensure that workers are aware of health and safety issues, and receive induction, training and instruction
- allow only authorised persons on site
- display the notification particulars required by HSE
- provide information to the planning supervisor for incorporation in the project health and safety file.

As with HSWA and MSHWR, infringement of the CDM Regulations becomes a matter for HSE and local authority inspectors to investigate with the potential to prosecute if serious offences are manifest.

An appropriate way forward for the principal contractor is the implementation of a health and safety management system (H&SMS) within which the planning function, demanded by CDM, becomes a key element.

The Construction (Health, Safety and Welfare) Regulations 1996

The Construction (Health, Safety and Welfare) Regulations 1996, or CHSWR (HSE, 1996), replaced the earlier provisions of the Construction (General Provisions) Regulations 1961, The Construction (Health and Welfare) Regulations 1966 and The Construction (Working Places) Regulations 1966 (HSE, 1961; 1966a; 1966b). The CHSWR apply to all construction work and although rigorous are less prescriptive in nature and application, allowing the principal contractor greater freedom in determining working methods and management practices.

The Regulations comprise 35 regulations and 10 appended schedules and apply to those who control the undertaking of construction works. CHSWR cover a wide range of health, safety and welfare provision applicable to almost all construction sites although there are situations, for example, DIY work, where exemptions apply. The Regulations place onerous responsibilities on employers, self-employed persons and employees to uphold their duty to co-operate in the interests of health, safety and welfare and to invoke safe working practices.

CHSWR encompass provisions applicable to the organisation and management of the site in addition to safe working practices. For example, they cover operation of site vehicles, traffic routes, signage and site lighting. They also embrace the provision of welfare facilities including temporary accommodation and services. The Regulations therefore are key to planning, organising and managing the construction site.

The Personal Protective Equipment at Work Regulations 1992

The Personal Protective Equipment at Work Regulations 1992, or PPE (HSE, 1992d), impose health and safety requirements on workers using personal protective equipment at the workplace. The Regulations place a duty on employers to ensure that where their employees could be exposed to health and safety risk then they are provided with appropriate protective equipment. Self-employed persons are duty bound to meet the provisions and provide themselves with protective equipment.

The PPE Regulations encompass all equipment and clothing intended to be worn or used at work which affords adequate protection against risks to the person's health or safety. By definition, personal protective equipment protects only the person wearing it or using it and gives no protection to others. PPE Regulations augment other legislation and the provision of protective equipment does not permit the people supplied with it to be any less vigilant. In fact it should be seen as a last resort in managing health, safety and welfare risk at the workplace.

Personal protective equipment – which includes, for example, helmets, footwear, eye protection, ear defenders and respirators – must be of a suitable standard to meet the PPE Regulations. As such it should carry a 'CE mark' of conformity to quality and safety standards.

The Reporting of Injuries, Diseases and Dangerous Occurrences Regulations 1995

The Reporting of Injuries, Diseases and Dangerous Occurrences Regulations 1995, or RIDDOR (HSE, 1995a), require that specific accidents occurring on construction sites are reported to the HSE. Such occurrences include: fatal and serious accidents; accidents where the person(s) concerned are unfit for three consecutive days; a dangerous occurrence where person(s) are placed at risk; and specified diseases associated with a person's job of work.

Under RIDDOR, it is the principal contractor's responsibility to inform the HSE of the occurrence of a reportable incident. The HSE provides companies with a set of report forms for this. Forms must be completed and submitted to HSE within specified time periods according to the nature of the occurrence. For example, a serious incident must be reported immediately by telephone followed by written confirmation within ten days. Employers are duty bound to maintain adequate records for RIDDOR. In addition, there is an obligation to maintain an auditable accident book under the Social Security (Claims as Payments) Regulations 1979.

Any accident, dangerous occurrence or disease should be investigated and a mechanism within the principal contractor's H&SMS established to facilitate this. The HSE is empowered by law to investigate cases and can request the principal contractor to provide documented evidence to assist that undertaking.

The Control of Substances Hazardous to Health (Amendment) Regulations 1999

The Control of Substances Hazardous to Health (Amendment) Regulations 1999, or COSHH (HSE, 1999b), replaced the 1988 and 1994 versions of the same regulations. The Regulations impose duties on the employer to:

- assess the risk of possible incidents where employees might be exposed to hazardous substances
- prevent, or control, such exposure
- monitor exposure
- ensure that information and training are provided to employees at risk from potential exposure.

Substances hazardous to health and safety are defined as:

- substances listed within The Chemicals (Hazard Information and Packaging) Regulations 1993, as: very toxic; toxic; harmful; corrosive; or irritant
- substances to which a maximum exposure limit is specified by HSE within EH 40: Occupational Exposure Limits (HSE, 1999c)
- biological agents capable of causing allergy, toxicity, infection or other hazard
- dust of any kind
- substances, other than those listed above, which create a health hazard to any person(s).

Health and Safety Management for Construction

Effective health and safety management is founded on the provision of a safe and healthy working environment with safe systems of work at its core. The key to success is to ensure that health and safety aspects are carefully considered – planned, organised, monitored, controlled and reviewed – and that the risk of danger and hazard to persons, as a result of site activities, is systematically safeguarded.

To address these requirements, many principal contractors have chosen to establish a formal health and safety management system (H&SMS) within the company and to implement project-specific health and safety plans (H & S plan). Through the adoption of formal management mechanisms, the principal contractor will be best equipped to encourage a positive health and safety following throughout its corporate staff and site workforce. The organisation will be able to develop the ethos, policies, plans, management procedures and safe working practices which are essential to delivering effective health and safety management at the project site. Equally important, a systems approach will add value to the organisation's culture and structure, enhancing its capabilities to maintain its core business and generate new business in the future. As with many aspects of construction management, health and safety management has become a holistic organisational consideration. It is now an integral part of an organisation's framework, structure and operations.

Health, safety and welfare legislation has a significant influence on the development and implementation of a H&SMS. As explained earlier in this chapter, particular Acts of Parliament and Regulations require that specific requirements are met. The HSWA 1974 require that employers develop clear health and safety policies. The MHSWR 1999 require that risk assessment is considered. The CDM Regulations 1994 demand pre-tender and construction-phase health and safety planning.

The CHSWR 1996, RIDDOR 1995 and COSHH 1999 impose duties on employers to maintain safe and welfare-conscious working and supervisory practices. All of these regulations, and others, place onerous responsibilities on the principal contractor to devise, implement and maintain an effective company H&SMS which extends its corporate culture into good health and safety practices on the construction projects that it undertakes.

In meeting the challenge to provide effective health and safety management, the principal contractor should consider two key dimensions to H&SMS development. These are:

(1) The Company H&SMS – the company, or corporate, system which devises policy, organisation and structure for health and safety implementation throughout the company and its business.
(2) The Project H & S Plan – the project H & S plan is a project-specific plan which implements the corporate H&SMS through: (i) health and safety 'planning' to meet the requirements of the CDM Regulations (i.e. the construction-phase H & S plan); and (ii) the implementation of 'procedures' to ensure safe systems of work during the siteworks.

The Company Health and Safety Management System (H&SMS)

Within the scope of the CDM Regulations the principal contractor has a duty and a responsibility not only to develop and implement a construction-phase health and safety plan but to establish and maintain effective H & S management. This must provide safe working conditions and the use of safe working practices by management and workforce throughout the siteworks.

The most apposite way for a principal contractor to address the onerous demands for health and safety and meet the requirements set by health, safety and welfare legislation is to develop, implement and maintain a structured and formal, or documented, H&SMS. An appropriate system may be dedicated to H & S or be part of other company and project organisational systems, for example, a quality management system (QMS). The standard for occupational H&SMS in the UK is BSI-OHSAS 18001. The standard provides a specification for H&SMS development to meet the requirements of UK certification schemes (BSI, 1999). Through such an approach, the company is likely to perpetuate the necessary corporate culture, policies, organisational structures and management functions for effective H & S management. It will also provide an enabling framework for the establishment of project-based management procedures and safe working practices on site. Together, a well-structured corporate approach allied to sound project mechanisms will provide effective H & S management throughout the company, contributing significantly to the company's holistic business.

Most organisations adopt a formal, or documented, approach to H&SMS development. The essence of such a system is that H & S procedures are implemented and documents kept which provide evidence that what is purported to be carried out is, in fact, carried out and can be confirmed by the documented records.

The features of effective H & S management are founded in sound corporate and project-based management practice. The general principles of such practice provide a useful basis for the conceptualisation, development and implementation of a H&SMS by the principal contracting organisation.

The key elements of all management systems are reflected in a H&SMS, and these are:

- policy – company policies on H & S management
- organisation – corporate and project organisation structures for H & S management deploying duties and responsibilities to management and supervisors
- risk assessment – management approach to identifying and mitigating hazards and dangers to persons employed or involved with the company's project sites
- planning – requirement under CDM Regulations to provide a construction-phase health and safety plan which ensures that health, safety and welfare are planned, monitored and controlled throughout the siteworks
- procedures implementation – site-based management and supervisory procedures and safety rules to ensure the use of safe working practices by the workforce during the siteworks
- auditing – approach and mechanisms used to gather, report, analyse and review H & S performance on site and provide useful feedback to the corporate organisation, and also to the client in the form of information to the H & S file, a requirement of the CDM Regulations.

The corporate elements of the H&SMS are the establishment of the H&SM Manual. This is configured in 3 parts: 1 – Systems; 2 – Procedures; and 3 – Working Instructions for Safe Working Practice. This is accompanied by the project-based extension of these in the project H & S plan. Together these elements combine to translate the corporate H&SMS into the project H&SMS enacted through the project H & S plan (see Figures 10.2 to 10.6). Associated documentation may include: organisation charts; method statements; construction programmes; and site layout plan. These will be presented as appendices to the H & S plan where appropriate. It is absolutely essential to highlight that different contracting organisations will have their own way of approaching H & S management. Each will develop its own systems to meet its individual needs given its own circumstances. Therefore, it is inappropriate to be overly prescriptive. However, an outline, or skeletal, framework can embrace the key elements of approach to developing the hierarchy of H & S documentation which forms the H&SMS.

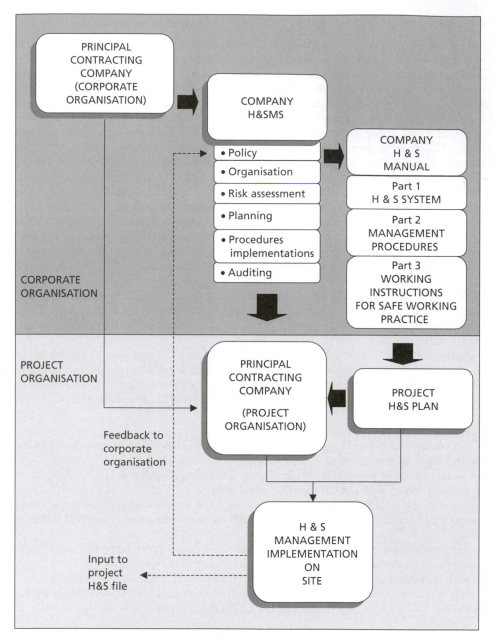

Figure 10.2 *Principal contractor's health and safety management system framework for corporate and project organisation*

Planning: The Construction-Phase Health and Safety Plan under the CDM Regulations

While all regulations impinge on the undertaking of any construction project, the CDM Regulations have a direct bearing on the project planning process. The CDM are directly concerned with planning and co-ordination of the management of project H & S. The Regulations recognise the traditional separation

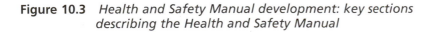

HEALTH AND SAFETY MANUAL

PART 1: HEALTH & SAFETY MANAGEMENT SYSTEM

- Frontispiece:
  - Title page
  - Contents
  - Preambles
  - Instructions for use
  - Version number
  - Circulation list

- Policy Statement:
  - Policies
  - Standards
  - Commitment
  - Duties of employer and employees
  - Relationship of manual to management procedures and working instructions

- Introduction:
  - Using the manual
  - Section descriptors and explanations
  - Management procedures
  - Working instructions
  - Project plan

- Definition of Terms:
  - Standards
  - Terms used in the documents

- System Documentation:
  - Manual
  - Procedures
  - Working instructions
  - Project plan

- Structure and Organisation:
  - Corporate structure
  - Board of Directors
  - Divisional/departmental structure
  - Management support services for H&SM
  - Company support services for H & S training
  - System management of H&SMS
  - Corporate–project management interface structures
  - Project organisation

Figure 10.3 *Health and Safety Manual development: key sections describing the Health and Safety Manual*

of the participants and the complexities, issues and demands that this can pose. The principal aim is to ensure that project H & S is consciously considered by all the project parties. Moreover, it seeks to embed H & S management principles and practices in the project from project evolution and development through to completion of the siteworks. The CDM Regulations require a two-phase approach to

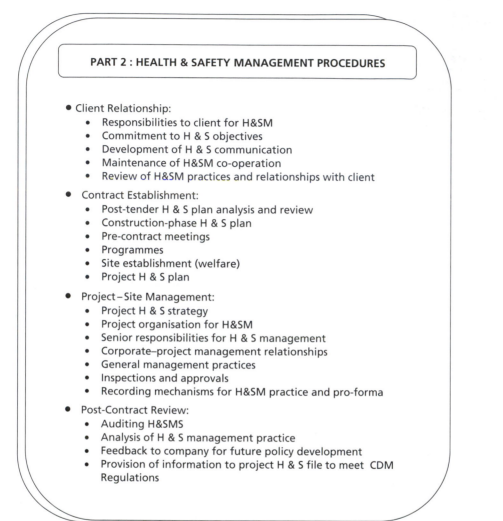

PART 2 : HEALTH & SAFETY MANAGEMENT PROCEDURES

- Client Relationship:
  - Responsibilities to client for H&SM
  - Commitment to H & S objectives
  - Development of H & S communication
  - Maintenance of H&SM co-operation
  - Review of H&SM practices and relationships with client
- Contract Establishment:
  - Post-tender H & S plan analysis and review
  - Construction-phase H & S plan
  - Pre-contract meetings
  - Programmes
  - Site establishment (welfare)
  - Project H & S plan
- Project – Site Management:
  - Project H & S strategy
  - Project organisation for H&SM
  - Senior responsibilities for H & S management
  - Corporate–project management relationships
  - General management practices
  - Inspections and approvals
  - Recording mechanisms for H&SM practice and pro-forma
- Post-Contract Review:
  - Auditing H&SMS
  - Analysis of H & S management practice
  - Feedback to company for future policy development
  - Provision of information to project H & S file to meet CDM Regulations

Figure 10.4 *Health and Safety Manual development: key sections describing management procedures*

H & S planning:

(i) Pre-Tender H & S Plan – the responsibility of the planning supervisor working on behalf of the client
(ii) Construction H & S Plan – the responsibility of the principal contractor.

Although the principal contractor has no involvement in the development of the pre-tender H & S plan, it does form a vital element in the principal contractor's tendering process. For this reason, the pre-tender H & S plan is explained.

The project H & S plan can be explained in two sections for simplicity. The first section is the requirement under the CDM Regulations for the principal contractor to develop and implement the construction-phase H & S plan. The second section is the extension of the company H&SMS into procedures to ensure safe working practices on site.

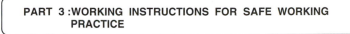

PART 3 : WORKING INSTRUCTIONS FOR SAFE WORKING PRACTICE

- Working Instructions

    - H & S induction, training, instruction, on–going education
    - hazard identification
    - risk assessment
    - mitigation and control mechanisms
    - site meetings for H & S discussion issues
    - toolbox talks
    - use of tools and equipment
    - working in hazardous situations (e.g. at heights)
    - site instruction for specific assignments
    - permits to work
    - working in confined spaces and use of respirators, PPE
    - hot works
    - handling hazardous substances
    - administration and record keeping
    - inspection procedures and signing-off

Figure 10.5 *Health and Safety Manual development: key sections describing working instructions for safe working practice*

There are obvious overlaps between the company H&SMS and the project H & S plan and indeed some common elements are present, for example, policies, risk assessment and auditing. This is so because such elements form a part of both levels of the system, being generic within the company H&SMS and project specific within the H & S plan. While the generic elements will apply to all the company's business and projects, each project will also have its own specific elements to be addressed for the particular characteristics of the site and its organisation.

The pre-tender plan is explained before examining the construction-phase H & S plan. One needs to be aware of the features and inputs to the pre-tender planning phase as information from this impinges directly on the construction H & S plan.

The Pre-Tender Health and Safety Plan

The CDM Regulations propound that the management of H & S on any project starts with planning, design and specification. Clients and designers, therefore, make a considerable contribution to the identification of hazards and assessing construction risks by formulating a clear and comprehensive health and safety plan early in the project development.

The pre-tender H & S plan is a key contributor to the identification and assessment of risk. The information gathered is essential to project evaluation and development. Moreover, the information is a prerequisite to providing the basis for a safe working environment and safe working practices on site.

The pre-tender H & S plan should contain information that enables prospective principal contractors who tender for the project to plan and cost for safety measures in their bid. Where a contract is awarded or negotiated rather than bid for, the same information should still be provided for this first stage of H & S planning.

PROJECT HEALTH AND SAFETY PLAN

- Project (Sitework) Health and Safety Plan:

  - H&SM legislation and obligations (HSWA, MHSWR, CDM, COSHH)
  - Company H & S policy
  - Management responsibilities
  - Company safety officer/adviser
  - Project (site) manager's responsibilities
  - Site safety supervision
  - H & S training
  - H & S method statements
  - Risk assessment
  - Permits to work
  - Site rules (linked to working instructions)
  - First-aid provision
  - Emergency procedures and contacts
  - Accident register and HSE notification under RIDDOR
  - Investigation procedures
  - Inspection routines
  - Record keeping and pro-forma

Figure 10.6   *Health and Safety Manual development: key sections describing the project health and safety plan*

The information needed for developing a pre-tender H & S plan is gathered within three broad elements of the project:

- environment
- design
- site.

Information is contributed by the client, designer and planning supervisor. The information is project specific and therefore will vary from one project to the next. However, information can be gathered under standard headings to simplify matters.

Managing Construction for Health and Safety, the HSE Approved Code of Practice, suggests nine elements under which information for the pre-tender H & S plan might be gathered:

(1) nature of the project
(2) existing environment
(3) existing drawings
(4) design
(5) materials
(6) site elements
(7) overlap with the client's activities
(8) site safety rules
(9) continuing liaison.

These elements are outlined as follows:

- *Nature of the project* – contains information to specify: the name and address of the client; the location of the project site; details of the construction; and time frame for the construction phase.
- *Existing environment* – describes the environment on and around the project site. Details of existing service utilities, for example, water, electricity and telephone, together with possible concealed services such as gas pipes or water mains will be included, as these may represent a potential safety risk. Existing buildings and structures will also be included in the plan as they constitute potential site hazards.
- *Existing drawings* – existing drawings may be available from landowners, building owners or occupiers, utility companies, or local authorities. These detail site layouts, structural details of existing or neighbouring properties and location of existing services.
- *Design* – findings of the designer's risk assessment should be considered. Hazardous works should be identified and detailed for the attention of the principal contractor in compiling its H & S method statement.
- *Materials* – details of any potentially hazardous materials specified in the design and for which specific precautions will be required should be given. This relates to particular materials noted as hazardous, for example, adhesives.
- *Site elements* – includes information which is important to any persons present on the site, for example: temporary accommodation; delivery points; unloading and storage; pedestrian and vehicular traffic routes; services and amenities; and site boundaries and security protection requirements.
- *Overlap with the client's activities* – where the project involves work to existing premises or populated areas there may be restrictions to working practices. For example, working hours may be limited, noise and pollution controls may be required beyond those usually encountered, or access may be limited to specific areas or times.
- *Site safety rules* – these are imposed to protect persons on site, for example, vehicular speed restrictions, material movement, or handling dangerous substances.
- *Continuing liaison* – the plan should contain details of key contacts for the client, designer and planning supervisor. Information will include the full name, address and telephone number of each contact.

The client has an essential contribution to make to the pre-tender H & S plan. The client provides information which is key to project development including:

- location and description of the siteworks
- time frame for project development and construction
- existing documentation and site plans
- details of site topography and ground conditions
- condition of existing buildings and structures
- location of temporary services
- arrangements for site access and traffic management
- need for site security
- facilities for material and plant storage and protection
- potential restrictions to work routines and siteworks
- findings from risk assessments.

These aspects cover the key information that can be provided by the client and the appointed consultants. Information may be available from existing sources while some might have to be acquired through undertaking specific surveys. The information that is provided by the client is project site specific. Details relating to the construction form are incorporated in the information provided by the designer.

The design consultant will provide information to the planning supervisor which considers both the site and the project, such as:

- details of existing buildings and internal layouts
- results from soil investigation and analysis
- phasing and sequencing of the works
- hazards presented by the proposed design
- specific requirements for method statement preparation, for example demolition works.

This information is vital to the potential principal contractor who must consider, incorporate and resource for any significant risks within the pre-tender health and safety plan.

The Construction-Phase Health and Safety Plan

The principal contractor has responsibility for the development of the construction-phase health and safety plan. On some projects there may be development and design works still being undertaken even though the principal contractor has been appointed. In this case these works remain the responsibility of the planning supervisor until they are at the appropriate stage to be passed to the principal contractor for inclusion in the construction-phase health and safety plan.

The information included in the pre-tender health and safety plan forms the basis of development for the construction-phase health and safety plan. The client has a responsibility to ensure that the health and safety plan is prepared subject to any outstanding aspects under the control of the planning supervisor, before construction commences on site.

When developing the construction-phase health and safety plan, the principal contractor will provide information which enables:

- the development of a framework for managing health and safety of all participants involved with and affected by the construction stage of the project
- the contributions of other organisations involved in the construction phase to be included
- the organisation of and action needed to investigate hazards, including informing all personnel who might be placed at risk
- good communication by the contractual parties
- the development of method statements, design details and specification for all project contractors (subcontractors to the principal contractor)
- the security of the site to ensure that only authorised personnel gain access.

To facilitate the detailed development of the construction-phase health and safety plan, the principal contractor will require inputs from other project participants. These include the following:

The *client* contributes to the development of the construction-phase health and safety plan through discussion concerning production matters throughout the construction stage.

The *designer* is duty bound by the CDM Regulations to avoid foreseeable risks, combat risk at source, protect the entire workforce and communicate appropriately on any risk to the project as a result of the design solution.

It is important that the designer has identified the potential hazards together with those that may occur during construction as a result of the design. Hazards must not only be recognised but be highlighted to other contractual parties and be priced for by tendering prospective principal contractors, subcontractors and suppliers.

The information collected by the designer during design review, presented in the hazard identification record and risk assessment record, will be passed to the planning supervisor. It will be included in the pre-tender health and safety plan and ultimately form part of the project health and safety file. Good information is vital because the information will be used in so many subsequent health and safety management stages – the construction health and safety plan, health and safety management on site and the generation of the health and safety file.

Subcontractors can contribute to the construction-phase health and safety plan by providing information concerning:

- identified hazards occurring directly from their participation
- the assessment of risks identified
- the potential measure of mitigating the risks.

Subcontractors are duty bound to make their personnel aware of all project risks and provide training for those employees where necessary.

A Guide to Managing Health and Safety in Construction (HSE, 1995b) suggests a number of specific elements which should be considered in compiling the construction-phase health and safety plan. These are:

- *Project overview* – develops further the information contained in the sub-section of the pre-tender health and safety plan entitled 'nature of the project'
- *Health and safety standards* – standards specified by the client, the designer or principal contractor
- *Management arrangements* – the management structure and organisation for the health and safety management of the project and the key responsibilities of the parties
- *Contractor information* – the protocol involved in advising contractors about project risks
- *Selection procedures* – procedures for assessing the competence and resources of contractors
- *Communication and co-operation* – communication methods for co-ordinating health and safety and ensuring co-operation between the parties
- *Activities with a risk to health and safety* – how hazards identified in the pre-tender health and safety plan are to be communicated to site personnel
- *Emergency procedures* – the notification of alarms, escape routes and assembly areas, and personnel checks
- *Accident recording* – how the contractor will comply with responsibilities under RIDDOR (HSE, 1995a)
- *Welfare facilities* – the arrangements for all temporary site welfare facilities
- *Training* – health and safety training, including induction training for employees
- *Site safety rules* – any site rules and restrictions identified in the pre-tender health and safety plan should be included in the construction-phase health and safety plan
- *Consultation* – the procedures for personnel to raise health and safety issues with their supervisors
- *H & S monitoring* – procedures for inspection and audit
- *Project H & S review* – the procedure for reporting on health and safety incidents
- *H & S file* – the procedures for the passing of information from the principal contractor to the planning supervisor.

Arrangement of Content for Construction-Phase Health and Safety Plan

Since different principal contractors will address the content, presentation and formatting of their H & S plans in different ways to meet their own needs it is sensible to avoid over-prescription. Nevertheless, common elements within all construction H & S plans does support an outline framework for presenting the documentation. Typical content and format might be as follows:

PROJECT DETAILS
- Project title
- Site address
- Client
- Planning supervisor
- Design consultants

- Contract manager
- Project description
- Principal contractors design responsibility (where applicable)
- Restrictions on the works

HEALTH AND SAFETY SYSTEMS
- Policies
- Organisational arrangements
- Standards (applicable regulations)

MANAGEMENT STRUCTURE AND SYSTEMS
- Management structure of site team
- Relationship to other parties
- Selection procedures of contractors
- Materials and plant procurement
- Management procedures
- Meetings of the parties
- Communication mechanisms

ASSESSMENT AND REPORTING
- Risk identification and assessment
- Management of risks
- Method statements
- Programmes
- Site layouts
- Reporting mechanisms (RIDDOR)

THE SITE
- Welfare provision
- Induction, training and instruction
- Management procedures and site rules
- Emergency procedures
- COSHH provisions

INPUT TO CLIENT'S HEALTH AND SAFETY FILE
- Responsibilities
- Reports
- Communication of feedback

HEALTH AND SAFETY REVIEW
- Review mechanisms
- Impact on H & S plan as implemented
- Safety amendments
- Policy review

MONITORING HEALTH AND SAFETY PERFORMANCE
- Instructions
- Recording mechanisms and pro-forma

APPENDICES
- Organisation charts (H & S management)
- Method statements (site safety)
- Programmes (master/short-term/section)
- Site layouts and floor plans (where relevant)
- Risk assessments (hazardous operations).

Procedures for Ensuring Safe Systems of Work

The elements of the construction-phase health and safety plan and the company's H&SMS implemented during the siteworks by the principal contractor are focused on ensuring *safe systems of work*. Safe systems of work are safety-conscious procedures which are formalised in the documentation and communicated to all employees, and are workable and implemented through effective management and supervision.

The plan consists of eight elements. These cover the principal contractor's duties and responsibilities for project health and safety under the HSWA 1974, the MHSWR 1992, the CDM Regulations 1994, the CHSWR 1996, RIDDOR 1995 and COSHH 1999.

Activities commence with information from the pre-tender health and safety plan and flow through a series of managerial implementation to auditing procedures. Feedback closes the system loop with information passed back to health and safety policy-making and the external flow of information provides H & S records for the health and safety file as shown in Figure 10.7.

Elements of the System Project H & S Plan to Establish Safe Systems of Work

There are key elements forming the framework for safe working practices. These elements, recognised in almost all construction-phase health and safety plans, were first discussed in detail by Griffith *et al.* (2000) and Griffith and Howarth (2001). Here, the originally identified elements have been further developed to meet the wider scope of current H&SMS and H & S plans. It should be remembered that as experience grows among principal contracting organisations, so the scope, extent and content of construction-phase H & S plans evolve. Likewise, the methods of establishing and maintaining site safety records, presented subsequently, are also evolving as greater knowledge is acquired through application. The significant aspect of these key elements is that they include all the core site-based matters identified previously in the content of the construction-phase health and safety plan. These elements are:

  (i) Policies
 (ii) Risk Assessment
(iii) Safety Method Statements
 (iv) Safety Induction, Training and Instruction
  (v) Permits-to-Work
 (vi) Management Procedures and Site Safety Rules
(vii) Safety Audits
(viii) Feedback (to policies and H & S file).

(i) Policies
Management policies from the company H&SMS set out the organisation's statement of commitment to health and safety management in the context of the specific project. They detail how organisational procedures are to be established to ensure that safe systems of work are developed and applied to the project. Policy is fundamental to the development of the H&SMS. It forms the basis for creating the managerial framework, operational elements and the assignment of responsibilities to both managers and employees. To encourage leadership and the motivation of employees, policies must be clearly stated and widely communicated. Policies must reflect the holistic organisational commitment at corporate levels. They should reach the whole organisation through the hierarchy of management to the workforce on site.

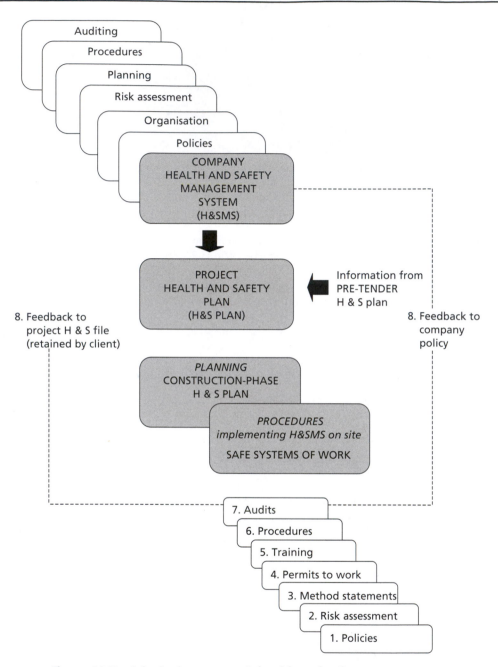

Figure 10.7 *Principal contractor's health and safety management system – project health and safety plan relationship*

To implement the H & S plan the principal contractor must establish a clearly determined set of policies which specify basic procedures and minimum standards of H & S performance. Although particular characteristics of health and safety policies will differ among principal contractors and across construction projects, a generic outline can be suggested. The H & S plan should address the wide range of features characteristic of construction siteworks. These include:

- boundaries, gates, fencing, hoardings; site access and registration procedures for employees and visitors
- plant use and vehicular access to site egress and associated public safety
- storage of materials and potentially hazardous substances
- waste management, site housekeeping and tidiness
- handling loads manually and with equipment
- accommodation, welfare facilities and services
- statutory notes and registers
- hazard and safety signage and site lighting
- working at heights: on platforms, scaffolds, ladders, hoists, using guard rails and harnesses
- working with formwork, steel erection, roofing groundworks and in other hazardous situations
- personal protective equipment and clothing
- working in confined spaces
- occupational health (noise, vibration, hazardous substances, pollution, exposure limits to specific conditions, manual handling)
- emergency procedures and first-aid provision
- fire prevention and incidents
- use of electrical supplies, work around services and portable hand tools
- appointment of safety supervisors
- third-party and public safety
- safety induction training and briefings
- safety method statements
- reporting and record-keeping of injury, ill-health, near miss incidents and damage.

(ii) Risk Assessment

The responsibility for devising safe systems of work rests with the principal contractor. Risk assessment is a fundamental aspect of establishing safe systems of work and is a key element in the principal contractor's H&SMS and project plan. The function of risk assessment is to provide management with details of potential hazards and risks, and to specify when safe systems of work are needed to control them. Detailed guidance to risk assessment for safety management is given in the MHSWR 1999.

Risk assessment for all sitework features as previously identified and any construction operation requires the consideration of three key tasks: (1) the identification of the potential hazard; (2) the evaluation of the degree of risk; and (3) the consideration of prevention and protection measures. In simplified terms, risk assessment is an analytical technique which determines the degree of risk through considering the likelihood of a safety incident occurring and the severity of harm that the incident could create. Assessment assigns a quantitative value to the degree of risk which is then translated into qualitative ratings of low, medium or high risk (see Figure 10.8). This outcome can then be transferred to the risk assessment form where each element of sitework, potential hazards, population at risk, risk rating and control measures to be implemented to remove or reduce the risk are specified.

(iii) Safety Method Statements

Safety method statements provide details of how safe systems of work relate to construction operations and tasks. The statements inform supervisors of the proposed method of carrying out the sitework tasks. For each task the health and safety implications are assessed and precautions taken to mitigate risk and ensure a safe working environment. The function of this within the project plan is to provide management with a systematic and thoughtful approach to linking the physical works to health and safety issues. During the construction phase, the safety method statements are an essential reference to guide on-going site activities.

The safety method statement allows operatives carrying out the work to:

- conduct tasks in a safe manner
- identify hazards associated with tasks
- use mechanisms to mitigate risk.

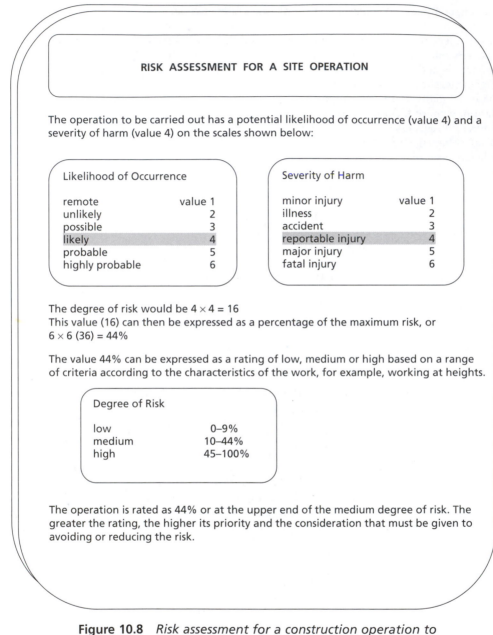

Figure 10.8 *Risk assessment for a construction operation to determine the degree of risk*

To enable this, the safety method statement should:

- accurately describe operations and tasks
- identify the location of the works
- specify supervisory arrangements
- identify the safety supervisor
- specify the plant needs for the operations
- consider likely occupational health implications
- specify precautions to be taken to minimise risk by those carrying out the works.

(iv) Safety Induction, Training and Instruction

The principal contractor should establish management procedures for safety induction, training and instruction for all personnel who are appointed to the project site. The mode of delivery will differ among organisations, yet common elements of safety induction include:

- specification of health and safety policy
- communication of the construction-phase health and safety plan and safety method statements
- specification of expected safety performance standards
- penalties and disciplinary issues for breaching standards
- clarification of any project aspect
- encouraging support and co-operation among site personnel.

Induction meetings will integrate technical aspects of the project with the specified health and safety procedures. Induction meetings will therefore:

- familiarise personnel with the construction requirements
- inform personnel about the safety aspects of undertaking the works
- determine supervisory management for the technical and health, safety and welfare aspects.

The MHSWR 1999 require a designated company safety officer to attend induction events where the works are complicated or where they involve high-risk characteristics. At such induction events, a typical agenda will require the safety officer to:

- remind personnel of the safety procedures and standards
- identify operation-specific safety aspects
- advise on, for example, designated access or egress points; traffic management; public welfare and safety; site boundaries and restricted access
- specify permit-to-work procedures
- highlight first-aid and emergency procedures
- specify safety inspection arrangements and identify safety supervisors present on site
- outline disciplinary measures for non-compliance with procedures
- specify modes of reporting safety incidents
- provide personnel with a safety management handbook.

Safety training and instruction should be seen as an on-going management procedure with follow-up events being held where, for example, there is a major change to the construction programme, a variation to operational procedures or new personnel taking up post on site.

(v) Permits-to-Work

A permit-to-work is an authorisation to carry out work which can be issued for all construction activities. It is usually applied to activities that give rise to a high risk of danger to those undertaking the work. The issue of a permit-to-work is a formal control procedure designed to ensure safe systems of work and as such it is issued by a designated individual. Essential aspects of a permit-to-work are that they always relate to tasks where:

- specific training is needed to carry out the task safely
- the work is considered to be high risk
- the works are complex or difficult, for example, carried out in a restricted or dangerous environment.

This element of the plan focuses on ensuring that procedures are in place to consider, issue, monitor and control permits for hazardous work.

The format and content of documentation in issuing a permit-to-work will differ across organisations and even from task to task. However, common aspects are:

- permit number
- date and time of issue

- duration of the permit
- location of the works
- description of the hazard and likely risk
- precautionary measures to be adopted
- any testing or validating procedures needed
- emergency procedures, signals and reporting mechanisms
- acknowledgement by the receiving operative in charge of the work
- signing-off the completed work
- cancellation of the permit.

(vi) Management Procedures and Site Safety Rules

Management procedures set out the specific methods to be used and the minimal standard requirements of all general activities on the project site. While any aspect of general project administration is common across most projects, the CDM Regulations require that site-specific safety rules are established which accommodate the particular characteristics of the project. These are termed site safety rules.

Management procedures can be developed to cover almost any aspect of project administration but, in general, procedures are set in place to cover those aspects of health and safety policy established in the construction-phase health and safety plan.

Site safety rules are project specific and exist to provide safe systems of work where works will take place in high-risk environments or where the works themselves are hazardous to those persons carrying them out.

Site safety rules will be detailed in the site health and safety handbook applicable to the project to ensure that all personnel are aware of and follow safe working procedures.

(vii) Safety Audits

An important aspect of the H & S plan is that it acts as a fundamental link to the company H&SMS safety management auditing. An audit is:

- a detailed review of how safety management is being applied
- an analysis of the degree to which the plan is being complied with
- an appraisal to determine the benefits being gained from implementation of the plan
- an identification of systems aspects where a change in procedure could make improvements to safety management.

A safety management audit may be conducted in-house in association with the organisation's company safety management or may be conducted by specialist consultants contracted for the task. The approach of the audit should be to:

- evaluate current safety policy
- approve the safe working procedures specified
- appraise the construction-phase health and safety plan
- identify the control mechanisms in place to ensure compliance with procedures
- review the current construction programme, site structure and organisation, and management personnel responsible for safety aspects
- appraise safety management reports of incidents
- provide a detailed report on the findings of the audit.

Activities to be audited can cover almost any aspect of site procedure and activity. General items which will be audited include standards of:

- administration
- welfare
- accommodation

- access control
- waste management
- plant and materials management
- groundworks
- temporary works
- occupational health controls
- fire prevention
- hazardous substance protection
- public safety.

(viii) Feedback

The principal contractor should provide feedback in two specific directions: (i) to the company which will assist in updating company H & S policies and H&SMS procedures; and (ii) to the project H & S file in accordance with the requirements of the CDM Regulations and which is retained by the client organisation.

Establishing and Maintaining Site Safety Monitoring and Records

The achievement of the H&SMS implemented is evidenced in the effectiveness of information gathering through monitoring and the methods by which it is recorded. Different organisations will have their own procedures for documenting their H & S activities on site. Typical approaches include the use of a set of forms which are completed by supervisors and operatives as elements of the H & S plan are implemented. This ensures that operations with attendant risk are considered, planned, completely monitored and controlled to mitigate hazards and dangers. Thus, as far as can be reasonably anticipated, operatives are protected against foreseeable risk.

Figures 10.9 to 10.19 show a set of site safety record forms which can be used for H & S activities throughout the siteworks. This set is not exhaustive but does cover the elements involved in implementing safe working practices on site.

The forms can be formatted in any way to suit the requirements of the principal contractor. The important point with their use is that each should be issued and signed off by a designated supervisor as each element is undertaken and completed. The range of documents covers the designation of management, or supervisory, responsibilities through to H & S incident and investigative reports. In addition to the documentation developed and implemented by the principal contractor for its own use there will be various documents to be completed in the event of H & S incidents to meet the requirements of RIDDOR.

Management and Supervisory Designated Responsibilities Form		
Project:		
Document Ref No:		
Date issued:		Page of
Compiled by:	Signed:	Date:
Authorised by:	Signed:	Date:
Name	Function	Responsibilities

Figure 10.9 *Management and supervisory designated responsibilities form*

Health and Safety Document Register Form					
Project:					
Document Reference No	Title	Issue Date	Amendment Date	Amended Reference	Authorised by

Figure 10.10 *Health and Safety document register form*

Risk Assessment Form

Project:				Document Ref No:	
Contractor:				Specialist Discipline:	
Assessor:			Signed:	Date:	
Activity/ Element	Potential Hazards	Population at Risk	Risk Rating		Priority	Control Measures Specified
			L	S	R		

Sources of information:

Legend:
L – Likelihood
S – Severity
R – Risk (Severity × Likelihood)

Figure 10.11 *Risk assessment form*

Safety Method Statement Review Form					
Project:					
Document Ref No:					
Principal Contractor:					
Evaluation of:					
The method statement is returned for reconsideration*/accepted* (*delete as applicable)					
Next action:					
Assessed by:					
Date:					
	TEST	YES	NO	IN PART	N/A
a.	Task/process and area of the work				
b.	Sequence of work				
c.	Supervisory arrangements				
d.	Monitoring arrangements				
e.	Schedule of plant				
f.	Reference to occupational health standards				
g.	First aid				
h.	Schedule for personal protective equipment				
i.	Schedule of arrangements for demarcation and security				
j.	Controls for the safety of third parties				
k.	Are the assessed high-risk or safety-critical phases identified with controls specified?				
l.	Emergency procedures				

Figure 10.12 *Safety method statement review form*

Site Safety Co-ordination Record Form

Project:						Document Ref No:					
Activity	Contractor	Start Date	Pre-qualification Questionnaire		Initial Safety Meeting	Safety Method Statement Evaluation	Risk Assessment			Safety Information given	Authorised to Start
			Sent	Received and accepted			General	COSHH	Noise	Manual Handling		

Figure 10.13 *Site safety co-ordination record form*

Induction and Training Form		
Project:		
Document Ref No:		
Contractor:		
Type of training:		
Name of trainee	Trainer	Date of training

Figure 10.14 *Induction and training form*

Safe Working Procedures Form
Project:
Document Ref No:
Task or Work Operation:
This Safe Working Procedure has been prepared for the following work
Location of work:
Description of work:
Safe methods to be implemented:
Prepared by:
Name:
Designation:
Signature:

Figure 10.15 *Safe working procedures form*

Permit-to-Work Form	
Project:	
Document Ref No:	
Task or Work Operation:	Duration of permit:
This Permit-to-Work is issued for the following:	
Is work to be carried out when plant, equipment or systems are in operation?	Yes/No
Location of work:	
Description of work (specific hazards):	
Precautions to be taken:	
Extra precautions to be taken if plant and equipment are being used:	
Additional permits: • Hot work • Electrical • Confined space • Other	
Authorisation	
Name of person issuing Permit:	
Designation:	
Signature:	
Time:	Date:
Receipt	
Name:	
Designation:	
Signature:	
Company:	
Clearance	
The work stated above has/has not been completed. Details if not completed:	
Cancellation	
Permit to work is cancelled	
Name:	
Designation:	
Signature:	
Date:	Time:

Figure 10.16 *Permit-to-work form*

Site Safety Inspection Form			
Project:			
Document Ref No:	Date:		Time:
Location:			
Any unsafe conditions or work:			
Remedial action:			
Further action to be taken	By (named person)	Date	Complete
Inspected by (Safety Inspector): Action authorised by:	Date:		

Figure 10.17 *Site safety inspection form*

Accident Report Form		
Project:		
Document Ref No:		Page of
Injured person:	**Accident**:	**Person reporting accident**
Name:	Date: Time:	Name:
Home address:	Location:	Home address:
	Work process involved:	
	Cause (if known):	
Occupation:		Occupation:
	Details of injury:	Signature:
		Date of report:

Figure 10.18   *Accident report form*

Health and Safety Incident Investigation Report Form			
Project:			
Document Ref No:			
Parties involved:			
Location of incident:			
Date of incident:		Time of incident: AM/PM	
Type of incident:			
Potential severity:	Major	Serious	Minor
Probability of recurrence:	High	Medium	Low
Description of how incident occurred:			
Immediate causes: what unsafe acts or conditions caused the event?			
Secondary causes: what human, organisational or job factors caused the event?			
Remedial actions: recommendations to prevent recurrence:			
Signature of investigator:	Date:		
Follow−up action/Review of recommendations and progress:			
Name of reviewer:			
Position/Title of reviewer:			
Signature of reviewer:	Date:		

Figure 10.19 *Health and Safety incident investigation form*

Case Study – Construction-Phase Health and Safety Plan

The set of documents at the end of this chapter presents an extensive and comprehensive construction-phase health and safety plan compiled by a major principal contracting organisation undertaking a large city centre building development project.

It will be seen that the H & S plan is divided into sub-sections following closely the description of H & S plan development given earlier in this chapter. The siteworks are broken down into key construction aspects and, for each, detailed information is provided to promote efficient and effective safe working practices on site. Of particular importance within the documents is the comprehensive coverage of health and safety risk and its assessment together with instructions to mitigate hazards and dangers which might be encountered during the siteworks.

It should be remembered that the construction-phase health and safety plan in application will have personalised details contained within the document as implemented on the project. For the purposes of presentation here, all references have been deleted, otherwise the H & S plan is in the form applied to the actual project.

Summary

A carefully considered and developed construction-phase health and safety plan is a prerequisite to effective health and safety in the course of managing the project. Central to that plan is an all-embracing and comprehensive risk assessment which anticipates potential health, safety and welfare hazards and dangers, and provides the basis for developing appropriate mitigation measures. Also crucial is the successful development and implementation of a health and safety management system within the corporate organisation of the principal contractor. This provides not only the company focus, ethos and culture, but moreover establishes the sound base for translating company policies into effective management and supervisory procedures, and safe working practices on site.

References

British Standards Institution (1999) *BSI-OHSAS 18001: Specification for Occupational Health and Safety Management Systems*. BSI, Milton Keynes.

Department for Social Security (1979) *The Social Security (Claims and Payments) Regulations 1979*. HMSO, London.

European Commission (EC) (1989) *Directive 89/391/EEC on the Introduction of Measures to Encourage Improvements in the Health and Safety of Workers at Work*. HMSO, London.

Griffith A. and Howarth T. (2001) *Construction Health and Safety Management*. Addison Wesley Longman, Harlow.

Griffith A., Stephenson P. and Watson P. (2000) *Management Systems for Construction*. Addison Wesley Longman, Harlow.

Health and Safety Executive (HSE) (1961) *The Construction (General Provisions) Regulations 1961*. HMSO, London.

Health and Safety Executive (HSE) (1966a) *The Construction (Health and Welfare) Regulations 1966*. HMSO, London.

Health and Safety Executive (HSE) (1966b) *The Construction (Working Places) Regulations 1966*. HMSO, London.

Health and Safety Executive (HSE) (1974) *The Health and Safety at Work, etc. Act 1974*. HMSO, London.

Health and Safety Executive (HSE) (1992a) *The Management of Health and Safety at Work Regulations 1992*. HMSO, London.

Health and Safety Executive (HSE) (1992b) *The Workplace (Health, Safety and Welfare) Regulations 1992*. HMSO, London.

Health and Safety Executive (HSE) (1992c) *The Provision and Use of Work Equipment Regulations 1992*. HMSO, London.

Health and Safety Executive (HSE) (1992d) *The Personal Protective Equipment at Work Regulations 1992*. HMSO, London.

Health and Safety Executive (HSE) (1992e) *The Manual Handling Operations Regulations 1992*. HMSO, London.

Health and Safety Executive (HSE) (1992f) *The Health and Safety (Display Screen Equipment) Regulations 1992*. HMSO, London.

Health and Safety Executive (HSE) (1993) *The Chemicals (Hazard Information and Packaging) Regulations 1993*. HMSO, London.

Health and Safety Executive (HSE) (1994) *The Construction (Design and Management) Regulations 1994*. HMSO, London.

Health and Safety Executive (HSE) (1995a) *The Reporting of Injuries, Diseases and Dangerous Occurrences Regulations 1995*. HMSO, London.

Health and Safety Executive (HSE) (1995b) *A Guide to Managing Health and Safety in Construction*. HMSO, London.

Health and Safety Executive (HSE) (1996) *The Construction (Health, Safety and Welfare) Regulations 1996*. HMSO, London.

Health and Safety Executive (HSE) (1999a) *The Management of Health and Safety at Work Regulations 1999* (replaced 1992 version). HMSO, London.

Health and Safety Executive (HSE) (1999b) *The Control of Substances Hazardous to Health (Amendment) Regulations 1999*. HMSO, London.

Health and Safety Executive (HSE) (1999c) EH40: *Occupational Exposure Limits*. HMSO, London.

Principal Contractor's Construction-Phase Health and Safety Plan

CONTENTS

1. INTRODUCTION

This Construction Health and Safety Plan has been prepared in accordance with the requirements of the Construction (Design and Management) Regulations 1994 and is based on the tender documentation and the pre-tender Health and Safety Plan prepared and provided by the Planning Supervisor.

This Health and Safety Plan is intended to set out the general framework and arrangements for securing the health and safety during the construction phase and it deals with the key tasks during the initial stages of the works where the design has been completed or substantially completed.

This Construction Health and Safety Plan will be developed further as the design is progressed and/or the specialist subcontract packages are procured via the addition of detailed Method Statements. Any additional hazards will be identified and risk assessments will be carried out as the works proceed.

Unforeseen circumstances or variations will also be taken into account if and when they arise.

Prior to commencing any works it is a requirement that the client is to confirm in writing that he has satisfied himself that this document complies with regulation 15(4) and has been sufficiently developed to allow works to proceed (regulation 10). Possession of site will be deemed to be dependent on this requirement.

The level of detail provided by this plan will be determined by the health and safety risks of the particular project. Hence a project involving minimal risk will call for a fairly simple and straightforward plan. Larger projects or those involving significant risks will require more detail.

2. PROJECT DETAILS

2.1 Project Title

2.2 Site Address

2.3 Client

2.4 Contract Administrator

2.5 Planning Supervisor

2.6 Design Consultants

Architect	Name Company Address	Tel Fax
Structural Engineer	Name Company Address	Tel Fax
Mechanical Engineer	Name Company Address	Tel Fax
Electrical Engineer	Name Company Address	Tel Fax
Planning Supervisor	Name Company Address	Tel Fax
Fire Engineers	Name Company Address	Tel Fax

2.7 *Project Description and Scope of Works*

The works involve the construction of a six-storey office building with a gross floor area of approximately 10,800 m² above ground level, and a two-storey split level basement car park of approximately 3200 m² for 100 cars. The main plantrooms are at roof level.

The site was once occupied by industrial buildings which have been demolished and it is currently a pay and display car park.

The substructure works include grouting of the old mine workings, contiguous piled wall, deep excavations, reinforced concrete basement walls and floors and structural steel frame. The superstructure is a steel frame with composite floors and traditional cavity wall construction. There are some glazed curtain walling areas to the ground and fifth floors, and at the ends of the building. There is also a section of metal cladding above the main entrance.

The mechanical installation comprises LPHW heating, mechanical ventilation with chilled beam cooling system, hot and cold water systems. A separate ventilation system serves the basement car park. There are 4 no. passenger lifts, 1 dedicated goods lift, and a separate lift for the transfer of cash to the secure delivery area. Electrical installations include lighting, power, fire alarm, security/CCTV, public address and telephone/data infrastructure.

Fitting out of the building with furniture and fittings is included as part of the works. Access for the occupier's own contractors will be needed during the last 6 weeks of the works to install the IT and telephone hardware.

The drawings and outline specifications are included within the Contractor's Proposals dated

The contract will be a bespoke Building Contract written to reflect the PFI Project Agreement.

2.8 *Extent of Principal Contractor's Design Responsibility*

The works are being procured under a PFI Project Agreement.
            has full design responsibility for the whole of the works.

2.9 *Programme of Works Including any Phasing*

The works will be carried out in accordance with the Construction Programme which is included in Appendix 10.3.

   Key dates are as follows:

Contract Commencement
Site Possession
Completion of Building Works
Ready for Occupation

2.10 Restrictions Affecting the Works

Restriction	Details and effect of the particular interface with the works	Management and Control Responsibility
Service Diversions	The following services have been identified as being affected by the works and require to be diverted before any work takes place below the ground level in the location of these services • Low voltage cable at junction of St and St • Data/communication cables in pavement on St	Electricity Cable
Existing Services	The site is used as a pay and display car park. Power supplies to the machines on site are to be made safe and the machines removed.	City Council
Pedestrians	The works will require the closure of footpaths on Street and Street. Pedestrians will be directed across these roads to maintain safe routes.	
Cyclists	Cyclists use the roads around the site. A cycle lane is to be provided along Street and Street	City Council
Mine Grouting	A worked coal seam underlies the site. There may also be buried mineshafts. This seam and associated mineshafts are to be grouted before piling and groundworks are commenced.	
Access to	There is a loading bay to House in Lane serving The delivery route currently uses Street and Lane. The road outside is to be widened before Lane is to be closed as part of the works.	
Turning Head	A turning head (part of the existing highway) encroaches within the building line opposite House. The building will oversail the turning head. Safe access is to be maintained, or temporary closure agreed with the Council and occupants of House.	
Access to Phase 1B Car Park	Car park is in use at all times. Lane roadworks to be phased.	
Site Access	Temporary accesses will be required from Lane during the reconstruction of Lane. After the roadworks access will be via the existing turning head.	

Deliveries	Street and part of Street is one way from Gate. Traffic movements to be maintained during the works. A delivery area is to be included within the hoarding on Street. Deliveries to be managed to ensure no restriction of traffic movements.
Tower Crane	A tower crane will be erected to service the site. Banksmen to ensure that no loads are suspended over surrounding roads and properties.
Noise	The site is bounded by offices, shops and public houses. Noise to be within the statutory required levels.
Contamination	Ground contamination is generally assessed as posing a low hazard, though pockets of low to medium, and medium hazard are present. Dispose to licensed tips.
Pollution	Ground water is polluted with sewage from the surrounding system. Branches onto the site to be capped. Pumping to be controlled and into the foul drainage system only.
Working Hours	7.30–6.00 weekdays 7.30–6.00 Saturday 7.30–6.00 Sunday in exceptional circumstances (i.e. phased roadworks)

3. HEALTH AND SAFETY SYSTEMS

3.1 Health and Safety Policy Statement

STATEMENT OF POLICY

The continuing policy of the Company is to conduct its operations in such a manner as will ensure, so far as is reasonably practical, the safety, health and welfare at work of its Team and the prevention of risks, to the safety or health of others, and harm to the natural environment, to comply willingly with all relevant laws and regulations and to co-operate with those responsible for enforcing them.

Particular importance is given to the provision of information, instruction, training and supervision necessary for the implementation of this Policy.

The Board of Directors will appoint one member to perform the duties of the Company Safety Director.

All directors, managers and supervisors have a responsibility for implementation of the policy in accordance with the directions given in this document and which may be given from time to time in supplementary documents.

All Team Members have a part to play in the implementation of the Policy and in particular have a duty to take reasonable care for the safety and health of themselves, their fellow Team Members and anyone else who may be affected by their acts or omissions and a duty to co-operate with others in the discharge of their duties.

The effectiveness of this Policy and its implementation will be monitored and this document will be reviewed from time to time.

IMPLEMENTATION OF THE POLICY

At tendering and work planning stages, full account is to be taken of those factors which help to eliminate injury, damage and waste, and decisions about other priorities (e.g. production and finance) are to take proper account of health and safety requirements.

Specific and precise arrangements are to be agreed with each contractor or subcontractor, as the case may be, for the Policy to be implemented. All such contractors and subcontractors are to be properly supervised to ensure compliance.

Systems of work, incorporating risk assessments, are to be established, implemented and monitored so as to ensure appropriate standards of safety at all times.

High standards are to be applied in complying with legislation regarding the health and safety of Team Members and others, with proper attention also being paid to environmental conditions.

High standards of cleanliness, hygiene and housekeeping are to be maintained at all times.

Safe, adequate and clear means of access to and egress from places of work are to be provided and maintained.

Adequate and suitable protective clothing and equipment, appropriate to the work undertaken are to be provided.

Accidents are to be recorded, reported and investigated promptly, with preventive measures being implemented as appropriate.

Effective liaison will be maintained through Group Safety Department, with external accident prevention organisations. Safety information will be disseminated as appropriate.

Safety training programmes are to be promoted with the object of achieving high standards of personal awareness or risks and hazards, and knowledge of personal responsibility.

Responsibility and accountability in relation to the prevention of accidents, ill health, injuries and damage are to be specified clearly and in writing for each category of Team Member.

Facilities for joint consultation on matters of safety, health and welfare will be available through safety representatives and safety committees, where appointed, in accordance with the relevant Regulations,

Approved Code of Practice and Guidance Notes, and through Team Consultative Committees. The agreements reached through these consultants will be taken into account, when the policy is reviewed, periodically or as required.

Further arrangements for the implementation of the Policy on site are contained in the Site Safety Manual, copies of which are held on every contract.

The policy is to be explained to all new Team Members as part of their induction training, before they start work.

All reviews of the Policy are to be given wide circulation to update all of the Company Team Members.

RESPONSIBILITIES: DIRECTORS
REGIONAL MANAGERS

To so direct management under their control that the Policy (supplemented by special addenda prepared as considered necessary by the Board of Directors) is fully implemented at all times.

To promote good liaison between site or office management and Group Safety Department to ensure the fullest and most effective use of the services available.

To encourage good safety practices on their contracts and to discourage indifferent management by such means as are deemed necessary.

To show understanding of their personal responsibilities under the Act to treat the health and safety of persons under their control as a matter of importance equal to the other functions of a Director.

To demonstrate their commitment to safety, by making sporadic site safety tours, where health and safety are the sole topic of discussion.

To make recommendations to the Board for improving the safety performance of the company and its Team.

To appoint in writing to each contract on commencement (or in the event of management changes) a Safety Supervisor. This will normally be the senior manager, i.e. Project Manager, Site Manager or Site Supervisor.

To set up and maintain an efficient and adequate system for first-aid attention, good reporting procedures, assimilation and use of 'feed back' information received from Group Safety Department.

To co-operate with Group Safety Department and Team Training Department in arranging safety training for management and operatives.

When the size of the contract site warrants, to appoint a full-time Safety Adviser.

RESPONSIBILITIES: SAFETY DIRECTOR

To overall co-ordinate and implement the safety policy.

To co-ordinate the review of the safety policy, as required, after consultation with Team Members at all levels.

To ensure that the reviewed policy is communicated to all projects and Team Members.

To review the safety performance of the Company and report the findings to the board.

To review the investigation reports of all serious accidents.

To deal with all Health and Safety Executive or other enforcement authority, correspondence, involving the Company's activities.

To overall co-ordinate the Company's safety training programme, including induction training.

To ensure that all changes to legislation, new Codes of Practice, new official guidance, newly identified hazards and new safety products are communicated to all new projects and Team Members.

RESPONSIBILITIES: **CONTRACT MANAGERS**
                          PROJECT MANAGERS
                          SITE AGENTS

The above Team Members when in charge of contracts or sites will normally be appointed Safety Supervisors by written notification from the Director in Charge. When so appointed they will have specific responsibility for ensuring compliance with all statutory and Company regulations governing the safe conduct of the operations, and for promoting high standards of safety, health and welfare generally.

They will be required:

To appoint in writing one or more Safety Assistants of staff managerial status.

To organise the contract or site to fully comply with the Policy.

To take over the project Health and Safety Plan from the planning supervisor and maintain it.

To ensure that all relevant information, documents, drawings, Operations and Maintenance Manuals are collected and transferred to the Planning Supervisor for inclusion in the Health and Safety file.

To be familiar with and so observe all regulations applicable to the construction and related industries.

To control all contractors and/or subcontractors in order to ensure safe and healthy systems of work, low fire risk, proper use of shared facilities and co-operation between companies and/or trades in an unselfish and co-operative manner.

To set up and maintain an efficient and adequate system for first-aid, good reporting procedures, and assimilation and use of information from Group Safety Department.

To ensure that the needs for personal protective equipment are assessed and that the appropriate selection to be worn/used for the operations is made, and that all contractors/subcontractors do the same.

To perform the duties of, or appoint in writing, a local COSHH co-ordinator.

To perform the duties of, or appoint in writing, a Crane Co-ordinator, and a Plant Co-ordinator where required.

To arrange noise assessments and the implementation of control measures to protect Team Members and others from hearing damage, where required.

To ensure that all 'manual handling' operations are assessed and the assessments and training are recorded. All such operations which are at risk to health and safety are to be avoided by means of mechanisation and other measures.

To appoint a Site Fire Safety Co-ordinator to ensure compliance with Site Safety Manual, section 17.

To establish, where appropriate, site emergency plans, and appoint persons to be team leaders to manage the plans.

To accompany any visiting HM Inspector on his/her tour of inspection and to take prompt action concerning any complaint or advice received.

To co-operate with their management associates in providing and maintaining a good working relationship with any appointed safety representative and/or safety committee.

RESPONSIBILITIES: **SUB-AGENTS**
                          ENGINEERS
                          PLANT MANAGERS
                          GENERAL FOREMEN
                          SUPERVISORS

The above Team Members may be appointed Safety Assistants by written notification.

To ensure that work operations are only carried out by suitable and competent operatives, especially where mechanical appliances are concerned.

To instruct subordinates in precise terms as to work methods and by supervision to ensure compliance.

To reprimand or discipline any person who is persistently careless in regard to his/her own or others' safety. To encourage those persons who consistently show awareness and attention to safety matters.

By personal example and instruction to subordinates, to encourage the use of protective equipment provided by the Company in all circumstances where risk to health and/or safety is likely.

To investigate all reports made by subordinate Team Members alleging shortfalls of the Company's preventive and/or protective measures in the local workplace and to take action to remedy the situation, as required. Where this cannot be done immediately, to report the facts to senior management together with any suggestions for appropriate action.

To be familiar with and to observe all Regulations applicable to the construction industry. To co-operate with senior management in all opportunities available for safety training.

To prepare and deliver Toolbox Safety Talks, on relevant topics, as directed by the Site Manager.

At all times to take full heed of verbal and written advice from Group Safety Department and/or any Safety Adviser of the Company.

RESPONSIBILITIES: OTHER TEAM MEMBERS

To be familiar with the Policy and to co-operate in its implementation at all times.

To carry out instructions given by managers and supervisors.

To observe all safety regulations at all times.

To take reasonable care for the safety and health of themselves, fellow Team Members and anyone else who may be affected by their acts or omissions and to co-operate with others in the discharge of their duties.

To wear appropriate protective clothing and use appropriate safety equipment at all times as instructed.

To report all hazards to their immediate supervisor.

To report any shortcomings in the site preventive and protective measures, as would affect those persons, to their immediate supervisor.

To report all accidents whether persons are injured or not, and any damage, to their immediate supervisor.

Note 1

It is an offence under the Health and Safety at Work, etc. Act 1974 to intentionally or recklessly interfere with or misuse anything provided in the interests of health, safety or welfare in pursuance of any of the relevant statutory provisions.

Note 2

All Team Members may obtain further information about health and safety by consulting their managers, the site safety manual, the site safety adviser or by contacting Group Safety Department.

RESPONSIBILITIES: GROUP SAFETY DEPARTMENT

To advise all Group Companies, on all matters relating to the safety and health of Team Members and others as appropriate; relevant legislation, codes of practice and guidance material; fire precautions, the suitability of safety equipment; and accident reporting procedures.

To monitor, by the inspection of sites, workshops and premises, safety and health performance and to report on such inspections.

To prepare statistical analysis on accidents and causation classification, with recommendations on preventive measures.

To investigate and report on major injuries, notifiable dangerous occurrences, other accidents and incidents at the direction of the Group Safety Manager, and to attend and report on legal proceedings in which Group Companies may be involved.

To arrange Safety Training for Team Members at all levels.

To promote good working relations with the HSE and other Enforcing Authorities and to strive at all times to achieve, with the co-operation of site management, compliance with current legislation.

To keep under periodic review the published Group and Company Policy documents, making recommendations to the Board of Directors for revisions to meet changes in legislation or Company circumstances.

To represent Group Companies at external safety organisations.

ADDENDUM NO. 1
SAFETY IN OFFICES

A. GENERALLY

1. This supplementary document gives guidance on specific aspects of safety in offices. It in no way absolves any Team Members of their responsibilities under the Company Policy Statement notes nor should it be taken as superseding any specific safety notices or instructions displayed in offices or attached to equipment etc. It applies specifically to the Head Office and in general to all site offices.
2. Responsibility for ensuring compliance with the procedure and taking any remedial action necessary will be undertaken by the office Safety Supervisor.

B. FIRE EQUIPMENT AND EMERGENCY DRILL

1. Fire Fighting equipment must not be tampered with nor must it be used for any purpose other than that for which it is provided.
2. If any Team Member has any reason to believe that a fault has developed in the fire alarm system or an item of fire fighting equipment, or sees an item of fire fighting equipment being used for a purpose other than that for which it is provided, they should report the matter immediately to the safety supervisor who will take appropriate action.
3. Emergency exits must be kept clear at all times and fire break doors must not, in any way, be fixed in the open position.
4. Signs indicating fire escape routes must not be obscured, obliterated or otherwise defaced.
5. Notices indicating fire emergency drill will be displayed in accordance with the appropriate instructions and must be read and adhered to by every Team Member and by every other occupant of the building.
6. Fire drills will be organised at appropriate intervals.

C. FIRST-AID AND MEDICAL

1. All accidents, irrespective of how minor they may appear at the time, must be reported to the Safety Supervisor.
2. Any illness must be reported in accordance with the instructions in the Team handbook.
3. In the event of any emergency arising from an accident or illness, the safety supervisor and first-aid assistant should be advised immediately.

4. Team Members responsible for first-aid boxes must ensure that they are fully and correctly stocked, that the contents are not used for purposes other than first-aid and that they are accessible at all times.

D. LIFTS

1. Lifts must not be loaded in excess of the manufacturer's instructions displayed in each lift car.
2. Lifts must not be used by Team Members when alone in the building.
3. Lift controls must not be tampered with or in any way misused.
4. If any Team Member has any reason to believe that a fault has developed with the lift or its operating system, they must report this immediately to the safety supervisor who will take the appropriate action.
5. In the event of a lift being faulty while moving passengers, an emergency telephone has been provided. Instructions for use are adjacent to the telephone.

E. EQUIPMENT AND APPLIANCES

1. All equipment and appliances must be operated in accordance with the manufacturer's instructions.
2. All typewriters/computer work stations are to be set up in accordance with the company's guidance (see Addendum No. 2).
3. All portable electrical equipment – typewriters/print machines/calculators etc. must be switched off when not in use and the plug removed from the socket when vacating the office.
4. Any frayed leads, missing screws or loose connections must be reported, immediately they are discovered, to the safety supervisor. The equipment must not be used until the faults have been repaired.
5. In the event of an item of equipment or an appliance developing a fault this must be reported immediately to the safety supervisor or his deputy. It must not be used again until the fault has been repaired by a qualified mechanic.
6. Any item of equipment or any appliances, which in the course of operation could cause injury, must be fitted with a safety guard or device. On no account should the safety guard be removed or the equipment be operated without the safety guard or device being properly fitted. (Particular examples are guillotines, paper punches and drills, electric stapling machines, paper shredders, envelope opening machines etc.) The safety supervisor will advise on the suitability of safety devices or guards.
7. The flexible covers on photocopying machines must always be used to protect the operator from glare.

F. LIGHTING, HEATING, VENTILATION AND POWER SUPPLIES

1. No Team Member should in any way interfere or tamper with any of the above systems.
2. Adjustments to the control systems must only be made by a suitably qualified and authorised person.
3. Any faults or damage must be reported immediately to the safety supervisor.

G. FURNITURE

1. Furniture must only be used for the purposes and in the manner for which it is intended.
2. On no account should furniture be moved into such a position as to block or obstruct corridors passageways, entrances or exits.
3. Any damage to furniture which could impede its proper use or which might cause injury to any person using the furniture must be reported to the safety supervisor.

H. STORAGE OF MATERIALS, FILES, DRAWINGS ETC.

1. Where materials, files etc. are stored on shelves or in racks above arms' reach, then only proper steps or ladders must be used when placing or removing the materials, files etc.
2. Filing cabinet drawers must not be opened more than one at a time and must be closed immediately after use.

J.   CORRIDORS, STAIRCASES AND PASSAGES

1.   No person must leave anything in a corridor or passageway or on staircases which could in any way obstruct their safe and proper use.
2.   When moving through corridors, passageways or on staircases, whether as an individual, in groups, or with plant or equipment, care must be taken not to endanger the safety of others.
3.   Corridors, staircases and passageways will be properly lit at all times.

K.   CLEANLINESS AND DISPOSAL OF WASTE

1.   Any waste must be disposed of in an appropriate container or in accordance with any special instructions.
2.   Cigarettes and matches must not be placed in waste paper containers, but in ash trays.
3.   Toilets, washing facilities and sanitary appliances must be properly used and left clean so as not to endanger the health of others.
4.   All offices and facilities will be regularly cleaned but it is the responsibility of every Team Member to co-operate to ensure that they are not left in such a state as would endanger the health and safety of others.

L.   MAINTENANCE

1.   Regular inspection will be made for the purposes of maintaining buildings, roads, car parks, fixtures, fittings, furniture, plant, equipment and appliances.
2.   It is the responsibility of every Team Member to report any damage or fault which might endanger the health and safety of themselves or others. Equally, to regularly inspect any equipment, appliances or furniture in their care to ensure that it can be operated effectively and used safely.

ADDENDUM NO. 2
WORK WITH VISUAL DISPLAY SCREEN EQUIPMENT (VDE)

Arrangements for implementation of the Policy in respect of the safe use of VDE are contained in the following supplementary documents, copies of which are available from the

1.   User's Guide to Working with Visual Display Equipment.
2.   Working with VDE Workstations – Review Checklist.
3.   Guidelines for Conducting and Review of Workstations.

3.2 *Health and Safety Organisation Chart and Directory*

Note: Details of the Client, the Planning Supervisor and Design Consultants are given in sections 2.3, 2.5 and 2.6.

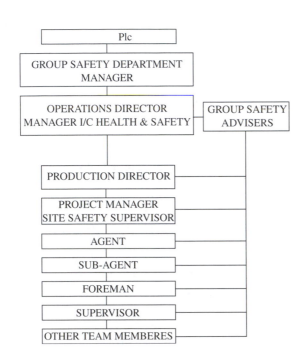

SITE-SPECIFIC DIRECTORY

Title:	Name:	Tel:
Director I/C Health and Safety		
Group Safety Manager		
Operations Director		
Production Director		
Group Safety Adviser		
Project Manager		
Construction Manager		
Safety Supervisor		
Fire Safety Co-ordinator		
Site Safety Assistant(s)		
Crane Co-ordinator		
Plant Co-ordinator (as appropriate)		
First-aid Attendant(s)		
Others		

3.3 *Health and Safety Standards*

Legislation/guidance to be complied with shall include but not be restricted to

The Factories Act 1961
The Offices, Shops and Railway Premises Act 1963
The Building Regulations
Health and Safety at Work Act 1974
Management of HASAW Regulations 1992
The Workplace (Health, Safety and Welfare) Regulations 1992
The Provision and Use of Work Equipment Regulations 1992
The Manual Handling Operations Regulations 1992
The Personal Protective Equipment at Work Regulations 1992
Construction (Health and Welfare) Regulations 1966
Food Hygiene (General and Amendment) Regulations 1970 and 1990 respectively
Offices, Shops and Railway Premises Act 1963
The Fire Precautions Act 1971
Fire Certificates (Special Premises) Regulations 1976
The Electricity at Work Regulations 1989
The Noise at Work Regulations 1989
The Abrasive Wheels Regulations 1970
COSHH Regulations 1988
Construction (Lifting Operations) Regulations 1961
Construction (General Provisions) Regulations 1961
Construction (Working Places) Regulations 1966
Construction (Health and Welfare) Regulations 1966
Construction (Head Protection) Regulations 1989
Control of Pollution (Special Waste) Regulations 1980
Builders Skips (Markings) Regulations 1984
The Highly Flammable Liquids and Liquefied Petroleum Gases Regulations 1972
Health and Safety (First Aid) Regulations 1981
First Aid at Work ACoP (42)
RIDDOR 1985
The Construction (Design and Management) Regulations 1994

British Standards
BS6187 1982 Code of Practice for Demolition

HSE Guidance Notes
GS29 Health and Safety – Demolition Works
Part 1 Preparation and Planning
Part 2 Legislation
Part 3 Techniques
Part 4 Health Hazards

4. MANAGEMENT SYSTEMS AND COMMUNICATION

4.1 Management Structure of Team

4.1.1 Client/Design and Team

Refer to the Project Organisation charts included in the **Appendix 10.1**

4.2 Direction and Co-ordination of Other Contractors

Overall responsibility for the direction and co-ordination of other contractors rests with the designated Project Manager. These responsibilities and the detailed procedures are laid down in the Project Quality Plan.

Once the order documentation is in place specifying the scope of works and responsibilities of the subcontractor, regular meetings will be held with the subcontractor to review progress.

These meetings will be arranged and chaired by the Project Manager or his Nominated Deputy and will be held at least once a month.

The Agenda of every meeting will include but not be limited to the following

– Safety and Welfare (including Risk Assessments and Method Statements)
– Industrial Relations
– Planning and Progress
– Labour
– Technical Queries

Specific actions will be agreed, monitored and recorded.

Where Method Statements are required/produced the typical Guide Documents within the Project Quality Plan will be used as a checklist of requirements together with any site specific items.

These items are:

Temporary Works, Protection of Services, Protection of Adjacent Property, Sequence of Work, Method of Erection, Safety Procedures, Safety of Persons, Safety of Plant, Local Environment Attendance, Compliance with Health and Safety, Method of Monitoring, Testing, Certificate of Tests, Plant, Materials, Samples, Mock-ups, Fixing, Systems, Plant Proposed, Transport, Deliveries, Storage, Setting Out, Preparation of Work Area, Services Required, Hoisting, Lifting, Access, Chain of Responsibility, Attendance Required, Subcontractors' Interface, Quality Assurance, Project Quality, Plan, Production Schedules, Co-ordination Procedures, Out of Sequence Working, Adverse Weather Precautions and Duty of Care.

Particular reference will also be made to section 10 of the   Site Safety Manual regarding safe systems of work and 10.2 regarding the preparation of Method Statements. If deemed necessary by the Project Manager, the Group Safety Adviser will be consulted during the preparation of any Method Statements.

Works will not proceed until the Method Statement has been received and checked by

As necessary, clear and specific written directions (instructions) shall be issued to the subcontractor. Subcontract site instructions, sheets and standard forms as defined by the Project Quality Plan are to be used for this purpose. Such instructions will be revised and signed by the Project Manager or his authorised nominee prior to issue.

The above mentioned Quality systems will be implemented by the Project Manager and will be audited by the Quality Manager. Production and Safety matters will also be monitored by the Production Director and Group Safety Adviser via regular meetings and safety audit reporting.

Site-Specific Measures

No specific measures at this stage.

4.3 Selection Procedures

4.3.1 Contractors

SUBCONTRACTORS

RESPONSIBILITIES

The Director-in-Charge

The director-in-charge has responsibility for ensuring compliance with this section, so far as it relates to the assessment and selection of subcontractors, and to the wording of health and safety clauses in subcontract orders.

The Project Manager

The project manager has overall responsibility for ensuring compliance with this section so far as it relates to subcontractors working under his direction or control.

THE ASSESSMENT AND SELECTION OF SUBCONTRACTORS

The awarding of contracts to subcontractors must be determined, not only on the grounds of price and technical ability, but also on past safety record and present ability to carry out work safely and without risk to health. The Health and Safety Questionnaire (at Appendix A to section 5 of the   Site Safety Manual) must, therefore, be sent to every subcontractor who wishes to be considered for work.

On receipt of the completed Health and Safety Questionnaire, an appraisal of the answers, and any supporting documentation must be undertaken using the subcontractor's Review Sheet (at Appendix B to section 5 of the   Site Safety Manual) with points being awarded in accordance with the instructions therein. In cases of doubt as to the awarding of points, a   safety adviser should be consulted.

Note: The subcontractor's Review Sheet must not be sent to the subcontractor.

Subcontractors scoring an average of 4 or 5 points may be considered acceptable. Those scoring an average of 3 may, at the discretion of the director-in-charge, be invited to rectify their shortcomings and submit a further completed Health and Safety Questionnaire for review. Those subcontractors scoring an average of less than 3 must be considered not acceptable.

THE SUBCONTRACT ORDER

Every subcontract order should contain the following clause:
  'Health and Safety'
  In accordance with the terms and conditions of the Subcontract Agreement covering work described in this Subcontract Order, you are required strictly to comply with the Health and Safety at Work, etc. Act 1974; the Management of Health and Safety at Work Regulations 1992; all other statutory acts and regulations; Safety Policy (the provision of which the subcontractor shall be deemed to have notice of); and local authority by-laws, so far as they relate to safety, health, welfare and the environment, and are for the time being in force.
  Assessments will be required of the risk to the Health and Safety of your employees to which they may be exposed while they are at work, and the risks to the health and safety of non-employees arising out of or in connection with your undertaking.
  Prior to work commencing on site, a senior and appropriately empowered representative of your company will be required to attend a meeting to determine and agree safe working methods and also to identify any high risk activities following a risk assessment in respect of which activities and appropriate method statements will be demanded.

PRE-START SAFETY MEETING

The Meeting

In order that specific and precise arrangements may be agreed with regard to safe systems of work in accordance with the Company Safety Policy, and also to ensure adequate co-operation and co-ordination in compliance with the MHSW Regulations, each subcontractor must be required to attend an initial safety meeting before he starts work on the site.

Checklist for Initial Safety Meeting with Subcontractors

A checklist for use at the initial safety meeting, together with a specimen letter to the subcontractor, confirming the matters agreed, is given at Appendix C of section 5 to the   Site Safety Manual. This checklist is for the use of the   Personnel conducting the meeting, and is not to be sent or given to the subcontractor, either before or after the meeting.

High Risk Activities

In cases where subcontractors are to be engaged in high risk activities, a safety adviser should attend the initial safety meeting.

Further Safety Meetings

Following the initial safety meeting, safety should normally be on the agenda at site production meetings. Further safety meetings may be called, however, involving one or more subcontractors, should the site manager, in consultation with his safety adviser, consider them necessary.

CLIENT'S RESPONSIBILITIES

Every subcontractor must be carefully briefed as to any requirements of the client which may relate to his work on site.

4.3.2 Designers

RESPONSIBILITIES

The Director-in-Charge

The Director-in-Charge has responsibility for approving the identification and appointment of any Design Consultants.

The Project Manager

The Project Manager is responsible for ensuring that all information requested from the Design Consultants by the Planning Supervisor is supplied.

The Assessment and Selection of Designers

Where a design service is required for a project, the selection of the Design Consultants will be carried out at tender stage in accordance with the procedures defined in the   Construction Ltd Quality System.
   As part of the selection procedure the Director-in-Charge will consider the following

- Their overall technical ability
- Their capability to deal with the particular project both in terms of resources and timescale
- Past safety record
- Their ability to implement and comply with the requirements of the CDM Regulations 1995
- Their commitment to Health and Safety training.

Potential Design Consultants will be requested to respond to a Pre-qualification Questionnaire. These will subsequently be assessed by the Director-in-Charge during the Selection process.

Site-Specific Measures

No specific measures at this stage.

4.3.3 Material/Suppliers

<u>**THE CONTROL OF SUBSTANCES HAZARDOUS TO HEALTH**</u>

**This section contains a resume of
the Control of Substances Hazardous to
Health Regulations 1988 (known as the
COSHH regulations), followed by
Group procedures for compliance.
It does not purport to be a definitive
work on the subject.**

<u>APPOINTMENTS AND RESPONSIBILITIES</u>

<u>Local COSHH Co-ordinator</u>

(i)   A person specifically appointed by a Group undertaking to be responsible for the co-ordination of all matters relating to COSHH. He or she should be the manager or a person specifically nominated by him or her who is responsible for specifying and/or ordering chemicals, substances and materials which could be subject to the COSHH Regulations. Alternatively, in appropriate cases, the person so appointed could be a full-time resident safety adviser.

(ii) Group undertaking includes every factory, workshop, depot, plant yard, site, quarry, laboratory, office or any other premises owned or operated by a   Group Company where persons are at work.

(iii) In the case of substances under the control of   and/or to which team members could be exposed, the local COSHH Co-ordinator will be responsible for maintaining an up-to-date list, on Form   /COSHH/01 (see Appendix 9.1 D of the   COSHH Manual), for ensuring that the relevant data is to hand, that assessments are undertaken and that appropriate control measures are identified and implemented.

(iv) For substances in common use, data, general assessments and advice on appropriate control measures will be available on Form   /COSHH/02 (see Appendix 9.1 E of the   COSHH Manual) in the   COSHH folder, a copy of which must be held on every site.

(v)   In the case of a substance for which no such information is available, data from the manufacturer or supplier will have to be consulted, an assessment undertaken and control measures identified on site. Blank forms   /COSHH/02 for this purpose are at the back of the   COSHH Manual. In cases of difficulty, the visiting Group Safety Adviser, or the Group COSHH co-ordinator, should be consulted.

(vi) It must be ensured that subcontractors maintain up-to-date lists of their substances, are in possession of all relevant data, undertake assessments and apply appropriate control measures. Copies of such lists and assessments should be obtained from subcontractors and filed separately from the   COSHH documentation.

<u>Group COSHH co-ordinator</u>

(i)   A Group Safety Adviser, trained in the principles of occupational hygiene, and charged with the duty of assisting local COSHH co-ordinators with routine assessments, and with advising generally on compliance with the COSHH Regulations.

(ii) Currently the Group COSHH co-ordinator is

<u>GROUP PROCEDURES – ASSESSMENTS</u>

<u>The Main Elements of an Assessment</u>

(i)   The identification of hazardous substances.
(ii)   Determination if the work presents risk of exposure.
(iii) Determination of the degree of exposure.
(iv)   Review existing precautions.
(v)   Decision on action to prevent or control risk and whether health surveillance is needed.

Substance Identification

 (i) Suppliers of substances are required to provide adequate information and precautions. Package labels and/or suppliers' data sheets must, therefore, be examined for relevant information (e.g. harmful by inhalation or skin irritant), approved use, noting in particular any limitations on use, and handling precautions (e.g. use in well-ventilated area or wear gloves).
 (ii) Every order, including every local purchase order, must contain clear instructions as to where the necessary information is to be sent. This will normally be to the delivery address for the attention of the Manager.
 (iii) Where the information provided is inadequate or non-existent, suppliers must be contacted. A specimen letter for this purpose is at Appendix 9.1 B of the   COSHH Manual.

Substance Data, Assessment, Control Measures Form   /COSHH/02

 (i) General (or generic) assessments for a range of materials in common use on site are in the   COSHH folder, a copy of which must be held on every site.
 (ii) For procedure to be adopted where no such general assessment has been undertaken, see 9.1.6 (v) of the   COSHH Manual.

SUBCONTRACTORS

 (a) A subcontractor has duties under the COSHH Regulations to his own employees and to others who may be affected by his work.
 (b) Proof that an assessment has been made in respect of every substance subject to the Regulations, stored, used or disposed of on site, by subcontractors, must be available on site.
 (c) An HSE Construction Summary sheet which explains the requirements of the COSHH Regulations is at Appendix 9.1 J of the   COSHH Manual. Copies of this Summary sheet may be issued to subcontractors as appropriate.

Site-Specific Measures

No specific measures at this stage.

4.3.4 Plant and Equipment Suppliers

PLANT SAFETY – GENERAL
Approved by the Executive Safety Committee 22 March 1995

PLANT SAFETY – GENERAL

PURPOSE AND SCOPE

This section contains the arrangements for the implementation of   Group Safety Policies with regard to the safe operation of plant. Its purpose is to ensure, so far as is reasonably practicable, that plant used on site, whether owned or hired, by   , package contractors or subcontractors, is operated, maintained, serviced, tested, examined and inspected, as is appropriate, in such a way as to ensure that manufacturer's/supplier's instructions, relevant legislation and related guidance are complied with, in order that employees and others are not exposed to risk to their health or safety on, or in the vicinity of,   Group construction sites.

RESPONSIBILITIES

 (a)   The Director-in-Charge

       The director-in-charge has responsibility for ensuring compliance with this section so far as it relates to the appointment, in appropriate circumstances, of a competent plant manager or plant co-ordinator.

(b) The Site Manager

The site manager has responsibility for

(i) ensuring compliance with this section so far as it relates to Health and Safety of persons working under his direction or control, and to third parties who may be at risk as a result of the Group undertaking, and

(ii) the appointment, in appropriate circumstances, of a competent plant co-ordinator.

(c) The Plant Co-ordinator

The appointed plant co-ordinator has responsibility for ensuring that

(i) no item of plant is introduced to site without his knowledge or approval

(ii) each item of plant introduced to the site is suitable for its intended purpose, and serviceable

(iii) all relevant statutory tests, examinations and inspections have been carried out and recorded prior to use on site

(iv) arrangements are in place for any necessary further tests to be carried out and recorded while the plant is on site

(v) plant operators, machinery attendants and banksmen are trained, competent and authorised

(vi) plant operators, machinery attendants and banksmen are given adequate information about site hazards, site rules, overhead and buried services, permit-to-work procedures, emergency plans, as is appropriate, and any other operation which could affect, or be affected by, the use of the plant

(vii) risk assessments concerning the use of the plant are undertaken and appropriate control measures implemented

(viii) arrangements are in place to provide for maintenance in accordance with manufacturer's/supplier's instructions

(ix) plant is properly maintained, and that maintenance and repair work is properly planned and executed, with full use being made of guards, cages, props and special tools as necessary, and that detailed records are kept

(x) fuel storage and refuelling arrangements are properly set up with regard to fire prevention and control of pollution

(xi) any discharge of fumes, fuel, oil or waste is properly dealt with to avoid harm to persons or the environment

(xii) measures are taken to reduce noise levels to as low as is reasonably practicable, and that signs and barriers are erected as required to define hearing protection zones

(xiii) electricity supplies are not adapted or modified by anyone other than an authorised electrician

(xiv) all plant accidents and near misses are reported promptly to the site manager.

FACTORS COMMON TO THE USE OF ALL PLANT

(a) The plant must be suitable for the purpose required.

(b) Plant must be clearly marked with appropriate safety information e.g. SWL, max RPM, or SWP.

(c) Appropriate warnings or warning devices must be fitted and serviceable.

(d) Guards, switches, interlocks etc., where required, must be fitted and serviceable.

(e) The written text of signs, notices, instructions etc. must be in English.

(f) Control systems for starting, stopping, emergency stopping and for changing speed and direction must be clearly identified and serviceable.

(g) Means of isolating plant from its energy source must be provided.

(h) Inspections and, in appropriate circumstances, thorough examinations and tests must be undertaken before use.

(i) Operators, machinery attendants, banksmen, maintenance fitters and supervisors must be given appropriate instruction, information and training.

(j) Operators, machinery attendants and banksmen must be over 18 years of age, competent and specifically authorised.

(k) Dangerous moving parts, and parts which give off high or low temperatures, must be effectively guarded.
(l) Supervision must ensure a safe place for the operations intended, including access and egress for the plant.
(m) Plant must be set up and used in stable condition.
(n) Sufficient lighting, space, and a safe environment must be provided.
(o) The use of plant must be considered when undertaking risk assessments.
(p) Appropriate PPE must be provided and used by the operator and others who may be at risk.
(q) Records must be kept of inspections, examinations and tests, and of authorisations and training.

Site-Specific Measures

No specific measures at this stage.

4.4 Information for Contractors

Contractors will have attended initial interviews for the subcontract works where Safety Matters were on the Agenda. Additionally, Contractors will have received subcontract documentation defining the scope of their works and the environment prior to commencing on site.

Prior to work commencing on site, a senior and appropriately empowered representative of the Contractor will be required to attend a meeting to determine and agree safe working methods and also to identify any high risk activities following a risk assessment in respect of which activities and appropriate method statements will be demanded.

Regular progress meeting will be held with Contractors and these are used as a vehicle to continually monitor and update contractors about risks to Health and Safety arising on the site as well as evaluating the need for further Risk Assessments etc.

Other means used to advise Contractors about risks to Health and Safety are

• Toolbox talks
• Posting of information on general site notice boards
• Posting of information on special Health and Safety notice boards

Site-Specific Measures

(To be reviewed and updated)

4.5 Communication and Co-operation

4.5.1 Communication and Information Issue Procedures

The ISO 9001 Quality Management System operated by   contains procedures for the receipt, registration, storage and distribution of design and technical documents generated by subcontractors, design consultants, site and works subcontractors.

These design and technical documents comprise:

Drawings, Sketches, Specifications, Schedules, Testing Procedures, Calculations, Operating and Maintenance Instructions, Installation Details, and Work Control and Administrative Procedures.

The detailed procedures are laid down in the project-specific Project Quality Plan and are available for inspection with the Project Manager.

Site-Specific Measures

When there is a need for co-ordination of two or more subcontractors, the principal means of communication and co-operation will be provided for at weekly Subcontractor Co-ordination Meetings.

4.5.2 Health and Safety Co-operation between Contractors

Safety matters are to be discussed at subcontract progress meetings and will be subject to formal and informal inspections. Where particular operations require, joint meetings will be held. Copies of risk assessments will be provided to subcontractors as applicable.

4.5.3 Management Meetings

operate a system of regular internal progress meetings at site level at which health and safety issues in relation to the work and the working environment are reviewed.

The Site Safety Supervisor, or his delegated assistant, is responsible for reviewing health and safety provisions on site, taking appropriate action where necessary on a weekly basis and submitting a written report to the Production Director responsible.

Group Safety Department has permanent Safety advisers who regularly and independently inspect and audit health and safety provisions on site and report to our Parent Company in order to monitor our Company Safety Policy. This report is also submitted to the Production Director responsible.

Site Managers and Supervisors receive safety training, which is updated on a regular basis, which is appropriate to their responsibilities as part of the company's on-going Safety Training Programme.

Induction safety training is provided for directly employed personnel and on-site training of subcontract personnel is provided if deemed necessary

Site-Specific Measures

Nothing further envisaged.

4.5.4 Arrangements for Contractor's Design Elements

Contractor design elements envisaged at this stage are

Normal Temporary Works
Structural Steelwork Connections
Lift Installation
Curtain Walling
Metal Cladding
Roof Lights
Proprietary Roofing

will manage contractor's design generally in accordance with those measures that are applicable to the professional design team – refer to section 4.3.2.

---

5. ASSESSMENT AND REPORTING

5.1 Activities with Risk to Health and Safety

5.1.1 Arrangements for Identification and Management of Risks

---

<u>**RISK**</u>
Approved by the Executive Safety Committee

<u>RISK ASSESSMENT</u>

<u>PURPOSE AND SCOPE</u>

This section contains the arrangements for the implementation of   Group Safety Policies with regard to risk assessment. Its purpose is to ensure, so far as is reasonably practicable, that, in appropriate circumstances, risk assessments are undertaken in accordance with the relevant legislation and guidance, in respect of employees and others who may be exposed to risk to their health or safety on, or in the vicinity of, Group construction sites.

<u>REFERENCES</u>

(a) <u>Legislation</u>

 (i)   The Health and Safety at Work Act 1974, Sections 2 and 3.
 (ii) The Management of Health and Safety at Work Regulations 1992, Regulations 3 and 4.

(b) <u>Guidance</u>

 HSE Approved Code of Practice relating to the Management of Health and Safety at Work Regulations 1992. Copies available from Group Safety Dept.

<u>EXCEPTIONS</u>

This section does not apply to risk assessment in relation to the following processes or activities which are dealt with elsewhere in the   Site Safety Manual:

(a) COSHH (section 9.1 refers)
(b) Noise at Work (section 9.8 refers)
(c) Manual Handling (section 25 refers).

<u>DEFINITIONS</u>

(a) The 'site manager' means the senior   Group employee who is for the time being in charge of the site or construction related activity.
(b) 'HSE' means Health and Safety Executive.
(c) A 'hazard' is a source of possible harm.
(d) A 'risk' is the likelihood of harm being realised.
(e) A 'risk assessment' is an evaluation of the chance that a hazard will cause harm which identifies and takes account of all significant factors, and which reaches a conclusion on whether and how the management of such factors needs to be improved to eliminate or lessen the chance.
(f) 'Control measures' means measures taken to control, minimise, or eliminate a risk. They may take the form of physical safeguards, e.g. guard rails and toeboards, special procedures, e.g. detailed method statements or permit-to-work, the provision of additional instruction, training or supervision, or, as a last resort, the issue and use of appropriate PPE.

(g) 'PPE' means personal protective equipment.
(h) 'MHSW Regulations' means the Management of Health and Safety at Work Regulations 1992.

RESPONSIBILITIES

(a) The Site Manager

The   Group site manager has overall responsibility for ensuring compliance with this section so far as it relates to persons working under his direction or control and to third parties who may be at risk as a result of the Group undertaking.

(b) Subcontractors

See 10.6.6 (a) of   Site Safety Manual.

(c) The self-employed
See 10.6.6 (b) of   Site Safety Manual.

ACTION SUMMARY

(a) Employers, including subcontractors must

(i)   assess the risks to employees and any others who may be affected by their undertaking
(ii) record the significant findings of the assessment where five or more persons are employed
(iii) review and revise each assessment as often as is necessary
(iv) make arrangements for implementing the health and safety measures identified as being required by the risk assessment, and
(v)   record the arrangements where five or more persons are employed.

(b) Self-employed persons must

(i)   assess the risk to themselves and any others who may be affected by their undertaking
(ii) review and revise each assessment as often as is necessary.

GENERAL PRINCIPLES OF RISK ASSESSMENT

(a) Risk reflects both the likelihood that harm will occur, and its severity. It is essentially a three-stage process, namely

(i)   identification of all hazards
(ii) evaluation of the risks, and
(iii) measures to control the risks.

(b) Non-routine activities such as maintenance and breakdowns must also be considered.

THE GENERAL PRINCIPLES OF PROTECTION

In order of priority:

(a) Avoid the risk completely.
(b) Combat the risk at source.
(c) Give priority to measures which protect:
 (i) the whole workforce
 (ii) the largest number of people.

(d) Provide protection specific to the individual, e.g. PPE, only as a last resort.

RECORDING THE ASSESSMENT

Arrangements for recording risk assessments are contained in the   Risk Assessment Manual, copies of which are available from the Group Safety Dept.

RISK ASSESSMENT REVIEW

Risk assessments, and any associated method statements, or permit-to-work procedures, must be reviewed should there be a substantial change in the matters to which they relate, or should it be suspected that they are no longer valid, and updated accordingly.

RECORDS

Copies of the risk assessments and associated method statements or permits-to-work must be kept available for inspection by those persons who may need to see them, e.g. factory inspectors, environmental health officers, safety representatives and safety advisers.

5.1.2 Hazards/Matters Including Those Identified in Tender Health and Safety Plan

The principal/significant hazards identified to date and which will require further and detailed consideration are as follows.

Where possible, further information and risk assessments have been included in **Appendix 10.7**, or will be added when a detailed assessment is undertaken with a subcontractor.

Hazard identified	In Tender H&S Plan Y/N	Risk Assessment done
Local traffic	Y	N
Pedestrians	Y	Y
Contaminated ground	Y	Part
Leaking sewers – contaminated groundwater	Y	Part
Existing services	Y	Part
Buried mineshafts	Y	N
Collapse of deep excavations	Y	N
Noise (pilling equipment and breakers)	Y	N
Unloading materials in highway	N	N
Falls, and falling objects	Y	N
Installation of large glazing panels	Y	N
Cutting and welding on site	Y	N

Widening of existing highway	Y	N
COSHH	Y	N

5.1.3 Contractor's Suggestions where Design and/or Materials could be Modified

> No matters apparent at this stage.
>
>
>
>
>
>
>
>
>
> Safety in design will be continually kept under review as detail design progresses. Subcontractors and site personnel will also be encouraged at interview/progress meetings to put forward suggestions either directly or via their own company representative.

5.2 Project Method Statement

> Refer to the Project Method Statement included in **Appendix 10.2**.

5.3 *Construction Programme*

Refer to the Construction Programme included in **Appendix 10.3.**

5.4 *Site Layout Sketch*

Refer to the Site Layout sketch included in **Appendix 10.4.**

5.5 *Reporting of RIDDOR Information*

**THE NOTIFICATION AND REPORTING OF ACCIDENTS,
DISEASES AND DANGEROUS OCCURRENCES**

**This section contains the legal and the
Group requirements for the notification
and reporting of accidents, diseases and
dangerous occurrences, and the Group
requirements in relation to accident
investigation**.

THE NOTIFICATION AND REPORTING OF ACCIDENTS, DISEASES AND
DANGEROUS OCCURRENCES

5.5.1 INTRODUCTION

The Reporting of Injuries, Diseases and Dangerous Occurrences Regulations (RIDDOR) 1985, effective from 1 April 1986, replace the Notification of Accidents and Dangerous Occurrences Regulations (NADOR) 1980. They also replace the provisions for the notification of industrial diseases in the Factories Act 1961.

This section, which is based on the Company Procedures Circular C1/5, covers the requirements of the RIDDOR, so far as they relate to the principal operations of the Group, together with the Group Procedures for the internal notification, reporting and investigation of accidents, dangerous occurrences and cases of disease contracted at work.

Once Group Safety Department has been notified of a fatality, major injury or notifiable dangerous occurrence, a Group Safety Adviser will endeavour to attend the scene of the accident as soon as possible for the purpose of carrying out an investigation, liaising with the relevant authorities and producing a report.

Following a fatality, major injury or notifiable dangerous occurrence, Group Safety Department will liaise with the Director in Charge, Company Secretary, Insurance and Publicity Departments, as appropriate, and will keep them advised of developments.

5.5.2 RESPONSIBILITIES

The responsible person shall be the Manager in implementing the procedure.

5.5.3 NOTIFICATION AND REPORTING

1. Injuries Causing Incapacity for More Than Three Days (i.e. at least four days' absence from work)

    (a) To an employee, trainee or self-employed person working on Company premises

Complete 'Notice of Accident to a   Group Employee' (coloured blue) and distribute within seven days as follows:

(i)   Insurance Ltd
(ii) Insurance Dept
(iii) File

   and <u>Complete Form</u>   and distribute within seven days as follows:

   (i) the local office of the relevant Enforcing Authority
   (ii) Group Safety Dept
   (iii)   Insurance Ltd
   (iv) Insurance Dept
   (v) File.

(b) <u>To an employee of a subcontractor</u>

Complete '   Group Third Party Accident Report' (coloured green) and distribute within seven days as follows:

(i) Insurance Dept
(ii) File

and

<u>Obtain a copy of the completed Form</u>   from the subcontractor and distribute within seven days as follows:

(i) Group Safety Dept
(ii) Insurance Dept
(iii) File.

Note: It is the responsibility of the subcontractor to send the completed form to the relevant Enforcing Authority.

2. <u>Major Injuries and Fatalities (for definition of Major Injury, see 3)</u>

(a) <u>To an employee, trainee, or self-employed person working on Company premises</u>

<u>Notify the following by telephone forthwith:</u>

(i) Group Safety Dept
(ii) the local office of the relevant Enforcing Authority (see 5)

and

<u>Follow the complete procedure outlined in 1(a).</u>

(b) <u>To an employee of a subcontractor</u>

(i) notify Group Safety Dept, by telephone forthwith
(ii) ensure that the accident has been reported by telephone to the local office of the relevant Enforcing Authority (see 4.5)

and

<u>Follow the complete procedure outlined in 1(b).</u>

(c) <u>To any other person (e.g. a visitor, RE, Clerk of Works, or member of the public) who was either on Company premises or was otherwise involved in an accident arising from Company work activities</u>

<u>Notify the following by telephone forthwith:</u>

(i) Group Safety Dept
(ii) the local office of the relevant Enforcing Authority (see 4.5)

and

Complete 'Group Third Party Accident Report Form' (coloured green) and distribute within seven days as follows:

(i) the local office of the relevant Enforcing Authority (see 5)
(ii) Group Safety Dept
(iii) Insurance Dept
(iv) File.

3. Definition of Major Injury

(a) fracture of the skull, spine or pelvis
(b) fracture of any bone
  (i) in the arm or wrist, but not a bone in the hand or
  (ii) in the leg or ankle, but not a bone in the foot
(c) amputation of
  (i) a hand or foot or
  (ii) a finger, thumb or toe, or any part therefor if the joint or bone is completely severed
(d) the loss of sight of an eye, a penetrating injury to an eye, or a chemical or hot metal burn to an eye
(e) either injury (including burns) requiring immediate medical treatment, or loss of consciousness resulting in either case from an electric shock from any electrical circuit or equipment, whether or not due to direct contact
(f) loss of consciousness resulting from lack of oxygen
(g) decompression sickness (unless suffered during an operation to which the Diving Operations at Work Regulations 1981 apply) requiring immediate medical treatment
(h) either acute illness requiring medical treatment, or loss of consciousness, resulting in either case from the absorption of any substance by inhalation, ingestion or through the skin
(i) acute illness requiring medical treatment where there is reason to believe that this resulted from exposure to a pathogen or infected material
(j) any other injury which results in the person injured being admitted immediately into hospital for more than 24 hours.

4. Delayed Fatalities

Where an employee dies after some delay as a result of an injury at work, the employer must inform the Enforcing Authority about the death, in writing, provided that it occurs within a year of the date of the accident.

5. Enforcing Authorities

(a) For construction sites, factories and workshops – the Health and Safety Executive
(b) For offices, shops, restaurants etc. – the Environmental Health Dept of the Local Authority
(c) For quarries and open cast coal sites – the Mines and Quarries Inspectorate.

6. Notifiable Diseases

(a) Certain diseases which are linked to specified work activities are notifiable to the relevant Enforcing Authority (see 5) on   Notification is to be sent only if
  (i) the employer receives in respect of an employee, a written diagnosis (for example, on a medical certificate made out by a doctor) of one of the prescribed diseases; and
  (ii) the ill employee's current job involves the corresponding specified work activity.

(b) Notifiable diseases include (in relation to construction and related work):
  (i) diseases arising from poisonings
  (ii) various skin diseases
  (iii) various lung diseases, including cancer linked to a work condition, e.g. hepatitis, leptospirosis, anthrax etc., and
  (iv) identified other conditions, especially in relation to work with

For further information and advice, and for copies of   please contact Group Safety Dept.

7. Accidents on the Public Road

Major injuries (see 3) and fatalities arising out of or in connection with a vehicle on a public road are notifiable and reportable only if they occur

(a) as a result of exposure to a substance being conveyed by the vehicle
(b) in connection with the loading or unloading of any article or substance onto or off the vehicle
(c) to a person who was either himself engaged in, or was injured or killed as a result of the activities of another person who was engaged in, work on or alongside a road being work connected with the construction, demolition, alteration, repair or maintenance of
   (i) the road or the markings or equipment thereon
   (ii) the verges, fences, hedges, or other boundaries of the road
   (iii) pipes or cables on, under, over or adjacent to the road, or
   (iv) buildings or structures adjacent to or over the road.

8. Dangerous Occurrences – Notification and Reporting (for definition see 9).

(a) Notify the following by telephone forthwith:

   (i) Group Safety Dept
   (ii) the local office of the relevant Enforcing Authority (see 5).

It is most important that Group Safety Dept be consulted before any dangerous occurrence is notified to the relevant Enforcing Authority.

Complete _____ and distribute within seven days as follows:

   (i) the local office or the relevant Enforcing Authority
   (ii) Group Safety Dept
   (iii) Insurance Dept
   (iv) File

and

Where damage has been caused to contract works, temporary works, material or plant (whether owned or hired) complete '_____ Group Loss and/or Damage Report Form' (coloured pink) and distribute within seven days as follows:

   (i) Insurance Dept
   (ii) File

or

Where a subcontractor is involved or where damage is caused to property not belonging to or in the custody or control of the Company complete '_____ Group Third Party Accident Report Form' (coloured green) and distribute within seven days as follows:

   (i) Insurance Dept
   (ii) File.

9. Definition – Notifiable Dangerous Occurrences

(a) Lifting Machinery etc.

   The collapse, overturning of, or the failure of any load bearing part of

   (i) any lift, hoist, crane, derrick or mobile powered access platform, but not any winch, teagle, pulley block, gin wheel, transporter or runway
   (ii) any excavator, or
   (iii) any pile driving frame or rig having an overall height when operating of more than 7 m.

(b) Pressure Vessels

Explosion, collapse or bursting of any closed vessel, including a boiler or boiler tube, in which the internal pressure was above or below atmospheric pressure which might have been liable to cause the death of, major injury (see 3) or notifiable disease (see 6) to, any person.

(c) Electrical Short-circuit

Electrical short-circuit or overload attended by fire or explosion which resulted in the stoppage of the plant involved for more than 24 hours and which, taking into account the circumstances of the occurrence might have been liable to cause the death of, major injury (see 3) or notifiable disease (see 6) to, any person.

(d) Explosion or Fire

Any explosion or fire occurring in any plant or place which resulted in the stoppage of that plant or suspension of normal work in that place for more than 24 hours where such explosion or fire was due to the ignition of process materials, their by-products (including waste) or finished products.

(e) Escape of Flammable Substances

The sudden, uncontrolled release of one tonne or more of highly flammable liquid, flammable gas or flammable liquid above its boiling point from any system or plant or pipeline.

(f) Collapse of Scaffolding

A collapse or partial collapse of any scaffold which is more than 5 m high which results in a substantial part of the scaffold falling or overturning, and where the scaffold is slung or suspended, a collapse or part collapse of the suspension arrangement (including any outrigger) which causes a working platform or cradle to fall more than 5 m.

(g) Collapse of Building or Structure

Any unintended collapse or partial collapse of

  (i) any building or structure under construction, reconstruction, alteration or demolition, or any falsework, involving a fall of more than 5 tonnes of material or
  (ii) any floor or wall of any building being used as a place of work, not being a building under construction, reconstruction, alteration or demolition.

(h) Escape of Substance or Pathogen

The uncontrolled or accidental release or the escape of any substance or pathogen from any apparatus, equipment, pipework, pipeline, process plant, storage vessel, tank, in-works conveyance tanker, land-fill site, or exploratory land drilling site, which having regard to the nature of the substance or pathogen and the extent and location of the release or escape, might have been liable to cause the death, major injury (see 3) or notifiable disease (see 6) to any person.

(i) Explosives

Any ignition or explosion of explosives, where the ignition or explosion was not intentional.

(j) Freight Containers

Failure of any freight container or failure of any load bearing part therefor while it is being raised, lowered or suspended.

(k) Breathing Apparatus

Any incident in which breathing apparatus malfunctions, other than while being maintained or tested, in such a way as to be likely to deprive the wearer of oxygen, expose him to a contaminated atmosphere, or otherwise pose a danger to his health.

(l) Overhead Electric Lines

Any incident in which plant or equipment either comes into contact with an uninsulated overhead electric line in which the voltage exceeds 200 V, or causes an electrical discharge from such a line by coming into close proximity to it.

(m) <u>Locomotives</u>

> Any case of an accidental collision between a locomotive or a train and any other vehicle at a factory or at dock premises, which might have been liable to cause the death or major injury or notifiable disease to any person.

NEAR MISSES

A 'near miss' is an incident, other than a notifiable dangerous occurrence, sometimes know as a 'no injury accident' which may or may not cause damage, but does not on this occasion result in personal injury.

Any such incident must be investigated in order that the cause may be established and eliminated.

Where the potential for serious personal injury is significant, the Group Safety Dept must be notified forthwith in order that an official investigation may be undertaken, a report produced and in appropriate circumstances, remedial action recommended.

KEEPING RECORDS

A record must be kept of every accident, case of disease, and dangerous occurrence reported to the Enforcing Authorities. For this purpose a copy of every record must be filed at the site or premises where the incident occurred. All such records (except records which reveal personal health information) must be readily accessible to Inspectors from the Enforcing Authority, and to appointed Safety Representatives.

FURTHER INFORMATION

The HSE may from time to time, with the approval of the HAC, require further information about reportable and notifiable accidents, notifiable dangerous occurrences, or notifiable diseases. In such cases, the HSE will despatch a further form which would ask the recipient to furnish more detailed specific information about a particular incident.

SCENE OF AN ACCIDENT NOT TO BE DISTURBED

Notwithstanding the requirements of section 3.4.4 of the   Site Safety Manual which relates specifically to quarries and open cast coal sites, the scene of any fatality, major injury accident or notifiable dangerous occurrence must not be disturbed until a Group Safety Adviser, Enforcing Authority Inspector, appointed Safety Representative and a representative of our insurers have had the opportunity to inspect it. This requirement shall not apply to any action necessary to safeguard against further hazards.

WITNESSES

The names and addresses of witnesses and other persons concerned with the accident must be obtained together with a brief account of the accident as it appeared to them.

PHOTOGRAPHS AND DRAWINGS

Arrangements should be made at once for photographs to be taken. In the case of a fatal accident, a professional photographer should be employed. In other cases the employment of a professional photographer will be at the discretion of the manager and/or safety adviser, dependent on the severity and circumstances of the accident.

Where drawings would be helpful to an investigation, they should be prepared preferably by an engineer, and should be signed and dated.

INVESTIGATION

(a) <u>Enforcing Authority and Police</u>

> Close liaison must be maintained with any Enforcing Authority Inspector or Police Officer who may have cause to investigate the accident. Such facilities and assistance as they require must be provided.

(b) Appointed Safety Representatives

Appointed Safety Representatives may investigate notifiable accidents where the interests of employees whom they are appointed to represent might be involved. Such facilities and assistance as they may require must be provided.

   For restriction on the information which must be made available to an appointed safety representative see Regulation 7(2) of the Safety Representatives and Safety Committees Regulations 1977.

(c) Other Interested Parties

Before any solicitor, engineer or investigator acting on behalf of any other interested party is allowed at the scene of the accident, approval must be obtained from Insurance Dept.

(d) Persons Conducting an Investigation to be Accompanied by a Company Representative

Any person other than Group   employee, who is entitled, or has been given permission to conduct an investigation, must be accompanied by a Company representative, save that Enforcing Authority Inspectors, Police Officers and appointed Safety Representatives are entitled to have private discussions with employees.

CORRESPONDENCE

(a) The Enforcing Authority

Any correspondence from the Enforcing Authority relating to an accident must be forwarded unanswered to Group Safety Dept.

(b) Other Interested Parties

Any correspondence from solicitors or other interested parties relating to an accident must be forwarded unanswered to Insurance Dept.

CORONER'S INQUEST

It is essential that the Company be represented at an Inquest. Group Safety Dept must be advised of date, time and venue as soon as they are known.

6. SITE

6.1 Welfare Facilities

Initial Construction Phase (Mine Grouting, Bored Piling and Bulk Excavation)

Toilets – portable chemical toilets with washing facilities to be maintained by hire company
Changing/Drying Room provided by subcontractor
Canteen – not provided
Storage of PPE by subcontractors
Subcontractors to advise on storage of oils and fuels.

Main Works

Estimated that at peak there will be 150 to 170 operatives
Wash facilities to include 6 basins
Toilets to include 6 WCs and 6 urinals; separate WC and basin for females
Canteen to sit 50 persons – breaks to be phased
Drying rooms for   employees with storage for personal PPE. Subcontractors to provide separate facilities for their own operatives
Full site layout and facilities to be developed.

6.2 Information and Training for People on Site

All personnel working within the confines of the site boundary will be required to undergo induction/orientation training. The following procedure will be adopted.

The Construction Manager will carry out induction for the senior site management of the Subcontractor. This will follow a pre-set syllabus which will cover all issues relevant to the nature of the works and the current site conditions. The induction records will be retained by the Construction Manager and a certificate of training will be issued to each person as appropriate.

The Subcontractor's site personnel will be inducted by the Subcontractor's senior management (previously inducted by the Construction Manager) using the same approach but adding any company specific information not covered by the Construction Manager's syllabus.

Records of the induction will be maintained by the respective Subcontractors or the Construction Manager as appropriate. These records will be subject to audit by the Construction Manager at regular intervals.

The Subcontractor will be expected to maintain contemporary records of all safety training and toolbox talks together with a log of the names of all operatives who have received the training, which must be available for inspection by the Construction Manager on request. These items will be discussed as part of the regular progress meeting process and audited as appropriate.

A project information sheet containing the following information will be issued to each Subcontractor at the Pre-Start Meeting and he is to ensure this information is passed to the site personnel under his control. This is to be recorded on the individual's site record induction sheet.

Project Information sheet to include:

1. Client's name and project representative.
2. Planning Supervisor's name.
3. Principal Contractor's name.
4. Subcontractors' names and brief description of their works.

5. Emergency procedures and Fire Safety procedures.
6. Site rules.
7. Location of Statutory Notices.

In addition the Subcontractor is to include details of his intended 'on-site' toolbox and method statement talks schedules at the pre-start meeting.

6.3 Health and Safety Consultation

It is intended that the Construction Manager will encourage an open policy with regard to Health and Safety issues.

Subcontractors will be expected to provide feedback to the Construction Manager at the regular site progress meetings, the site co-ordination meeting and at any other time deemed appropriate.

The Construction Manager by discussion with the site personnel will also obtain views from the workplace direct.

The Subcontractors will be expected to carry out daily inspections of the workplace either by their senior management or visiting safety adviser, record any non-conformance and report to the Construction Manager any areas of concern. The Construction Manager will carry out regular audit of these inspections to confirm compliance with these requirements.

The Construction Manager's safety adviser will also visit site on a regular basis to comment and advise on the site conditions/issues and provide instructions and information direct to work face or site. A copy of his report will be sent to any subcontractors mentioned for their action.

If certain safety issues keep on reoccurring (i.e. safe use of lightweight towers), an emergency site meeting will be held with those persons involved in this activity to discuss and review the issues and explain the safety requirements and standards required.

6.4 Site Rules and Procedures

General

1. No smoking on site.
2. Hard hats and safety boots to be worn.
3. Access and egress to adjoining owners' offloading and car parking areas to remain clear at all times.
4. Wheel cleaning facilities to be used, as necessary, during excavation works.
5. As far as is reasonably practicable, the site must be kept tidy at all times to avoid hazards arising out of the build-up of debris, particularly of a flammable nature, on a confined site.
6. Because of the number of office and retail buildings in the vicinity of the site, the subcontractors must minimise noise generated when carrying out the works.
7. Eating and drinking will be restricted to mess facilities only.

Safety

Subcontractors shall carry out documented general risk assessments:

(a) for all significant risks which are foreseeable within the contractor's work activities on the project and to implement and maintain suitable preventative and/or protective measures necessary to protect employees and others from those risks, and

(b) to cost in terms of resources, time and finance the preventative and/or protective measures as part of the pre-qualification or tendering process of the contract.

If, during the risk assessments, any significant risks are identified which are relevant to the work activities and are not identified in the Project Health and Safety Plan, those risks shall be assessed and brought to the attention of the Planning Supervisor.

Subcontractors shall ensure that employees designated to work on the project have, as appropriate, sufficient skills, training and experience or knowledge to properly undertake the work activities and to comply with the requirements and prohibitions imposed by or under the relevant statutory provisions applicable to the project.

Subcontractors shall ensure that adequate resources, including the necessary plant, machinery, technical facilities, trained personnel and time, are provided to fulfil their health, safety and welfare obligations within the project.

Subcontractors shall ensure, in relation to the project, that they have:

(a) the arrangements in place to actively manage health and safety
(b) the procedures for developing and implementing the health and safety plan (or any relevant part of it)
(c) the appropriate measures in place to deal with high risk areas identified in the health and safety plan
(d) the arrangements in place for monitoring compliance with health and safety legislation
(e) the people to carry out or manage the work, with the required skills and training
(f) allocated adequate time to complete the various stages of the construction work without risk to health and safety, and
(g) employment procedures in place to ensure the way people are employed complies with health and safety law.

The Construction Manager (assisted by the Health and Safety Adviser) shall monitor work activities to ensure that they consistently comply with the requirements and prohibitions of this Project Health and Safety Plan and current health and safety legislation.

Subcontractors shall carry out all work activities in compliance with any rules applicable to the work activities contained in the Project Health and Safety Plan and in accordance with current health and safety legislation.

Where appropriate, subcontractors shall carry out construction work in accordance with BS-EN standards and codes of working practice or other published authoritative standards.

All subcontractors' work activities and workplaces (including means of access and egress) shall, as appropriate, be maintained in a safe condition, efficient state, efficient working order and in good repair.

All subcontractors shall in relation to the project:

(a) co-operate with the Planning Supervisor and the Principal Contractor so far as is necessary to enable each to comply with their duties under the relevant statutory provisions
(b) promptly provide the Planning Supervisor and the Principal Contractor with any information (including any or any relevant part of any risk assessment) which might affect the health or safety of any person carrying out construction work or of any person who may be affected by the construction work or which might justify a review of the Health and Safety Plan
(c) comply with any reasonable direction given by the Planning Supervisor and the Principal Contractor so far as is necessary to enable each of them to comply with their duties under the relevant statutory provisions
(d) comply with any rules in this Project Health and Safety Plan and any rules for the management of construction work which are required for the purposes of health and safety
(e) promptly provide the Planning Supervisor and the Principal Contractor with that information relating to any death, injury, condition or dangerous occurrence which the subcontractor is required to notify or report under the Reporting of Injuries, Diseases and Dangerous Occurrences Regulations 1995
(f) take reasonable steps to control access by unauthorised persons to danger areas where construction work has started but has not been completed
(g) shall not cause or permit any employee to work on construction work unless the employee has been provided with the following information:
    (i) the name of the Planning Supervisor for the project, and
    (ii) the contents of the health and safety plan or such parts of it as are relevant to the construction work which any such employee is to carry out

(h) ensure that their employees at work on the construction work are able to consult with, and offer advice to the Construction Manager and/or the Planning Supervisor, on matters connected with the project which will affect their health and safety

(i) ensure that there are arrangements made for co-ordinating the views of employees at work on construction work, where necessary for reasons of health and safety

(j) ensure that every employee carrying out construction work is provided with comprehensible and relevant information on:
   (i) the risks to their health and safety identified by risk assessments
   (ii) the preventative and protective measures
   (iii) the procedures to be followed for serious and imminent danger and danger areas and the names of persons nominated to implement evacuation procedures in the event of an emergency
   (iv) areas occupied by other subcontractors to which it is necessary to restrict access on the grounds of health and safety unless the employee concerned has received adequate health and safety instruction
   (v) the risks to their health and safety arising out of or in connection with the conduct of other workers at work on the project

(k) ensure that their employees are provided with adequate health and safety training on their being employed on the project and being exposed to new or increased risk because of:
   (i) being transferred or given a change of responsibilities
   (ii) the introduction of new work equipment into or a change of work equipment already in use
   (iii) the introduction of new technology
   (iv) the introduction of a new system of work into or a change in a system of work already in use

(l) provide the Planning Supervisor and the Principal Contractor with any information which is necessary for inclusion into the Health and Safety File.

6.5 Emergency Procedures

6.5.1 Fire

Preliminary Works (Grouting, piling and excavation)

In case of fire:

1. Warn all other personnel within the area and ensure evacuation is under way.
2. Only tackle the fire if you have the correct fire fighting equipment, you have been trained in its correct use and you do not expose yourself to any unnecessary risk.
3. Advise   or your own supervisor of occurrence and ensure emergency service is called if required.
4. Ensure area of emergency is sealed from further access.
5. Give specific directions to the emergency service vehicles on arrival at the site entrance.

Main Works

The Fire Safety Plan will be developed in due course.

6.5.2 Injury

If you are in the vicinity of the accident carry out the following:

- Stop work
- Attend accident and assess situation
- If casualties, immediately contact and advise nature and extent
- Check with the Construction Manager that the Emergency Service has been called

- Apply first-aid if it is safe and if you are trained to do so. *Do not* move casualties unless under extreme emergency
- If no casualties – barrier area off to prevent persons straying into area until help arrives
- Report any observations to your site supervisor as soon as possible.

All contractors are to ensure that they have adequate first-aiders and equipment on site and these arrangements are to be detailed in the Subcontractor's Safety Plans.

All accidents on site are to be recorded in the Subcontractor's Site Accident Book and reported to the Construction Manager.

A member of the site-based team will be trained in First-Aid, and name and location of First-Aid Box and Accident Book will be displayed and form part of the site induction.

6.5.3 Dangerous Occurrences

Actions as 6.5.2 above plus
- Under no circumstances enter the area – wait for the emergency services.

7. HEALTH AND SAFETY FILE

7.1 Programme for Provision of Information

All relevant information for insertion into the project Health and Safety File, provided by   will be continuously reviewed, recorded and collated for final issue to the Client at or before Building Completion.
   A Programme for information release is to be developed.

8. PLAN REVIEW

8.1 Safety Plan Issue

Recipients of copies of the Construction-Phase Health and Safety Plan are:

Recipient	Date Issued	Revision
		O
		O
		O

8.2 Project Review

This Construction Health and Safety Plan will be developed further as the design is progressed and/or the specialist subcontract packages are procured via the addition of detailed Method Statements. Any additional hazards will be identified and risk assessments will be carried out as the works proceed.

Site-Specific Measures

Nothing to add at this stage.

8.3 Safety Amendments

Amendments and additions to the plan will be made only by The Project Manager once they have been reviewed by   Significant changes to the plan are to be brought to the attention of the appropriate Director responsible for checking/approval.

Site-Specific Measures

Nothing to add at this stage.

9. ARRANGEMENTS FOR MONITORING PERFORMANCE

Monitoring of Health and Safety performance will be carried out by virtue of the procedures and systems described in sections:

3.0   Health and Safety systems
4.0   Management Systems and Communication
5.0   Assessment and Reporting.

<u>Site-Specific Measures</u>

No further measures envisaged.

10. APPENDICES

10.1 Project Organisation Charts

10.1.1 Phase 1A Project Organisation Chart

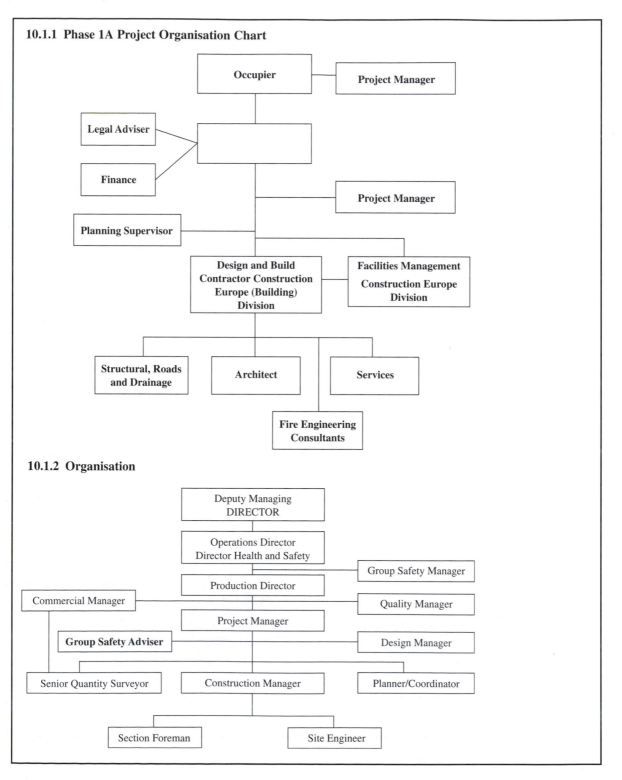

10.1.2 Organisation

10.2 Method Statement

CONSTRUCTION METHOD STATEMENT

10.2.1 Introduction

10.2.2 Site Establishment

10.2.3 Site Layout and Accommodation

10.2.4 Pedestrian and Traffic Management

10.2.5 Site Logistics

10.2.6 Enabling Works

10.2.7 Main Construction Works

10.2.1 Introduction

The replacement offices for   Council are the first phase of the   redevelopment that will enable the existing town hall extension to be demolished providing the space for the future development phases. It is vital that the Phase 1A construction works take due consideration of the redevelopment as a whole.

The Phase 1A site is bordered by public roads and footpaths. The need to minimise the impact of our operations on the general public, and the normal operation of the city centre, is of paramount importance. The physical measures that we will put in place and the management procedures that we will implement will be crucial to the success of the project.

The construction of the new offices is challenging, particularly in terms of providing a facility that has to be far superior in comparison with the existing town hall extension that it is to replace. Constraints imposed by the location and ground conditions of the site present further difficulties. To meet this challenge, we have allocated a number of experienced managers, supervisors and support staff. They will act as a forward thinking, coherent team providing good management, effective planning and problem-solving skills.

The subcontractors and suppliers that we will employ will be experienced and suited to the needs of the project. We have thoroughly reviewed the building fabric, services and internal finishes to carefully determine the plant, equipment and systems to facilitate the construction works safely, speedily, within budget and to the required quality.

The following sections outline our approach to the project. In due course, further detailed method statements will be produced for specific elements of the works.

10.2.2 Site Establishment

The site will be established in four phases. The first will involve localised works to divert the existing underground services in   Street,   Street and   Lane.

The second involves the complete closure of the surface car park that currently occupies the Phase 1A site. This will enable the perimeter underground obstructions to be removed, the redundant sewer connections capped and the commencement of grout stabilisation of the mine workings.

The third phase will be the closure of a portion of the Phase 1B car park to enable early construction of the new turning head to facilitate the closure of   Lane. With   using the turning head and the grouting works suitably advanced, the main piling works will be commenced. The proposed location of the hoarding and the provision of the traffic and pedestrian management system in shown in Figures 1 to 8.

Access to the site will be via   Street and   Street to the main site entrance in   Lane. Construction traffic routing through   will generally follow and share the permanent arrangements for the vehicles servicing   House and other surrounding buildings.

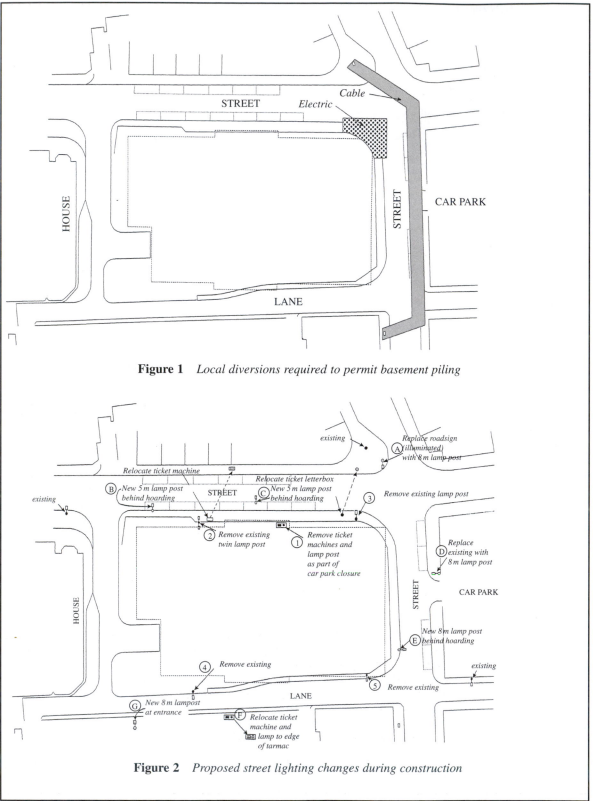

Figure 1 *Local diversions required to permit basement piling*

Figure 2 *Proposed street lighting changes during construction*

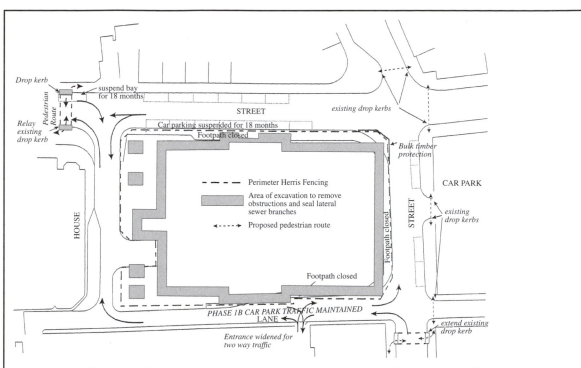

Figure 3 *Underground obstruction removal and sewer sealing to permit piling*

The footprint of the building occupies much of the Phase 1A site and being a city centre development, we recognise space limitations and confined working arrangements. Site working hours will be 7.30am to 6.00pm Monday to Friday. Weekend working will be carried out as required or deemed necessary.

The site will be protected by a secure 2.4 m high hoarding and fencing. During the enabling works, substructure and main frame construction, we will provide a site-based security presence during the normal working hours with a visiting patrol covering nights and weekends. A full 24 hour site-based security presence will be in place during the later stages of the construction works and fit out prior to occupation. We see a seamless transfer of site security to the FM provider at completion.

Levels of noise, dust and mud will be monitored and controlled by site personnel. On-site precautions will include secure hoarding and monoflex sheeting to scaffold elevations and the provision of a wheel cleaning facility to reduce the impact of muckaway traffic on the normal traffic routes.

A liaison manager will be appointed at the commencement of the site works. He will be the point of contact to promote a good working environment and deal with any nuisance issues that may arise during the construction period.

10.2.3 Site Layout and Accommodation

Space or site accommodation is very limited. On site accommodation will consist of a temporary site office and security hut. Welfare facilities, including male and female toilets, drying room and canteen, will be provided.

Subcontractors' site accommodation will be limited to small mobile or temporary offices constructed and contained within the building footprint.

It is our intention to rent local office accommodation for our site management team.

There will be no parking available on site.

10.2.4 Pedestrian and Traffic Management

It is our intention to close the footpath on the east side of   Lane and on the south side of   Street once the Phase 1A car park has been closed and the main construction works commenced on site.

Pedestrian routes with suitable signage will be provided in   Street and   Street. A cycle lane will be marked out on the west and north sides of   Street and   Street respectively.

These pedestrian routes and cycle lanes will be maintained during the site construction period until the permanent footpaths and cycle routes are available. The proposed routing is shown in Figure 5.

     will ensure that all suppliers and hauliers will notify the highway authorities in respect to loads over 38 tonnes and of abnormal width. Construction traffic entering or leaving the site will be actively managed.

     Street and   Street will be subjected to light and heavy day to day traffic such as vans, excavation tippers, concrete delivery lorries, steel delivery lorries, brick and block delivery lorries, skip lorries and delivery vehicles generally. All deliveries will be booked into a delivery schedule system controlled by site personnel under the overall guidance of the Construction Manager.

Site traffic will conform to local requirement guidelines and permanent routing changes as advised.

10.2.5 Site Logistics

It is good construction policy to facilitate efficient movement of materials to and from the workface.

A tower crane, providing good lifting capacity and complete hook coverage, will be erected within the building at basement level 4 following the excavation and local construction of the ground slab. It is our intention to utilise a conventional saddle jib tower crane mounted on a static base with kentledge reducing the need for deep foundations and expendable fixing angles. This crane will oversail adjacent properties in its out of service condition. If a suitable oversail licence proves difficult to obtain, then the use of a luffing jib crane will be reviewed.

Forklifts will be employed to offload and distribute materials at the ground floor level and in the basement areas. These forklifts will be licensed for use on the public roads and will be equipped with the mandatory warning provisions.

A materials goods hoist will be erected in the lightwell void to feed palletted construction materials from the ground floor to all levels above. Horizontal distribution within the building will be by electric pallet truck or similar. The hoist platform will be of sufficient size and capacity to accept pre-palletted materials and prefabricated modules. The location of the goods hoist will enable the platform to be loaded and unloaded directly by the forklifts.

Following construction of the structural frame, waste materials and rubbish will be marshalled to a dedicated area on each floor and removed in suitable containers to skips outside the building footprint.

It is the responsibility of everyone on site to maintain a clean, uncluttered and safe working environment.

The spatial constraints of the site offer little room for material storage. Material deliveries will be structured and managed on a 'just in time' basis. Daily deliveries will be scheduled and controlled by the site team using tried and tested procedures.

In the later stages of the construction, the permanent lifts will be used for material handling to enable the goods hoist in the lightwell to be dismantled to progress the One-Stop Shop fit out.

Preliminary meetings with representatives of   Design and Property Department   and   Water have established that the new surface water outfall will be constructed from a 4 m diameter drive shaft/chamber located in the Phase 1B car park near to the   public house on what was   Lane. The outfall works are planned to commence in the late summer of   for completion by Christmas   . Our proposals include for liaison with the SP&D contractor and the provision of fair and reasonable access for him to carry out his works.

10.2.6 Enabling Works

The site covers an area that has been subject to underground mine workings. From the site investigation borehole logs, trial excavations and archaeological surveys, it is clear that these old mine workings require suitable treatment to allow the construction of the proposed building on the site. The investigations also indicate basement walls and floor slabs left in place from previous occupation of the site and partially filled by demolition materials. Surveys of the local combined sewer system indicate lateral connections that have not been capped.

The enabling works will commence with the clearance of the oversite materials, removal of the previous building obstructions and the capping of the redundant sewer laterals to enable the grouting of the mine working to commence. Relocation and diversions of the underground services will be undertaken by the statutory authorities and utilities prior to this initial period.

Reconstruction of Union Lane and the provision of the new turning head will be undertaken following approval of   design, adoption by   , the tendering and mobilisation of the Highways Contractor.

The proposed method of construction ensuring continuous access for   deliveries is shown in Figures 4 to 6.

The grouting operation will commence with perimeter drilling and formation of a grout curtain outside the foundation area of the building to minimise leakage of pressurised grout outside the required treatment area. The drilling and grouting will commence at the lowest location of the coal seam which is the corner at the junction of   Lane and   Street.

The line of the piled retaining wall will have to be excavated to remove obstructions and backfilled with a suitable fill material to assist the piling operation and capping beam construction. Trial trenches need to be carried out to establish and locate underground services and the location of underground structures and their possible influence on the perimeter piling works.

10.2.7 Main Construction Works

Sub-structure

Bored piling will commence with the contiguous bored pile basement wall at the   Street/   Lane elevations following completion of the grouting in this the lowest area of the coal seam. Once the grouting is complete at the highest point, bored piling to form the column foundations on grid 1 will commence together with the contiguous piled wall and public stair access recess on grid 2. Generally the drilling/grouting and piling will follow the sequence of lowest point up using two bored piling rigs.

The contiguous piled wall and capping beam will be designed to support the basement excavation with additional propping. The depth of the excavation and the variability of the ground conditions deem a self-supporting retaining wall an uneconomical option. It is proposed to install temporary propping at the level of the in-situ capping beam in a triangular configuration supporting the corners. Excavation of the centre section of the basement and early construction of the ground slab will facilitate a base for raking props at the core locations as well as

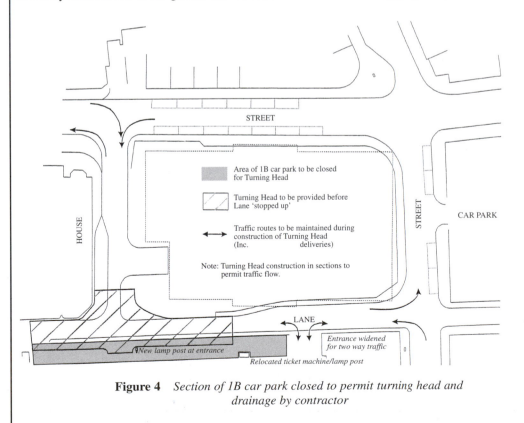

Figure 4 *Section of 1B car park closed to permit turning head and drainage by contractor*

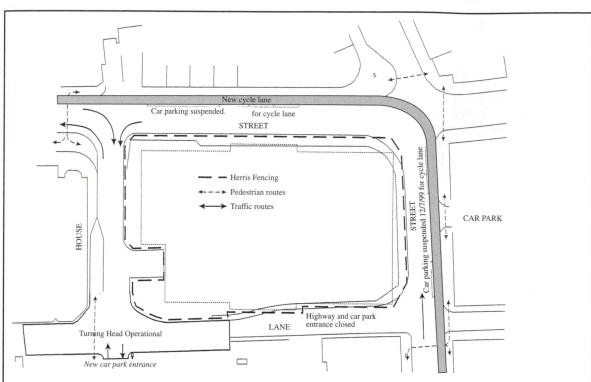

Figure 5 *Site boundary and traffic routes at union lane stopping up 19 July 99*

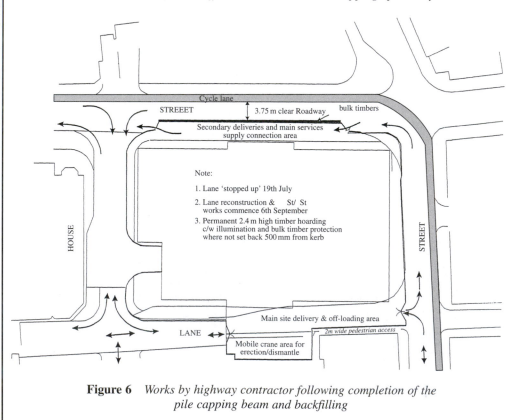

Figure 6 *Works by highway contractor following completion of the
pile capping beam and backfilling*

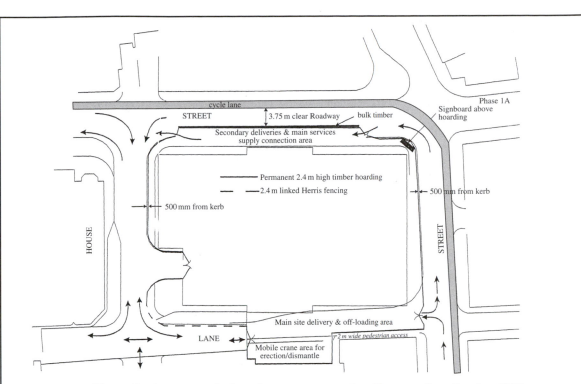

Figure 7 *Permanent site hoarding, entrances and traffic routes from October 1999*

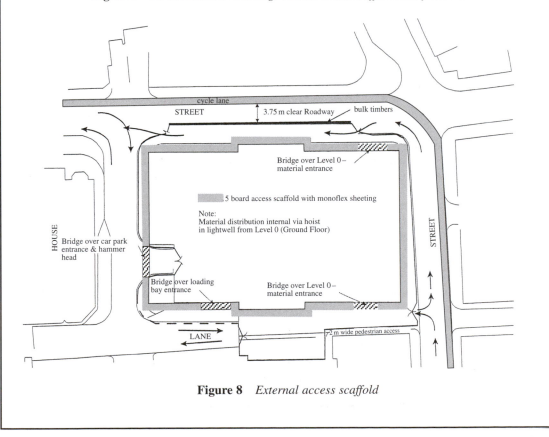

Figure 8 *External access scaffold*

providing a permanent base for erection of the tower crane. Progressive installation of the props and excavation of the perimeter beams will enable the construction of the column foundations to proceed.

The basement retaining sub-structure requires permanent support from the diaphragm action of the suspended floor slabs at levels 1 and 2 and support from the core walls. It is our intention to erect the steel frame from the lower basement to ground floor level and construct the precast plank floor and in-situ structural topping to provide a working platform. This will enable the superstructure to progress in parallel with the basement works. The proposed location of the corner props will not restrict the space for construction of the in-situ concrete lining to the piled basement walls. Similarly, the positioning of the racking props will enable unhindered construction of the core foundations, lift pits and in-situ reinforced concrete shear walls.

The steel columns in the basement car park will be concrete encased to provide support for the level 1 and 2 beams and slabs and for fire protection and durability.

Materials for the construction of the in-situ concrete beams and slabs will be off-loaded at the car park entrance ramp and distributed to the workface via a forklift running on the basement ground slabs.

Superstructure

Construction of the composite steel frame will commence when the ground floor slab is sufficiently far advanced to provide structural stiffness and a safe environment to the basement works below.

To reduce the beam/slab thickness and to provide good durability, the ground floor will utilise precast concrete planks with an in-situ concrete topping on UC beams with shelf angles. Level 1 to 5 construction will consist of in-situ concrete infill to metal decking on UB structural steel frame. The cores will be erected in advance of the main structure utilising the permanent bracing to provide rigidity during the erection process. Columns will span 2 floors, beams and metal decking fixed and concrete floors cast using mobile concrete pumps serviced by readymix truckmixers at ground level.

The steel frame will be fabricated with all bracing, windpost brackets, metal cladding support columns and sheeting rail fixings and services fixings to assist the overall site installation process.

It is our intention to utilise proprietary masonry fixing channels cast into the composite concrete deck edges for all external brickwork, metal cladding and glazing supports. The position of this fixing channel will be determined by the setting out requirements of the brickwork, blockwork, metal cladding and glazing and their relative construction interfaces.

The stairs and landings to the cores will be of precast concrete complete with all inserts for balustrades etc. and will be installed and cast in as the structural frame and composite decking advance.

Envelope

The main elevations are a combination of brick and blockwork with metal cladding to the feature corner elevation at the main entrance. The blockwork forming the cavity wall inner leaf will be advanced as closely as possible behind the main frame construction to provide a weatherproof working environment for the services and associated installations.

Completion of the metal cladding to the roof will coincide with completion of the rainwater drainage pipework in the cores and this with the inner cavity wall construction will achieve watertight status using the permanent roof drainage system at the earliest possible opportunity.

The tower crane will be utilised to erect the metal cladding to the roof and associated areas following completion of the structural frame. An external scaffold will be provided from which the brick and blockwork external cavity leaf and the vertical metal cladding and glazing will be constructed.

All bricks, blocks and mortar tubs will be distributed to the upper floors from within the building using the goods hoist in the lightwell. The external scaffold will be sheeted in monoflex to protect the elevations during construction and provide a weatherproof cocoon to the building prior to installation of the glazing and permanent watertightness completion. It will also act as a barrier to protect the pedestrians and traffic in the public streets below.

The tower crane will be used to install the roof plant modules. A temporary roof will be erected over the lightwell at roof level to maintain a waterproof environment to the ground floor and basement structure after completion following the removal of the tower crane.

The construction goods hoist will be kept as the sole means of vertical distribution until completion of the permanent goods lift and associated loading bay. Removal of the goods hoist will enable the completion of the glazed roof at the bottom of the lightwell, the glazing to the main entrance doors and the advancement of the One-Stop Shop fitting out works.

Services Installation

It is our intention to pursue the off-site manufacture of services modules and the phased installation of these vertical and horizontal pre-assembled units within the building envelope.

This will require a different approach to the production of design information but the benefits in terms of quality, productivity and cost predictability have been proved on previous projects.

The vertical and horizontal distribution of services are contained in the core risers and the primary circulation route bulkheads on each floor level. Details of the proposed modules, make-up and dimensions will be produced in due course.

It is our intention to complete all underground services diversions and connections in advance of the main construction works to enable the permanent connections to be made, reducing the need for any temporary or interim arrangements.

High level modular services installations will follow the construction of the external cavity wall blockwork inner skin and the core and riser blockwork walls. We will implement a Contractor Area Ownership procedure that enables Contractors to complete all their works before handing the area over to the next trade. We see the Services Contractor completing the high level services installation, with the exception of the final fix items, and handing over the area for the ceiling and bulkhead installation works. These being completed, the area is handed back to the Services Contractor for the low level LPHW pipework, containment and light power cable installation. Again, with this completed, the area is handed over to the Raised Flooring Contractor for him to fix the pedestals and fit the primary grid of floor tiles. The IT Infrastructure Contractor is then called in to install the copper IT cabling and floor box connections etc.

believe that this method of contractor ownership of working area, together with the implementation of a logistics strategy that delivers materials to the workface quickly and efficiently, is the fundamental recipe for a successful construction project.

The main items of plant will be supplied in pre-manufactured weatherproof modules for mounting on a galvanised structural steel grillage above the roof membrane. All services penetrations to the core risers will be through the vertical metal cladding panels, simplifying the weathering details. Permanent expanded metal walkways will be provided for roof access, services installation and maintenance.

Finishes and Fit-out

Internal blockwork walls and partitions installation will run in parallel with the external wall inner leaf construction.

In the cores, toilets and areas of solid floors, walls will be plastered and screeds laid prior to the commencement of the high level works. In the main office areas and One-Stop Shop, modular services installation, ceiling lining and high level works will be completed generally before the underfloor services and raised floors are installed.

The pre-manufactured services module approach will require a reduced M&E workforce and smaller storage areas on site, releasing greater areas for the finishing trades. In the office areas on the upper floors, the finishes to the central circulation bulkhead and associated ceiling installation will benefit from the elimination of a carcassing activity.

The hardwood skirtings and doors will be installed towards the end of the programme, together with the painting and general decoration.

As the area on the ground floor will be used to feed materials to the central goods hoist from off-loading areas accessed from   Street and   Lane, the fit-out of the One-Stop Shop will be the subject of an intense programme of activities prior to completion and occupation. We anticipate that these ground floor works will overlap the IT installation, commissioning and furniture installation of the office areas on the upper floors.

Hand-over and Occupation

The sequence and interface between the final stages of construction, commissioning and occupation will be subject to detailed discussion and common agreement.

We anticipate that the Council's IT and communications providers will be allowed access to commence their installation works 6 weeks before practical completion but this is subject to detailed discussions and legal agreement. Following the installation of the IT equipment will be the progressive assembly of the furniture and fittings and a two-week training and familiarisation period before the decanting and occupation by the first wave of Council staff.

10.3 Construction Programme

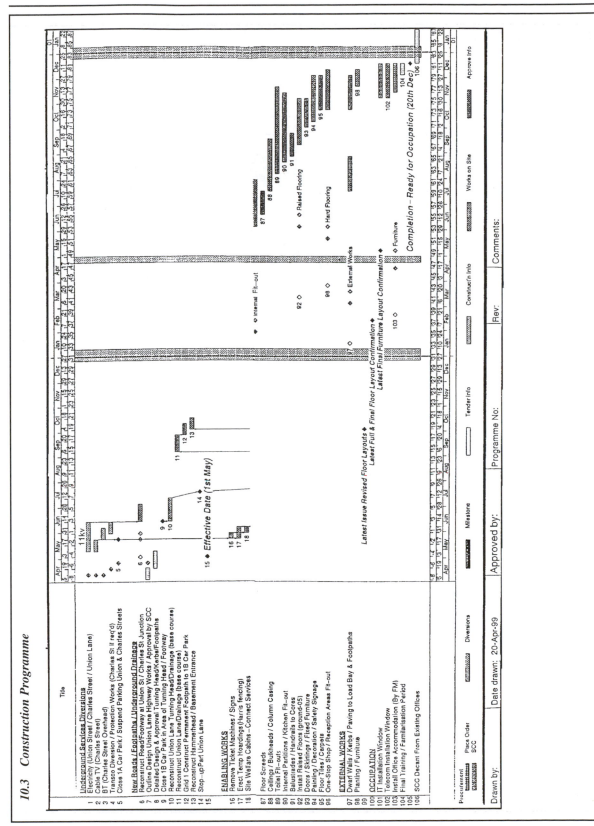

10.4 Site Layout Plan

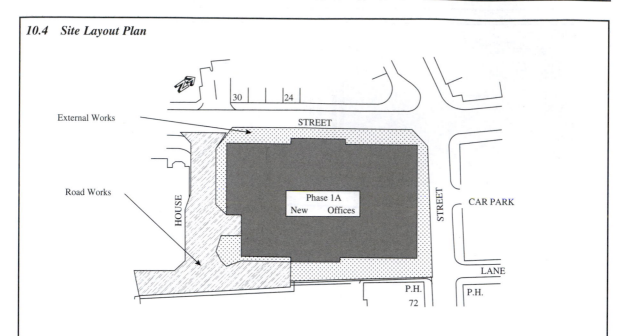

10.5 COSHH Assessments

10.6 Construction-Phase Design

10.7 Risk Assessments
STANDARD HAZARD AND RISK ASSESSMENT MATRIX COMPLETED FOR:

BUILDING OPERATION	RISK SEVERITY			LIKELIHOOD			RATING
	High	Medium	Low	High	Medium	Low	
1.0 UNDERGROUND SERVICES (EXISTING)							
.1 REMOVAL OF COVER TO FACILITATE CONSTRUCTION							
.a Contact with existing services		✓		✓			4
.b Existing sewer branches		✓		✓			4
2.0 DEMOLITIONS							
.1 BREAKOUT UNDERGROUND			✓	✓			3
3.0 SUBSTRUCTURE							
.1 EXCAVATIONS FOR AND CONSTRUCTION OF BASEMENT							
.a Trapped/hit by collapsing structure/earth		✓				✓	2
.b Adverse weather			✓	✓			3
.c Chemicals/substances		✓				✓	2
.d Use of machines/operation of vehicles	✓			✓			4
.e Noise		✓			✓		3
.f Falling object/materials from height		✓			✓		3
.g Asphyxiation	✓					✓	3
.h Existing/redundant services		✓		✓			3

STANDARD HAZARD AND RISK ASSESSMENT MATRIX COMPLETED FOR:

	BUILDING OPERATION	RISK SEVERITY			LIKELIHOOD			RATING
		High	Medium	Low	High	Medium	Low	
.2	EXCAVATIONS OVER UTILITIES (as item 3.1+)							
.a	Electricity/electric shock	✓					✓	3
4.0	**FOUNDATIONS**							
.1	EXCAVATIONS							
.a	Trapped/hit by collapsing structure/earth	✓					✓	3
.b	Adverse weather			✓	✓			3
.c	Chemicals/substances		✓				✓	2
.d	Use of machines/operation of vehicles	✓					✓	3
.e	Noise		✓			✓		3
.f	Falling object/materials from height		✓			✓		3
.g	Asphyxiation	✓				✓		4
.h	Existing/redundant services		✓				✓	3
5.0	**DRAINAGE**							
.1	EXCAVATIONS							
.a	Trapped/hit by collapsing structure/earth		✓			✓		3
.b	Adverse weather			✓	✓			3
.c	Chemicals/substances		✓				✓	2
.d	Use of machines/operation of vehicles	✓					✓	3

STANDARD HAZARD AND RISK ASSESSMENT MATRIX COMPLETED FOR:

	BUILDING OPERATION	RISK SEVERITY			LIKELIHOOD			RATING
		High	Medium	Low	High	Medium	Low	
.e	Noise		✓				✓	2
.f	Falling object/materials from height		✓				✓	2
.g	Asphyxiation	✓					✓	3
.h	Existing/redundant services		✓		✓			4
6.0	**SERVICES/UTILITIES**							
.1	**ROADWORKS**							
.a	Hit by passing vehicle on public road	✓				✓		4
.b	Adverse weather			✓	✓			3
.c	Compressed air		✓				✓	2
.d	Noise		✓			✓		3
.e	Operation of vehicles/use of machines	✓				✓		4
.f	Vibration		✓			✓		3
7.0	**STRUCTURAL FRAME**							
.1	**PRECAST STRUCTURE (GENERAL)**							
.a	Adverse weather			✓	✓			3
.b	Falling object/ material from height	✓				✓		4
.c	Fall of person from height/ on same level	✓				✓		4

STANDARD HAZARD AND RISK ASSESSMENT MATRIX COMPLETED FOR:

	BUILDING OPERATION	RISK SEVERITY			LIKELIHOOD			RATING
		High	Medium	Low	High	Medium	Low	
.d	Manual handling		✓			✓		3
.e	Mechanical lifting operations	✓					✓	3
.f	Operation of vehicles/use of machines	✓					✓	3
.2	STEEL FRAME (GENERAL) INCLUDING COMPOSITE STEEL / CONCRETE DECK (as item 7.1)							
8.0	**IN-SITU CONCRETE**							
.1	IN-SITU CONCRETE (GENERAL)							
.a	Adverse weather			✓	✓			3
.b	Chemicals/substances		✓				✓	2
.c	Compressed air		✓				✓	2
.d	Falling object/material from height	✓				✓		4
.e	Fall of person from height	✓					✓	3
.f	Mechanical lifting operations	✓					✓	3
.g	Operation of vehicles/use of machines	✓					✓	3
9.0	**EXTERNAL WALLS**							

STANDARD HAZARD AND RISK ASSESSMENT MATRIX COMPLETED FOR:

BUILDING OPERATION	RISK						RATING
	SEVERITY			LIKELIHOOD			
	High	Medium	Low	High	Medium	Low	
.1 BRICK/BLOCK							
.a Adverse weather			✓	✓			3
.b Confined spaces			✓		✓		2
.c Falling object/material from height	✓				✓		4
.d Fall of person from height	✓					✓	3
.e Hand tools		✓				✓	2
.f Manual handling		✓				✓	2
.g Mechanical lifting operations	✓					✓	3
.h Stacking		✓				✓	2
.2 CURTAIN WALLING (as item 9.1)							
.3 METAL CLADDING PANELS							
.a Adverse weather			✓	✓			3
.b Falling object/material from height	✓				✓		4
.c Falling person from height	✓					✓	3
.d Hand tools		✓				✓	2
.e Manual handling		✓				✓	2
.f Mechanical lifting operations	✓					✓	3
10.0 INTERNAL WALLS							

STANDARD HAZARD AND RISK ASSESSMENT MATRIX COMPLETED FOR:

BUILDING OPERATION	RISK						RATING
	SEVERITY			LIKELIHOOD			
	High	Medium	Low	High	Medium	Low	
.1 BLOCK (as item 9.1)							
.2 LIGHTWEIGHT PARTITIONS							
No site/project-specific risks identified							
11.0 ROOFS							
.1 ASPHALT							
.a Adverse weather			✓	✓			3
.b Contact with hot surfaces/materials		✓			✓		3
.c Falling object/material from height	✓					✓	3
.d Fall of person from height	✓					✓	3
.e Risk of fire	✓					✓	3
.f Hand tools		✓				✓	2
.2 GLAZING							
.a Adverse weather			✓	✓			3
.b Cuts and abrasions	✓					✓	3
.c Falling object/material from height	✓				✓		4
.d Fall of person from height	✓				✓		4
.e Manual handling		✓				✓	2
.f Mechanical lifting operations	✓					✓	3

STANDARD HAZARD AND RISK ASSESSMENT MATRIX COMPLETED FOR:

408 • Construction Management

BUILDING OPERATION	RISK SEVERITY			LIKELIHOOD			RATING
	High	Medium	Low	High	Medium	Low	
.g Stacking		✓				✓	2
12.0 FINISHES – FLOOR (including screeds)							
.1 No site/project-specific risks identified							
13.0 FINISHES – WALLS							
.1 FIRE PROTECTION TO STEEL							
.a Confined spaces			✓		✓		2
.b Dust	✓				✓		4
.c Falling object/material from height		✓			✓		3
.d Fall of person from height	✓					✓	3
14.0 FINISHES – CEILINGS							
.1 No site/project-specific risks identified							
15.0 FABRIC CANOPIES							
.1 FABRIC CANOPIES/ SUPPORTS							

STANDARD HAZARD AND RISK ASSESSMENT MATRIX COMPLETED FOR:

	BUILDING OPERATION	RISK SEVERITY			LIKELIHOOD			RATING
		High	Medium	Low	High	Medium	Low	
.a	Adverse weather			✓	✓			3
.b	Falling object/material from height	✓				✓		4
.c	Fall of person from height	✓					✓	3
.d	Operation of vehicles	✓					✓	3
.e	Vibration		✓				✓	2
16.0	**SIGNAGE (EXTERNAL AND INTERNAL)**							
.1	INSTALLATION AND MAINTENANCE							
.a	Adverse weather			✓	✓			3
.b	Electricity/electric shock	✓					✓	3
.c	Falling object/material from height	✓				✓		4
.d	Fall of person from height	✓					✓	3
.e	Mechanical lifting operations	✓					✓	3
.f	Operation of vehicles	✓					✓	3
.g	Use of machines	✓					✓	3
17.0	**MAINTENANCE**							
.1	CLEANING GLAZING							
.a	Falls from a height	✓					✓	3
.b	Operation of vehicles	✓					✓	3
.c	Chemical/ substances		✓			✓		3
.2	CLEARING/CLEANING GUTTERS							

	BUILDING OPERATION	RISK SEVERITY			LIKELIHOOD			RATING
		High	Medium	Low	High	Medium	Low	
.2	CLEARING/CLEANING GUTTERS							
.a	Falls of persons or objects from height	✓					✓	3

RISK ASSESSMENTS

Risk Assessment Schedule

No.	Risk	Comment
001	Risk to Public	
002	Probing Trench/Capping of Sewer Branched	
	Mine Grouting	Develop with Subcontractor
	Bored Piling/Plant	Develop with Subcontractor
	Deep Excavations/Plant/Temporary Works	Develop with Subcontractor
	Roadworks	Develop with Subcontractor
	Working at Heights	
	Disposal of Waste Materials	
	Temporary Electrical Supplies	
	Mobile Towers	
	Scaffolds	
	Precast Concrete Units	
	Tower Crane Erection	
	Tower Crane – Unloading/Distribution	
	Structural Steelwork Erection	Develop with Subcontractor
	Confined Spaces (Basement)	

RISK ASSESSMENTS

A. RISK ASSESSMENT NO. 001

ACTIVITY.   General Construction Works

SIGNIFICANT HAZARDS	ASSESSMENT OF RISK		
	Low	Medium	High
1: Risk to public from vehicles and general works		x	
2:			
3:			
4:			
5:			
6:			
7:			
8:			
9:			
10:			

RELEVANT ___ GROUP PROCEDURES	RELEVANT GUIDANCE

PREVENTIVE & PROTECTIVE MEASURES

Planning & Organisation:

Car parking bays in   Street and   Street to be suspended.
Footpaths along the site boundary are to be closed.
Pedestrian routes are to be diverted to the opposite side of the road

Management Control:

Each construction operation is to be assessed with respect to environmental risk to neighbours.

Ensure that all notices warning and diverting pedestrians/public are clear and maintained.

Site boundary to be secure at all times.

Systems for deliveries to site to be established to control traffic around the works.

TO MAKE THIS ASSESSMENT SITE SPECIFIC, COMPLETE SIDE B

B. SITE-SPECIFIC ASSESSMENT

Whenever this task is undertaken, the Generic Risk Assessment must be reviewed to ensure that all significant hazards and their risks are identified and controlled. Completion of this side will ensure that your assessment is both appropriate and complete.

SITE LOCATION: Street/   Street.

MAXIMUM NUMBER OF PEOPLE INVOLVED IN ACTIVITY: Indeterminate

FREQUENCY AND DURATION OF ACTIVITY: 20 months

ADDITIONAL SIGNIFICANT HAZARDS IDENTIFIED:

ASSESSMENT OF ADDITIONAL RISKS

	Low	Medium	High
Noise		x	
Dust		x	
Pollution (groundwater)		x	

ADDITIONAL CONTROL MEASURES

ASSESSMENT OF REMAINING RISKS

	Low	Medium	High
See activity risk assessments			

EMERGENCY PROCEDURES (in addition to those already provided for in the site safety plan):

None

METHOD STATEMENT REQUIRED Yes/No (Attach Copy) See section 10.2	PERMIT TO WORK REQUIRED Yes/No (Attach copy)

CIRCUMSTANCES WHICH MAY REQUIRE ADDITIONAL ASSESSMENT

Change in method statement

CIRCULATION OF RISK ASSESSMENT•

All   staff and subcontractors' management.

NAME:•   DATE:

SIGNATURE:

Risk Assessments

A. RISK ASSESSMENT NO.　　　002

ACTIVITY.　　Pile Probing and Capping Redundant Sewer Branches

SIGNIFICANT HAZARDS		ASSESSMENT OF RISK		
		Low	Medium	High
1:	Live services			x
2:	Bacterial infection		x	
3:	Presence of toxic gas		x	
4:	Presence of flammable gas		x	
5:	Lack of oxygen	x		
6:	Collapse of sides of trench			x
7:	Persons falling in trench		x	
8:	Vehicles falling in trench		x	
9:	Soil or material falling in trench		x	
10:				

RELEVANT ___ GROUP PROCEDURES
Site manual 8.8 and 10.4

RELEVANT GUIDANCE
BS 6031: 1981; CIRIA guide — Trenching Practice

PREVENTIVE & PROTECTIVE MEASURES

Planning & Organisation:

Underground services to be located and diverted or made safe.
Plant to be used and method of trench support to be determined in advance and discussed fully with supervision.
Establish system of work and provide all necessary equipment including PPE, first-aid and hygiene facilities before work starts.
Operatives to be trained in procedures to combat forseeable hazards, importance and use of PPE, personal hygiene and emergency procedures.

Management Control:
Supervision to ensure that work does not start until the necessary support for the trench sides is available on site, that underground services have been located, marked and exposed where necessary, and that persons do not enter any excavation more than 1.2 m deep unless the sides are properly supported. Plant and transport to be routed away from excavations, stop blocks to be provided to prevent vehicles tipping from over-running the edge. Secure barriers to be erected where trench more than 2 m deep. Securely fix ladders to provide safe access and egress.
Ensure correct use of PPE to include overalls, hard hats, waterproof gloves and safety boots or wellingtons.
Persons in trench to be supervised from personnel outside the trench.
Apply site rules — no smoking on site.
Inspection examination to include daily inspections by a competent person of all excavations more than 2 m deep, and thorough examination before entry.

TO MAKE THIS ASSESSMENT SITE SPECIFIC, COMPLETE SIDE B

B. SITE-SPECIFIC ASSESSMENT

Whenever this task is undertaken, the Generic Risk Assessment must be reviewed to ensure that all significant hazards and their risks are identified and controlled. Completion of this side will ensure that your assessment is both appropriate and complete.

SITE LOCATION: St/   St.

MAXIMUM NUMBER OF PEOPLE INVOLVED IN ACTIVITY: 5

FREQUENCY AND DURATION OF ACTIVITY: At commencement of works, duration approx. 4 weeks

ADDITIONAL SIGNIFICANT HAZARDS IDENTIFIED:

ASSESSMENT OF ADDITIONAL RISKS

Low	Medium	High
x		

Unexploded bombs

ADDITIONAL CONTROL MEASURES

ASSESSMENT OF REMAINING RISKS

Low	Medium	High

EMERGENCY PROCEDURES (in addition to those already provided for in the site safety plan):

If someone collapses within the trench, call for assistance.
Do not enter the trench without supervision from outside the trench.

METHOD STATEMENT REQUIRED Yes/No̶ (Attach copy) To be developed with the subcontractor	PERMIT TO WORK REQUIRED Yes/No̶ (Attach copy) To be issued as services are located/ disconnected

CIRCUMSTANCES WHICH MAY REQUIRE ADDITIONAL ASSESSMENT

CIRCULATION OF RISK ASSESSMENT•

All   staff and all subcontractor operatives

NAME:•   DATE: _____

SIGNATURE:

Risk Assessments

A.   GENERIC RISK ASSESSMENT   No.

ACTIVITY.			

SIGNIFICANT HAZARDS	ASSESSMENT OF RISK		
	Low	Medium	High
1:			
2:			
3:			
4:			
5:			
6:			
7:			
8:			
9:			
10:			

RELEVANT ___ GROUP PROCEDURES	RELEVANT GUIDANCE

PREVENTIVE & PROTECTIVE MEASURES

Planning & Organisation:

Management Control:

TO MAKE THIS ASSESSMENT SITE SPECIFIC, COMPLETE SIDE B

B.   SITE SPECIFIC ASSESSMENT   NO.

Whenever this task is undertaken, the Generic Risk Assessment must be reviewed to ensure that all significant hazards and their risks are identified and controlled. Completion of this side will ensure that your assessment is both appropriate and complete.

SITE LOCATION:

MAXIMUM NUMBER OF PEOPLE INVOLVED IN ACTIVITY:

FREQUENCY AND DURATION OF ACTIVITY:

ADDITIONAL SIGNIFICANT HAZARDS IDENTIFIED:

ASSESSMENT OF ADDITIONAL RISKS

Low	Medium	High

ADDITIONAL CONTROL MEASURES

ASSESSMENT OF REMAINING RISKS

Low	Medium	High

EMERGENCY PROCEDURES (in addition to those already provided for in the site safety plan):

METHOD STATEMENT REQUIRED Yes/No (Attach copy)	PERMIT TO WORK REQUIRED Yes/No (Attach copy)

CIRCUMSTANCES WHICH MAY REQUIRE ADDITIONAL ASSESSMENT

CIRCULATION OF RISK ASSESSMENT•

NAME:•   DATE:

SIGNATURE:

11 Environment

Introduction

"The construction industry is coming under increasing pressure to make its activities more environmentally acceptable ... good practice on site to preserve our environment is now usually a high priority for clients, their professional advisors, contractors and regulators" (CIRIA, 1999a). Likewise, principal contractors are coming under increasing pressure to conduct their siteworks with greater responsibility towards the environment and towards persons on and around the site. All contractors' projects have unequivocal effects on the environment whenever and wherever they are undertaken. The demand for greater environmentally considerate and sustainable construction and increasingly stringent regulation also suggests that environmental management will become an important formal feature within the construction processes in the future, in the same way that quality management has evolved over the last thirty years. Effective environmental management, a product of good site management practice, can create improved environmental conditions on a project. It enables the principal contractor to fulfil its environmental responsibilities both on and around the site and ensure that environmental legislation is met. Effective environmental management focuses on ensuring that the siteworks are planned, organised and carried out with full awareness and understanding of the environmental effects that the works create; moreover, ensuring that the works are assessed for environmental risk and that, where necessary, mitigation measures are established. In this way, environmental safeguard becomes an intrinsic part of managing the project.

The purpose of this chapter is to outline the prominent environmental effects of the construction processes and the key environmental laws that impinge on the management and control of those effects. It then examines environmental management on site through the implementation of a project, or site, environmental management plan (EMP). Further, it examines the establishment of an environmental management system (EMS), commensurate with environmental management standard ISO 14001. It takes this route because many principal contractors, who are often small to medium enterprises, will have the need at this time to develop an EMP for site application, but may not yet have found the business justification to establish a formal EMS throughout their company organisation. It also provides a contrast to the approach to H&SMS, presented in the previous chapter, which followed the full systems route to establishing site management procedures. In addition, focusing on the EMP prior to the EMS dimension is useful where a principal contractor applies environmental management as an addition to its existing quality, or health and safety management system, rather than developing a dedicated environmental management approach within the corporate organisation.

The Construction Industry and the Environment

"Construction sites are often criticised for the damage that they cause to the surrounding environment and the adverse effects that they have on their neighbours" (CIRIA, 1999a). The environmental effects of the construction process can assume a variety of manifestations. They may cause damage to the natural environment affecting plant and animal life, cause disruption to local inhabitants and may create

pollution and excessive noise. Construction also creates waste products as a result of its siteworks. Some may be transformed to provide reusable materials and some may be recycled. Notwithstanding these useful reapplications, the majority of waste products from construction are disposed to landfill sites. Such practice creates knock-on environmental effects as more and more waste must be disposed of and transport is needed to convey this to handling sites. The effective management of construction works is important not just to the project's success but in contributing towards mitigating direct and indirect environmental effects, minimising waste products as a result of its processes and encouraging reuse, recycling and safe disposal.

Rising concern for the environmental performance of the construction industry, along with dissatisfaction for other industries which are seen to create significant environmental effects, comes at a time when clients expect better value for money and greater quality and sustainable construction are high on the government's agenda. Moreover, the quantity of waste products is increasing and the availability of landfill sites is decreasing. Against downward trends in other sectors of industry, environmental incidents giving rise to pollution virtually doubled in five years over the mid-1990s. Furthermore, the construction industry creates around 30% of landfilled waste with less than 20% being recycled (Porritt *et al.*, 1999). While environmental awareness within the UK is comparable to mainland European countries, it has been suggested that little available knowledge is translated into effective practice (CIRIA, 1995). The construction industry therefore has much to do in applying environmentally responsible site practices in a wider sense.

The environmental effects of construction are affected by three significant influences within the total development process. The first is extensive environmental legislation which seeks to protect both the natural environment and inhabitants neighbouring the project site. The principal piece of legislation is the Environmental Protection Act 1990 (Environment Agency, EA, 1990a). In addition, there are many regulations having significant influence over particular aspects of construction as they affect the environment. These will be detailed subsequently. The second significant influence is that of the responsibilities placed on the construction contract as a result of the governmental development planning and approval procedures at national and local levels. For example, some proposed construction projects will be subject to environmental impact assessment, or EIA, where the potential environmental effects must be identified and considered as part of the planning proposals. The third influence results from the interface between the environmental characteristics of the construction project and the workforce carrying out the work. Welfare considerations are central to this, and therefore statutory safety legislation is significant. The Health and Safety at Work, etc. Act 1974 and the Construction (Design and Management) Regulations 1994 play a key role in establishing a safe working environment (Health and Safety Executive, HSE, 1974; 1994).

Construction Works and their Effects on the Environment

All construction works must be carefully planned and conducted if serious environmental effects are to be avoided. Environmental effects occur whenever and wherever works are carried out. While all construction activities give rise to environmental effects through their general characteristics, some types of works can have a greater propensity to be disruptive or harmful. In alteration works, for example, modifying existing structures can be challenging because of the unknown nature of the structure and its material constitution which could contain hazardous elements such as asbestos. Demolition works prior to rebuild can be problematic as technically complex structures may require specifically sequenced dismantling. Potentially hazardous materials may have to be dealt with on site, and this will require particular consideration to ensure safe handling, removal, transportation and disposal. Even on new build projects of a routine nature, operations may be carried out in close proximity to existing buildings and people, and this will require environmentally controlled conditions to be maintained throughout the works. In addition, all projects are known to create some degree of disturbance and discomfort impacts such as noise and dust which, again, will require careful attention if such effects are not to constitute a nuisance both on and around the site. Works may be undertaken also on existing contaminated sites where measures need to be considered to handle situations where unexpected materials might be revealed as the works proceed. Seen in the context of varying project and site situations, the nature of construction works will always have some effect on its environs.

Statutory Environmental Legislation

The environment is protected by legislation at national and local levels. The Environmental Protection Act 1990, or EPA, mentioned previously, is the most prominent piece of national legislation. The EPA protects the natural environment and the construction project's surroundings and neighbours. In addition, other Acts or Parliament and local legislation, which are requirements imposed by local authorities under national laws, serve to protect particular features of the environment. For example, pollution restrictions are maintained under The Environment Act 1995 and tree preservation under The Wildlife and Countryside Act 1981 (EA, 1981; 1995).

The principal contractor has to consider most carefully its legal obligations under such Acts when configuring its environmental plans and management systems. The contractor is bound by a duty of care, infringement of which can lead to conviction under environmental laws, resulting in heavy fines and imprisonment. Common law also provides for civil claims for compensation where a plaintiff proves damage or loss as a result of the contractor's negligence or nuisance.

The principal environmental legislation affecting construction activities will be identified in the subsequent section focusing on the key environmental effects of construction siteworks.

Environmental Regulation

During all construction projects the principal contractor will, at some time, have to liaise closely with national and local environment regulatory bodies. Local authorities have the overall responsibility for regulating traffic, noise and air quality. The Environment Agency, the Scottish Environment Protection Agency and the Northern Ireland Environment and Heritage Service handle matters relating to contamination and waste management. English Nature, Scottish Natural Heritage and The Countryside Council for Wales control special sites and protected wildlife. In addition, the Health and Safety Executive oversees all matters related to health, safety and welfare on construction sites and there are clear overlaps where environmental matters impinge on employees' safe working practices. A principal contractor could become liable under environmental laws for infringements of regulations, resulting in notices to remediate site conditions for environmental safety or safeguard, or in conviction where a serious environmental breach of duty occurred.

Environmental Effects of Construction

Contamination

Contamination of the site and its surroundings is the principal manifestation of environmental effects created by construction works. Contamination is an intrinsic by-product of construction and therefore it cannot be completely avoided. However, proactive site management can help to reduce the potential for contamination during the siteworks. The key to proactive management is a clear understanding of the types and potential causes of contamination (see Figure 11.1). These are contamination:

- of the ground
- of ground and surface water
- by noise and vibration
- of the atmosphere through dust, dirt, emissions, odours and pollutants
- to plants and wildlife
- through disruption and damage to archaeological and historic features, building and structures.

The prominent characteristics of each type of contamination and the principal legislation which exists to control them are as follows:

Ground Contamination

Contamination of the ground can occur for a number of reasons. It may result from a previous use of the project site, for example leakage of chemicals from industrial processes. It might also be caused by

TYPE OF CONTAMINATION	POTENTIAL CAUSES OF CONTAMINATION
Ground contamination	• existing contamination of site • disturbing contaminated ground • stockpiling contaminated materials • spillage and leakage of hazardous substances • discharge of contaminated water • wind blown contamination
Water contamination	• discharge of contaminated water • spillage and leakage of contaminants • run-off from plant/vehicle washing • contaminated surface water run-off • pumping of water from siteworks • material spill to drains and watercourses
Plant and wildlife contamination	• habitat disruption/destruction • vegetation damage/contamination • removal of habitats (trees/hedges/grassland) • disruption to wildlife activity (feeding/breeding) • disruption to food/water sources • changes to natural environs (noise, dust, light)
Noise and vibration contamination	• particular methods of construction • use of heavy plant, equipment and haulage • pattern of working hours • haulage routes to/from site • anti-social activity and tonal noises
Dust, dirt, emissions, odours and pollutants contamination	• plant, equipment and vehicles • fuels, chemicals and dangerous substances • waste storage and decomposition • burning site waste/refuse • site toilet odours • surface conditions of ground on site
Historic/archaeological contamination	• disruption/destruction of natural features • impact damage to buildings/structures • vibration damage (cracking) • ground disturbance (subsidence) • disruption by temporary works (access roads) • disturbance of historic finds (burial sites/ structures

Figure 11.1 *Types of contamination and their potential causes*

the filtration of substances stockpiled on site during construction, such as fuels, oils and paints. Pollutants may be airborne, for example dust and dirt can be blown from stored rubble heaps during demolition works.

The main problems associated with ground contamination presents, first, the need for particular health, safety and welfare measures to protect both the project's participants and nearby inhabitants; second, the requirement to arrange specialised disposal arrangements, for example the safe handling of asbestos; third, the need, in some situations, to undertake remediation works to reinstate a site contaminated during the siteworks. These demands can frequently create delays to programme, additional works and their associated costs and legal liabilities where the contractor has been held to have infringed environmental legislation. The principal contractor is duty bound under the Environmental Protection Act (Prescribed Processes and Substances) Regulations 1991 (HSE, 1991) to ensure that the project site is safeguarded from the potential incidence of ground contamination. An essential element of mitigating against ground contamination is the commitment to undertake soil investigations, an ecological survey and to conduct environmental risk assessment as a composite part of pre-tender planning; further, to follow this with a detailed environmental site survey undertaken at the pre-contract stage. Together, these mechanisms should reveal the likelihood of ground contamination occurring and, moreover, assist site management in its task of contamination alleviation and control.

Water Contamination

Within the scope of The Environment Act 1995: "*It is an offence to cause or knowingly permit any poisonous, noxious or polluting matter or any solid waste matter (which includes silt, cement, concrete, oil, petroleum spirit, sewage or other polluting matter) to enter any controlled waters unless a discharge is authorised. Road drains and surface water gullies generally discharge into controlled waters and should be treated as such*" (CIRIA, 1999a). Construction siteworks harbour the potential to be destructive in any or all of these ways. It is possible for construction process sourced effluents to discharge into drains and sewers or leak into the ground; also for dust and particle-based debris to be blown into surface and ground water courses creating water contamination both on the site and beyond into the public sewage systems. The Environment Act 1995 may require a principal contractor found responsible for contamination of water courses to trace pollutants and decontaminate affected waters, and take action to safeguard against future incidents. Water contamination is a frequent environmental infringement with severe penalties invoked for conviction. Detailed risk assessment following a pre-contract site survey is essential to the principal contractor in avoiding such site contamination.

Noise and Vibration

Contamination of a project's environs due to high levels of noise and vibration is perhaps one of the most sensitive of all environmental effects. Excessive noise levels from normal site activities can present a serious health, welfare and safety hazard to the workforce. Furthermore, it can cause a nuisance to nearby inhabitants. Noise and vibration are types of contamination inherent to many operations within the construction processes. The involvement of heavy plant and equipment, and sometimes specialised machinery such as piling rigs, only serves to exacerbate an already difficult situation with which site management has to contend.

The management of the works may involve the noise abatement of plant and equipment, the provision of acoustic screening to site boundaries and the issue of personal protective equipment (PPE) to operatives working at noisy workfaces or when using noisy equipment. It may also involve the attenuation of vibration when using hand-held power tools and when working in reverberant environments.

A number of environmental laws serve to safeguard against excessive contamination from noise and vibration. The most significant of these are The Environmental Protection Act 1990, mentioned previously and throughout this chapter, The Control of Pollution Act 1974 (EA, 1974) and The Noise at Work Regulations 1989 (HSE, 1989). In addition to providing continuous monitoring and control of such contamination, site management should establish open communication with the neighbouring community. This is essential given the particular sensitivities to noise held generally by the public when exposed to the effects of construction projects.

Dust, Dirt, Emissions, Odours and Pollutants

Dust, dirt, emissions, odours and pollutants are types of contamination having the propensity to cause serious nuisance to local inhabitants and represent a serious health risk to site operatives, participants to the project when visiting the site and to passers-by. These contaminants are intrinsic to many site activities and particular operations, and therefore should be reduced and controlled. The environs may be safeguarded through the use of physical screening, pollution attenuation, for example damping-down loose surfaces, and the workforce protected with personal protective equipment. The Environmental Protection Act 1990 generally, and specifically The Clean Air Act 1993 (EA, 1993), are significant environmental legislation regulating these commonly encountered site contaminants.

Plants and Wildlife

Siteworks may have environmental effects on landforms, habitats, plants, trees and animal life. Although some disruption to the natural habitat and its wildlife from construction work will be both reasonable and expected, it is essential that measures are employed to safeguard against excessive and irreversible disruption and harm. The prominent piece of legislation protecting plants and wildlife is The Wildlife and Countryside Act 1981 (EA, 1981). Some natural locations are designated as Sites of Special Scientific Interest (SSSI), others are denoted as Special Areas of Conservation (SAC) and Special Protection Areas (SPA). These exist throughout the UK and also throughout mainland Europe, being protected within the scope of EC Directives. To minimise the disruption and damage from construction works the principal contractor may undertake a detailed ecological survey on plant and animal wildlife. This will be under-taken at both the pre-tender and pre-contract planning stages followed by environmental site manage-ment. Where this particular facet of management cannot be provided by the principal contractor then environmental consultants may be contracted.

Disruption and Damage to Archaeological and Historic Features, Buildings and Structures

Archaeological and historic features, buildings and structures are sometimes found on construction sites. Dismantling, alteration and renovation works are undertaken in and around buildings and struc-tures of historic significance. Archaeological finds can prove seriously problematic to the principal con-tractor as they may be encountered unexpectedly. They may well present a dilemma in terms of additional time and cost, and the eagerness to progress the works can conflict with treating such finds with due respect. Essentially, such instances must be treated fairly and justly, and measures taken to avoid unnecessary disruption and damage from the construction works. The Town and Country Planning Act 1990, The Planning (Listed Buildings and Conservation Areas) Act 1990, and the Ancient Monuments and Archaeological Areas Act 1979 are the prominent legislation protecting such sites and archaeological and historic finds (EA, 1979; 1990b; 1990c).

Construction Waste

Construction waste is also a significant environmental effect from the siteworks of many projects. Much construction waste is considered as just that – waste, and unsuitable for reuse after treatment or as a recyclable commodity. It is therefore simply disposed of, usually at regulated landfill waste sites. Nevertheless, some by-products of the construction site processes may be recycled into useful products or reused on the project itself or other projects, following suitable treatment. An example of this might be demolition waste in the form of masonry which can be sifted, cleaned and crushed to provide fill materials for external works. Some waste materials may be hazardous, with safe disposal being the only course of action.

"*Since the construction and demolition industries are the second largest producers of controlled waste in the UK, they have an important role to play in the adoption and implementation of waste minimisation and recycling programmes, which should essentially result in less construction and dem-olition wastes being disposed of to landfill*" (CIRIA, 1995). However, it was mentioned previously that only a small proportion of construction waste is recycled. With controls on disposal becoming more restrictive, landfill capacity reducing and costs increasing, the construction industry has to explore alternative methods of disposal and waste management and utilisation approaches.

There is considerable government commitment encouraging better waste management practices. A government paper, *Less Waste More Value* (Department of the Environment, Transport and the Regions, DETR, 1998), presented a 'waste hierarchy' of reduction, reuse, recovery and disposal, focusing on highlighting the environmental benefits. However, it is often perceived that the direct disposal of construction waste outweighs the practices of recycling and reuse. Only when it becomes financially attractive to the principal contractor will measures be taken to recycle and reuse construction waste. *"Much work is needed to promote good waste management practices on site in line with current health and safety initiatives ... education and training throughout the entire workforce is the key to making any waste minimisation and recycling programme on site fully effective"* (CIRIA, 1995).

CIRIA (1993a; 1993b; 1999b) has made recommendations to encourage recycling and reuse within construction. These are: to provide specifications for recycled and reclaimed materials and components; to establish information systems that connect potential users with available recycled materials; to review the potential for refurbishing whole buildings for reuse; to use excavation/demolition material as fill; and to encourage positive attitudes from specifiers, designers, users and occupants towards the use of recycled and reclaimed materials.

Construction waste is variable by type and quantity, yet can be categorised into three broad groups – those which have: (i) *potential and value for reuse* – such as concrete, masonry, bricks, blocks, asphalt, solid and aggregates; (ii) *potential and value for recycling* – such as timber, glass, paper, plastics, oils and metals; and (iii) *no potential and value for reuse and recycling* – such as paints, solvents, plaster and asbestos. *The Reclaimed and Recycled Construction Materials Handbook* (CIRIA, 1999c) suggests best practice which principal contractors can adopt. A selection of reclaimed and recycled materials from construction waste, together with their potential uses, in shown in Figure 11.2. A fundamental issue in construction waste management is the tendency to see waste purely as waste. There needs to be a focus in the future on waste as recyclable and reusable end products: *"This invites both the supply industries and manufacturers to become more actively involved in waste management"* (Griffith, 2001).

The recycling of construction waste materials is not currently widespread and tends toward providing low-grade products for limited markets. In one of the largest local authorities in the UK – the Greater London Authority (GLA) – there are some poor waste recycling practices with some boroughs recycling as little as 3% (Goode, 2000), while across the UK as a whole it is suggested that only up to 20% of construction waste is recycled (Porritt *et al.*, 1999). A particular problem within construction is the contamination of potentially recyclable and reusable materials. As a result, waste materials frequently go to make lower-grade products where a measure of contamination is allowable or they are treated as disposable waste because it is simpler and directly more cost-effective.

The use of construction waste in the applied transformation of materials for alternative uses is more widespread. It is more common on major projects for large-volume waste, from the demolition of, for example, masonry and concrete, to be sifted for contamination and crushed into hardcore for sub-base material. There can be dangers in reuse, in particular where sifting and cleaning are not thorough, and clients and specifiers should be wary of such practices unless they are carefully monitored and controlled: *"The reliability, durability and compatibility of reused materials can sometimes be of concern. Where the potential for reuse may exist, demolition materials may be judged on their perceived quality rather than on performance criteria and for this reason the design process is often reluctant to support their incorporation within a project specification"* (Griffith, 2001). The Building Research Establishment (BRE), CIRIA and RILEM – the International Union of Testing and Research Laboratories for Materials and Structures – have published many reports to assist with the specification of recycled materials, and these should be reviewed by principal contractors with a commitment to recycling and reuse.

The vast majority of construction waste is removed from site for disposal, irrespective of the high costs involved. *"The procedures involved in this depend upon the nature of the materials. There are three types of waste materials: (i) inactive – materials that do not undergo reactive change when deposited in landfill, for example uncontaminated rubble; (ii) active – materials which maintain activeness and remain hazardous after being deposited in landfill, for example oils; and (iii) special – materials which are dangerous to life, for example corrosives"* (Griffith, 2001). While waste disposal activities from demolition are broadly controlled by The Environmental Protection Act 1990, special wastes are considered under The Control of Pollution (Special Waste) Regulations 1996 with haulage covered by The Carriage of Dangerous Goods by Road Regulations 1996 (HSE, 1996c; 1996d).

RECLAIMED/RECYCLED MATERIALS	POTENTIAL USES (following treatments)
Demolition rubble	• hardcore • fill material • sub-base • drainage fill
Bricks and blocks	• reclaimed for walling • hardcore • fill material • aggregates
Concrete	• fill material • sub-base • aggregate • hardcore • sand/slag
Glass	• reclaimed for glazing • sand • glass-fibre • aggregate
Timber	• reclaimed for construction uses • shuttering • fencing • hoardings • piling • wood–based products (boards/sheets)
Plastics	• insulation • geosynthetics • doors/windows • pipes • claddings

Figure 11.2 *A selection of reclaimed/recycled materials, and their potential uses*

The Principal Contractor's Responsibility for the Project Environment

The principal contractor is bound generally by the Environmental Protection Act 1990 and also specifically by construction-related legislation: *"The Construction (Design and Management) Regulations 1994 (CDM) require that health and safety is taken into account and managed throughout all stages of a project from conception, design and planning through to site work and subsequent maintenance and repair of the structure"* (HSE, 1996a). Moreover, the Construction (Health, Safety and Welfare) Regulations 1996 dictate that all works must be *"carried out in a safe manner and with due planning, organisation and supervision"* (HSE, 1996b). Construction works can frequently give rise to the environmental effects identified and which impinge heavily on workers' health and welfare. For example, construction work is synonymous with noise, vibration, dust and contamination, all of which pose severe dangers to site operatives. The identification of physical dangers and hazards to health and welfare forms a crucial element in the principal contractor's approach to environmental management.

Successful environmental management of construction works is fundamentally dependent on effective environmental assessment and planning by the client organisation. It also requires that the site-works undertaken by the principal contractor are managed within an organisation of effective hazard identification, risk assessment, good supervision and close control. Moreover, everyone who works on site must know where danger might be encountered and what precautions they must take to protect themselves, to safeguard the welfare of others and to protect the environment when carrying out their tasks. Supervisors and managers must be cogniscant of the dangers inherent to some operations and know fully what actions they need to take to mitigate risks and control the hazards threatened. Furthermore, they must ensure this within a backdrop of stringent legislation and regulatory procedures.

Encouraging effective environmental management on all construction projects, the Construction Industry Research and Information Association (CIRIA) suggested that clients should give greater emphasis to the environmental proposals of contractors when procuring tenders and encourage principal contractors to adopt environmental management systems (EMS) and implement environmental management plans (EMP) as an intrinsic part of their approach to environmental management (CIRIA, 1999b).

Environmental Management

Environmental management throughout a construction project encompasses the consideration of environmental effects at regulatory, the company and project organisational levels (see Figure 11.3). Three key aspects make a prominent contribution (Griffith *et al.*, 2000):

(1) Environmental Impact Assessment (EIA) – the environmental evaluation, undertaken by the client organisation during project evaluation and development, of the potential effects of the proposed construction project.
(2) Environmental Management System (EMS) – the contribution made by the principal contractor to mitigating and managing the environmental effects of the construction project through the implementation of a formal company environmental management system.
(3) Environmental Management Plan (EMP) – the translation of the EMS into management procedures and working instructions for use on the site to ensure environmental control of the siteworks.

Although EIA and EMS/EMP are undertaken by different and separate contractual parties, the information generated is closely linked. Information from the EIA will form part of the documentation that the principal contractor will use in tendering for the contract. Similarly, information from EMS/EMP will be fed back into the client's project environmental file collated on completion of the project and used to provide information for future development projects.

Environmental Impact Assessment (EIA)

Environmental Impact Assessment, or EIA, is a detailed appraisal technique for ensuring that the potential environmental effects of any new development are identified and considered before any planning approval is granted. EIA is a legislative requirement for designated construction projects in many countries and throughout the EC, and the Environmental Assessment Directive underpins national EIA legislation within each member country. Within the UK, the requirements for EIA are specified through the various national regulations in England and Wales, Scotland and Northern Ireland. For construction projects requiring planning permission the Environmental Assessment Directive is given legal effect in:

• *England and Wales*: by The Town and Country Planning (Assessment of Environmental Effects) Regulations, 1988 (HM Government, 1988a)
• *Scotland*: by The Environmental Assessment (Scotland) Regulations, 1988 (HM Government, 1988b)
• *Northern Ireland*: by The Planning (Assessment of Environmental Effects) Regulations (Northern Ireland), 1989 (HM Government, 1989).

ENVIRONMENTAL MANAGEMENT

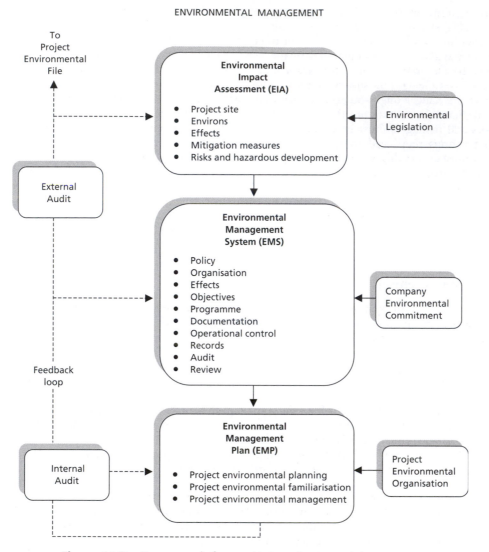

Figure 11.3 *Framework for project environmental management*

The principal function of EIA as an environmental management tool is to enable the lead consultant to systematically acquire as much information as possible on the potential environmental effects of the proposed development within the scope of applicable legislation. EIA should take into account: inhabitants, fauna, flora, soil, water, air, climate, landscape, material assets and cultural heritage. Furthermore, it should appreciate the interactions between these elements. The responsibility for EIA is assigned, within current regulations, to the client organisation which will play an important part in the assessment process along with the following participants: specialist consultants; the planning authority; statutory and other consultees; the public; and, depending on circumstance, central government departments.

The output from an EIA is the 'Environmental Statement' (ES). This is: "*a publicly available document setting out the developer's own assessment of the likely environmental effects of his proposed development, which he prepares and submits in conjunction with his planning application*" (Department of the Environment, DoE, 1989a).

An ES should provide information in five key sections (DoE, 1989b). These are:

 (i) information describing the project
 (ii) information describing the project site and its environs
(iii) information describing the assessment of environmental effects
(iv) information describing the measures to be taken to mitigate the effects
 (v) information describing the risks of accidents and hazardous development.

There are broadly five situations where the planning approval process will consider the relevance of environmental assessment (Griffith *et al.*, 2000). These are:

(1) A client, or developer, applies for planning permission with no reference to EIA because it is not necessary. Where the developer is unsure of its necessity, the planning authority will advise.
(2) A developer applies for planning permission and submits an environmental statement: because it is necessary under national planning regulations.
(3) A developer applies for planning permission and enquires with the planning authority if an EIA is necessary: because it is unsure if the proposal falls within the scope of regulations.
(4) The Secretary of State may independently exercise power to request on EIA where, for example, representations are made by third parties, for instance local communities, about a proposed development.
(5) The planning authority must undertake an EIA before granting itself permission for development: because the proposal falls within national planning regulations criteria.

There are eight elements in the process of EIA (Griffith, 1997). These are:

(1) Project description – a clear description of the project and its works
(2) Screening – the process of determining the need for environmental assessment
(3) Scoping – directing the assessment towards aspects of particular importance
(4) Baseline studies – the identification of significant environmental effects
(5) Impact prediction – the consideration of the degree of impact
(6) Mitigation assessment – the measures to be taken to mitigate the effects
(7) Environmental statement – a publicly available environmental effects statement
(8) Environmental monitoring – monitoring of environmental effects on the project.

While many construction projects do not fall within the legislative criteria for EIA, there is a growing trend among knowledgeable and visionary clients to undertake EIA on a voluntary basis. Many organisations see this as a positive projection of their commitment to and actions towards environmental protection. Such clients commission EIA to inform the briefing, design and construction processes and to ensure that the resulting development represents the best environmental option that is practically achievable. Organisations may find that such an approach fits in well with their current commitment to the environment. Moreover, the information that EIA generates greatly assists the principal contractor when formulating its EMP for the project.

Environmental Management Plan and System

The nature and characteristics of, and organisation for, a construction project will differ from site to site. However, ensuring that there is effective environmental management is no different from other key management functions, for example the management of time, cost, quality and health and safety. For effective environmental management it is necessary to compose a framework and structure which encapsulate the environmental ethos, policies and objectives of the principal contractor's corporate organisation. Furthermore, this must be translated into appropriate environmental method statements, management procedures and working instructions for implementation on site. Mechanisms need to be established which distribute information and instructions throughout the organisation, and which monitor to ensure that instructions are carried out and that checks for compliance are made. In addition, with all management systems and procedures it must be remembered that it is people who make them work. Awareness, induction and training for both management and workforce are, therefore, essential components in bringing about the necessary organisational change to improve practices on site (Griffith, 2001).

Principal contractors will adopt their own approach to environmental management to meet their own needs. Some organisations have developed simple procedures, for example a project Environmental Management Plan, or EMP, to meet client needs and project-specific legal requirements, while others have developed a company Environmental Management System, or EMS. There are two main international standards applicable to environmental system development – ISO 14001 (ISO, 1996) and EMAS [European Community Eco-Management and Audit Scheme] (European Union, EU, 1993). A well-conceived approach, ranging from an all encompassing EMS to a detailed EMP, can provide mechanisms by which the principal contractor can more effectively undertake environmental management: "*Most contractors now accept that everyone should follow environmental good practice on site to preserve our environment. Within the construction industry, all levels are trying to implement environmental improvements: those in the boardroom are demonstrating their commitment by preparing environmental policies for their company; clients are requesting evidence of environmental credentials from contractors before awarding them contracts; and those on site are already implementing a variety of environmental initiatives. However, despite all this action, work is still needed to improve the construction industry's environmental performance*" (CIRIA, 1999b).

Effective environmental management can only be delivered through greater awareness, action and commitment, supported by legislative and financial conditions which encourage widespread environmentally responsible site practices. For this to be achieved there needs to be a change in culture within the construction industry where cost benefits and market advantage are recognised and the expectations of both clients and contractors are clear (Griffith, 2001).

The principal contractor needs to ensure that key policy elements support its environmental undertaking. These are:

- integration of the project with the company's environmental management systems
- identification of the project's likely effects in terms of both environment and safety to workers and others
- implementation of environmental management plans for on-site management
- provision of environmental awareness and training.

An appropriate strategy to encourage the delivery of these elements is to adopt *environmental management*. "*Environmental management can be best described as those elements of the overall management of an organistion that deal with environmental issues, including control of environmental impacts*" (CIRIA, 2000). Environmental management presents the contractor with options to develop and implement various formal approaches when addressing environmental issues as part of its business. An organisation may establish an *Environmental Management System* (EMS) throughout the company, or develop a mechanism to manage the environmental management aspects of a specific project site – an *Environmental Management Plan* (EMP) (Griffith *et al.*, 2000).

The precise strategy adopted depends to a great extent on the operating business framework of the contracting organisation and the characteristics of the construction projects that it undertakes. One cannot be prescriptive or suggest best environmental management practice across all organisations or situations. All principal contractors must make up their own mind as to how they will manage environmental aspects at both corporate and project levels.

Generally and fundamentally, planning is at the core of environmental management. There must be a structured and committed approach to planning and therefore the principal contractor should seek to establish:

- a proactive planning process
- clear management structure
- robust procedures for site control
- sound supervisory monitoring mechanisms
- a process of review and improving practices.

Environmental Management Plan (EMP)

The *Environmental Management Plan* (EMP), sometimes referred to as a project environmental plan or site environmental plan, is the key reference document for any construction project. The EMP provides the necessary management framework for establishing good environmental site management control.

The EMP should contain the following information:

- project details
- management structure (organisation chart)
- key contacts/subcontractors/suppliers
- environmental policy
- significant anticipated environmental effects
- method statement (demolition/dismantling/environmental/special processes)
- contract programme (Gantt chart/network analysis)
- training (awareness/induction/on-site training)
- evaluation and review mechanisms.

The EMP is developed for each specific construction project. Although there are elements common to all construction projects which are reflected in the plan, other elements are specific and may require special construction environmental procedures. This is particularly true of, for example, demolition works where special procedures are often required to mitigate the specific environmental effects of, for example, noise, vibration and dust.

The EMP is prepared before management of the siteworks begin, although it may be prepared at the tender stage should environmental considerations with cost implications be identified. For the principal contractor, the key to developing the EMP lies in the compilation of the *Method Statement* and the consideration of special environmental procedures resulting from the methods adopted. As the precise methods of working and the necessary resources are determined, potential environmental effects can be anticipated and mitigation methods devised. The method statement is therefore a vital document within the EMP and provides the basis for site management and control.

A fundamental requirement of any EMP is that it must be accessible to project personnel, be easy to understand, be straightforward to implement, be used regularly and be kept up-to-date. Only in this way will the EMP be effective in application and lead to both environmental benefits and economic benefits to the contractor.

There are a number of key elements that require detailed consideration when developing the EMP (Griffith *et al.*, 2000; Griffith, 2001):

(1) environmental obligations
(2) environmental hazard identification
(3) environmental management structure and responsibilities
(4) routine and special environmental procedures
(5) awareness, induction and training
(6) environmental monitoring, review and improvement.

Environmental Obligations

There are four significant environmental obligations. First, there is legislation protecting the natural environment and the health, welfare and safety of those on and around the site. Second, there is local regulation devolved from national legislation. Third, *contract conditions* – specified obligations on the contractor in relation to the particular site or nature of the works – for example, the requirement to carry out environmental impact assessment (EIA) and to meet local community needs. Fourth, *company obligations* – resulting from corporate environmental policies that must be upheld to maintain the integrity and credibility of the company within its marketplace and community.

Environmental Hazard Identification

Identifying the environmental hazards specific to the site is fundamental for effective environmental planning. Aspects may have been highlighted in the EIA for which the contractor can consider mitigation and management measures. In addition, hazards may be identified in project documentation, from consultations with regulatory agencies, the client, consultants, designers and, of course, from drawing on the works planning and site team experience of previous contracts. Specialist surveys, for example ecological or soil surveys, are an essential part of this identification phase and should be carried out for all projects. Attention is focused on providing systematic and reliable identification and consideration of likely hazards together with those control measures necessary to mitigate or minimise their environmental effect. This can be achieved within a process of *risk assessment*. Risk assessment involves five key stages: (i) *risk identification* – where the potential on site for environmental effects to occur is identified; (ii) *risk assessment* – where the magnitude of each effect is determined; (iii) *risk mitigation* – where planned controls to prevent the effects occurring are devised; (iv) *risk monitoring* – where procedures are put in place to monitor the manifestation of effects; and (v) *risk management* – where a course of action is devised to manage the effects should they occur. A case study example of a project environmental audit, or assessment, follows later in this chapter. Assessment encompasses all those environmental aspects thought to impinge on the siteworks, together with a management approach to ensure effective environmental monitoring and control.

Environmental Management Structure and Assignment of Responsibilities

There must be an unambiguous structure for managing the siteworks. The definition of responsibilities is the key element and those personnel responsible for implementing, monitoring and controlling the works should be clearly identified. Lines of authority, accountability and communication should be made clear within the organisation on site. The interrelationship with third parties is also important as the works may involve subcontractors and the client's representatives and consultants will probably be encountered throughout the project's duration. The precise configuration of the management structure will be determined by the nature and scope of the construction project together with the types and number of parties involved. Within the environmental management plan the structure will be presented in the form of a simple organisation chart.

Routine and Special Environmental Procedures

The requirements for special environmental procedures will vary from site to site. As with any project, the procedures used to carry out the works are considered in the project's method statement. The method statement will consider all environmental effects, together with those actions needed to mitigate or reduce them, for each routine and special operation within the construction sequence. Environmental aspects will have been identified during the tendering and pre-contract stages, and the method statement will be developed during contract planning in readiness for management on site. Special environmental measures may be needed to alleviate contamination by noise and dust, or special processes may be needed to recycle debris into hardcore materials for use on the site. The operations identified in reducing or alleviating the environmental effect or special processes will be reflected in the contract programme, which usually takes the form of a Gantt chart or network diagram with the key tasks specified in the method statement. An example of an environmental method statement for dust mitigation is shown in Figure 11.4. This is then transferred into appropriate dust mitigation measures presented in the EMP as shown in Figure 11.5. Two other examples are shown in Figures 11.6 and 11.7 relating to contamination and dispersal procedures.

Awareness, Induction and Training of Project Personnel

Environmental responsibilities need to be designated throughout the organisation and among all personnel on site. It is essential that everyone carrying out tasks on or around a demolition site has a high level of awareness for the environmental dimensions to their work. Project induction briefings should take place, in particular where special environmental procedures have been identified with which personnel may be uncertain or unfamiliar. Training may be necessary to maintain general environmental awareness and to assist personnel to manage special processes.

ENVIRONMENTAL MANAGEMENT PLAN: Environmental Method Statement–Dust Mitigation									
ENVIRONMENTAL EFFECT	ENVIRONMENTAL IMPLICATION	HEALTH AND SAFETY IMPLICATIONS	PUBLIC SAFEGUARD IMPLICATION	SITE LOCATION	MAGNITUDE OF PROBLEM	LEVEL OF RISK (L/M/H)	FREQUENCY OF MONITORING	STATUTORY LEGISLATION APPLICABLE	MITIGATION ACTION AREAS
DUST (also, dirt/mud creating dust potential)	Yes	Yes	Yes	All	Widespread	High	Twice daily	• Environmental Protection Act 1990 • Clean Air Act 1993	See: EMP– contamination mitigation procedures; dust mitigation procedures Specific action areas: • Demolition • Haulage • Plant management • Materials handling

Figure 11.4 *Example section from Environmental Management Plan: Environmental Method Statement for dust mitigation (Griffith, 2001)*

ENVIRONMENTAL MANAGEMENT PLAN
DUST MITIGATION MEASURES

Demolition:

- Protect general environs from demolition areas on site, e.g. with screening or damping with water sprinklers
- Locate crushing equipment away from general work areas
- Use chutes for vertical movement of demolition debris

Haulage:

- Limit vehicle speeds on and around the site
- Damp down access roads
- Sweep public roads and accesses regularly
- Restrict haulage routes to surfaced roads where possible or protect base surfaces with temporary surfaces, e.g. boarding, textiles and hardcore
- Route haulage away from places of public disturbance

Plant Management:

- Clean plant and equipment when leaving site to avoid dust transfer to public areas

Materials Handling:

- Minimise the height-of-fall of debris materials
- Avoid spillage and clear where necessary
- Cover transported materials to avoid dust creation by movement
- Store materials away from sensitive site locations and protect
- Locate large volume loose materials out of the wind
- Damp down all loose site surfaces

Figure 11.5 *Example section from Environmental Management Plan outlining dust mitigation measures (Griffith, 2001)*

```
┌─────────────────────────────────────────────────────────────┐
│          ┌──────────────────────────────────────┐            │
│          │     ENVIRONMENTAL MANAGEMENT PLAN     │            │
│          │                                       │            │
│          │   CONTAMINATION MITIGATION PROCEDURES │            │
│          └──────────────────────────────────────┘            │
```

ENVIRONMENTAL MANAGEMENT PLAN

CONTAMINATION MITIGATION PROCEDURES

- Confine disposal materials to designated areas of site and protect from ground condition

- Protect water courses, drain and sewers from run off from debris materials

- Confine treatment of reusable materials to designated site location

- Stockpile hazardous materials in designated and clearly identifiable areas

- Protect site and environs from dust and dirt blowing off site

- Control backfill materials: investigate, analyse and treat before reuse

- Isolate potential contaminant liquids and substances to designated areas away from activity and store for controlled removal

- Secure specialised washing areas for plant and equipment for plant haulage leaving site

- Establish protection zones to limit site/neighbourhood traffic disturbance

- Establish bins and storage areas away from plant and wildlife habitats

- Provide protection to trees, hedges and retained vegetation by fences and hoarding

- Invoke checks to safeguard plants/wildlife before commencing work in previously undisturbed areas of site

Figure 11.6 *Example section from Environmental Management Plan outlining contamination mitigation procedures (Griffith, 2001)*

Environmental Monitoring, Review and Improvement

Monitoring is essential to determine if environmental responsibilities are being fulfilled and if environmental effects are being identified and managed effectively. The procedure and frequency of performance monitoring will differ from site to site, influenced by the characteristics and scope of the siteworks and the environmental effects anticipated.

The information gathered is used directly in compiling the EMP. The EMP document contains the necessary information to detail the management procedures and working instructions necessary to establish effective environmental control of the site based on the site characteristics identified in the key elements described. Furthermore, the EMP translates the project applicable aspects of the company's EMS and provides the effective interface between environmental management at site level with the corporate organisation. In this way the EMP responds directly to environmental policies and strategies set by the parent organisation. The key sections providing the framework for a comprehensive EMP are shown in Figure 11.8.

ENVIRONMENTAL MANAGEMENT PLAN

SITE RECYCLING, REUSE AND DISPOSAL PROCEDURES

- Identify demolition materials which can be retained and treated for use on site, e.g. concrete, bricks and blocks

- Stockpiles of demolition materials to be isolated from positions of potential pollution, e.g. away from water courses and drainage runs

- Contaminated demolition materials to be stored in sealed containers or wrapped in membrane sheeting to avoid leakage to ground

- Isolate potentially hazardous demolition materials for controlled disposal by registered contractor, e.g. asbestos sheeting

- Place waste receptacles close to work locations and ensure accessibility

- Segregate waste into labelled receptacles for clear identification and handling, e.g. paper, plastics and metals

- Isolate potentially hazardous waste for controlled storage, handling and collection

Figure 11.7 *Example section from Environmental Management Plan outlining recycling, reuse and disposal procedures (Griffith, 2001)*

Environmental Management System (EMS)

A small to medium enterprise (SME) may not have the need for more than the implementation of an EMP on its project sites. A large principal contractor however may have a need to structure environmental management within its corporate organisation in addition to application on specific project sites through the EMP. The principal mechanism to achieve this is the environmental management system (EMS). The EMS is *"The organisation's formal structure that implements environmental management"* (Griffith, 1994). The environmental management standard ISO 14001 *"requires that the organisation should develop, implement and maintain an EMS to ensure that its activities conform to the environmental policy, strategy, aims and objectives that it has set. Moreover, it should warrant that the system meets all current environmental legislation that regulates its business activities"* (Griffith, 1995). The key EMS elements and requirements expected under ISO 14001 are shown in Figure 11.9.

To establish an EMS the organisation should:

(i) develop an EMS manual – a documented set of procedures and working instructions that satisfy environmental standards
(ii) implement the procedures and working instructions in the course of its business
(iii) maintain procedures and instructions and upgrade them on a continual basis.

PROJECT (SITE) ENVIRONMENTAL MANAGEMENT PLAN (EMP)

- Project (Sitework) Environmental Management Plan

  - Company environmental policy
  - Management responsibilities
  - Company environmental management representative
  - Project environmental policy
  - Organisational chart and designated site responsibilities
  - Key subcontractors and suppliers' procurement programme
  - Environmental legislation and obligations (Environmental Protection Act 1990)
  - Significant environmental site aspects and risk assessment and mitigation
  - Project-specific environmental performance indicators
  - Environmental method statements
  - Site environmental rules
  - Emergency procedures
  - Incident reporting and HSE notification under COSHH
  - Investigation procedures
  - Record keeping

Figure 11.8 *Environmental Management Plan development: key sections describing the project (site) environmental management plan (EMP)*

ISO 14001 ENVIRONMENTAL MANAGEMENT SYSTEM (EMS) ELEMENTS AND REQUIREMENTS

Element	Requirements	Element	Requirements
Environmental policy	• to be available to the public • to provide continual improvement • to comply with environmental legislation • to meet organisational requirements	Document control	• to ensure that all documents in the EMS are controlled for version and distribution
Environmental aspects	• to identify environmental impact of activities • to identify those on which the company can impart influence	Operational control	• to develop documented procedures to control activities including subcontractors and suppliers
Legal and other requirements	• to identify applicable legislation and keep up-to-date	Emergency preparedness and response	• to develop mechanisms to handle emergency situations, followed by review and revision
Objectives and targets	• to establish documented objectives and targets and take in third-party considerations	Monitoring and measurement	• to develop documented procedures for monitoring: progress towards objectives; compliance with legislation; key characteristics of any significant environmental aspect identified
Environmental management programme	• to develop programmes (environmental management plans) that will deliver objectives and targets	Non-conformance and corrective and preventive action	• to develop procedures to handle any non-conformance and to take corrective and preventive action to avoid recurrence
Structure and responsibility	• to define roles and responsibilities to resource the EMS, including appointing an environmental management representative	Records	• to maintain accurate and complete records
Training, awareness and competence	• to identify training needs and implement them throughout the organisation and its projects	Auditing	• to develop mechanisms to carry out an audit of the EMS, and to review and report to company management
Communication	• to develop procedures for communicating internally and externally and to third-parties	Management review	• executive corporate management must review activities to determine changes needed to company policy and implementation of the EMS
EMS documentation	• to develop information to explain the EMS and related documentation		

Figure 11.9 *Requirements on the principal contracting organisation of the elements of the ISO 14001 Environmental Management System specification*

The fulfilment of the requirements set by ISO 14001 is fundamental to intra-organisational environmental management by the company. It is also beneficial in demonstrating to outside audiences that the company is upholding its commitment to its stated environmental policies and practices. This is essential where management system certification is sought or where the company is environmentally pre-qualified for tendering by the client or for external auditing by regulatory bodies. *"Increasingly, companies see the adoption of an accredited system as an important way to maintain a competitive advantage over their peers, as purchasers, clients and investors become more selective on an environmental basis"* (CIRIA, 1999b).

EMS standards in the UK influencing principal contracting organisations are: ISO 14001 and EMAS (European Community Eco-Management and Audit Scheme). ISO 14001 presents the international specification for any organisation in any sector of business seeking to establish an EMS. EMAS requires that participating organisations develop and implement an environmental protection system, predominantly for site applications. A similar scheme – SCEEMAS (Small Company Environmental and Energy Management Assistance Scheme) – has existed in the UK since 1995, making EMAS accessible to small and medium enterprises (SMEs). This may be of interest to smaller organisations within construction. Companies, large and small, will probably be familiar with the requirements for QMS – quality management systems (ISO, 2000). ISO 14001 shares key system elements with BS EN ISO 9000: 2000, the specification for quality systems. It is sensible therefore to configure an EMS much in the same way as a QMS, taking into account specific environmental elements of the standard (Griffith, 2001).

Relationship between the EMS and EMP

The EMS is, in practical terms, the mechanism by which a company sets, monitors and achieves its environmental objectives. It provides a framework within which management responsibilities are assigned throughout the corporate organisation. This framework is then translated into a framework of management for the site through the EMP. The EMP structures the management responsibilities on site in parallel with those of the EMS. For each project the contractor undertakes, the generic elements of system management are already established within the EMS thereby allowing the EMP to focus explicitly on the environmental issues of the project. The relationship between the EMS and the EMP can be seen in Figure 11.10. The considerations, tasks and actions required to develop an EMS and EMP by the principal contractor are considerable, as illustrated in Figures 11.11 and 11.12.

The EMS Manual

A key objective of the corporate organisation is to produce a set of documents that explains the purpose and implementation of the EMS. The principal document is the environmental management manual. The manual should describe: *"The organisation's documentation of procedures and working instructions for implementing environmental management throughout the company and on the projects it undertakes"* (Griffith, 1994). The manual is the organisation's reference point for the implementation and upkeep of the EMS. In its simplest format, the manual is similar to any company handbook of rules and general procedures but directed explicitly towards the environmental management of its business.

The document can take many forms depending on the principal contractor's requirements. It may encompass the whole organisation or specific parts and be a single manual or a suite of documents. It is usual to divide the EMS manual into three broad components: Part I – the system; Part II – management procedures; and Part III – working instructions. The key sections for each part are shown in Figures 11.13, 11.14 and 11.15.

It is essential that the facets of the EMS and EMP are effectively translated into applicable procedures and instructions, and moreover that these are effectively communicated to the operatives on site: *"Successful environmental management relies on communication. It is crucial that everyone is aware of the key issues, has relevant information to deal with them, understands their responsibilities, and provides feedback to those in charge. Site personnel must know whom they can contact for advice*

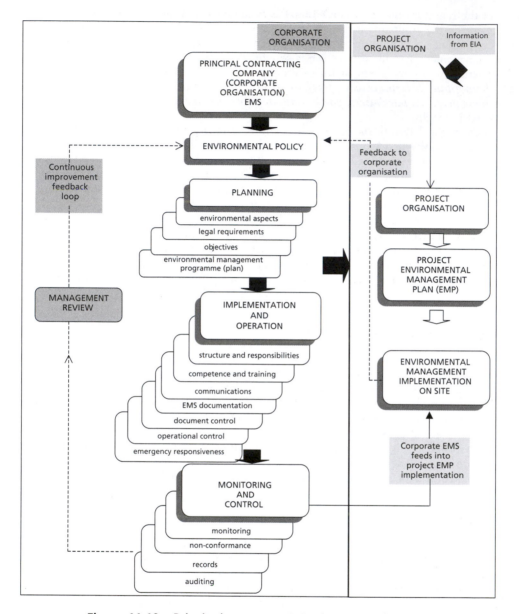

Figure 11.10 *Principal contractor's Environmental Management System framework for corporate and project organisation to meet ISO 14001 system elements*

on managing environmental issues and whom they can ask for training. Feedback down the chain is important in maintaining motivation and raising awareness" (CIRIA, 1999a).

An important element in ensuring good communication is the arrangement of management responsibilities. Designated managers must have clearly defined roles and they, together with everyone on the project, must take responsibility for their actions. To highlight this point the EMP will always have a site organisation chart contained in it. Within ISO 14001 there is a requirement for an organisation to appoint an environmental management representative. This could be a specialist brought in, but more usually the site manager assumes this role along with the many and onerous other duties that they perform.

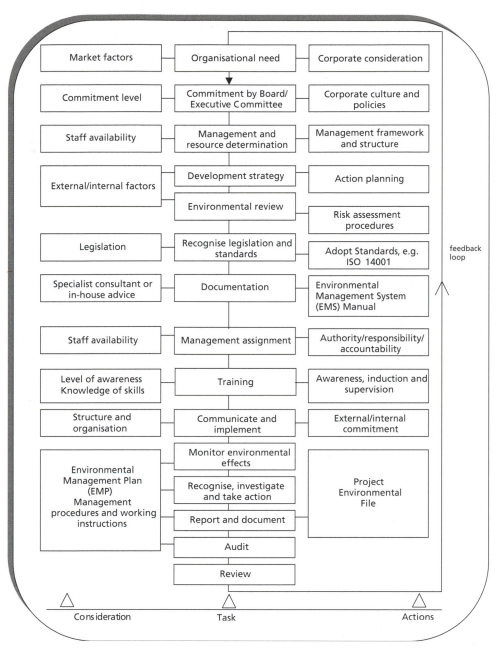

Figure 11.11 *Considerations, tasks and actions required to develop an Environmental Management System (EMS) and Environmental Management Plan (EMP) [adapted from Griffith et al., 2000]*

Environmental Site Control

"Although the majority of environmental concerns on site arise as a result of the actual construction works, the general manner in which the site is managed can influence considerably the implementation and success of control measures" (CIRIA, 1999a). *"Effective environmental management is based on good site management practice"* (Griffith et al., 2000). *"A well established EMS and carefully formulated EMP will provide the mechanisms to do this"* (Griffith, 2001).

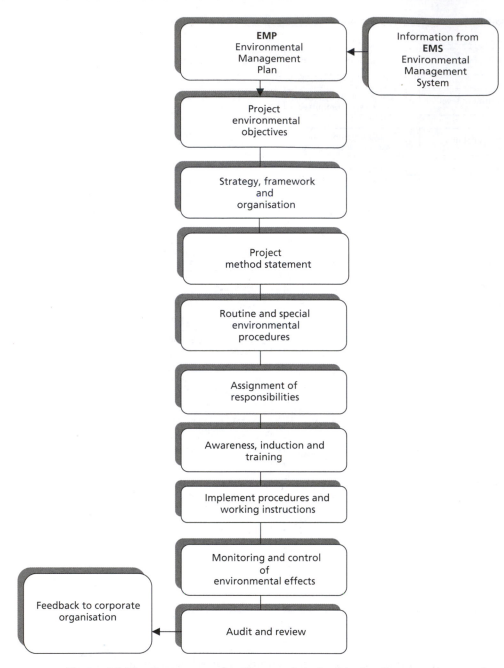

Figure 11.12 *Sequence of tasks associated with the development and implementation of an Environmental Management Plan (EMP) [adapted from Griffith et al., 2000]*

"*Environmental management can be assisted greatly by implementing a three-phase process of consideration which actually commences before start on site and proceeds through to site management and control: (i) environmental appraisal is concerned with the pre-contract period between contract notification and commencement of the demolition works on site. During this phase the contractor should carry out an environmental site survey to confirm and re-appraise the potential for environmental effects which were identified during the pre-tender site survey; (ii) environmental familiarisation is concerned*

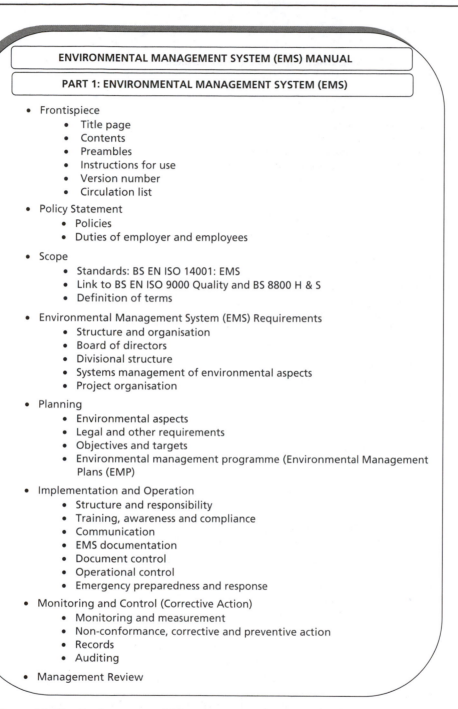

ENVIRONMENTAL MANAGEMENT SYSTEM (EMS) MANUAL

PART 1: ENVIRONMENTAL MANAGEMENT SYSTEM (EMS)

- Frontispiece
  - Title page
  - Contents
  - Preambles
  - Instructions for use
  - Version number
  - Circulation list
- Policy Statement
  - Policies
  - Duties of employer and employees
- Scope
  - Standards: BS EN ISO 14001: EMS
  - Link to BS EN ISO 9000 Quality and BS 8800 H & S
  - Definition of terms
- Environmental Management System (EMS) Requirements
  - Structure and organisation
  - Board of directors
  - Divisional structure
  - Systems management of environmental aspects
  - Project organisation
- Planning
  - Environmental aspects
  - Legal and other requirements
  - Objectives and targets
  - Environmental management programme (Environmental Management Plans (EMP)
- Implementation and Operation
  - Structure and responsibility
  - Training, awareness and compliance
  - Communication
  - EMS documentation
  - Document control
  - Operational control
  - Emergency preparedness and response
- Monitoring and Control (Corrective Action)
  - Monitoring and measurement
  - Non-conformance, corrective and preventive action
  - Records
  - Auditing
- Management Review

Figure 11.13 *Environmental Management System Manual development: key sections describing the environmental manual*

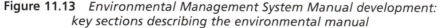

with commencement on site where the contractor should conduct an environmental site tour *to recap on the environmental aspects of the works and to induct management and the workforce to environmental site procedures; and (iii)* environmental monitoring *is concerned with the implementation of monitoring and control mechanisms within the EMP to ensure environmental performance during*

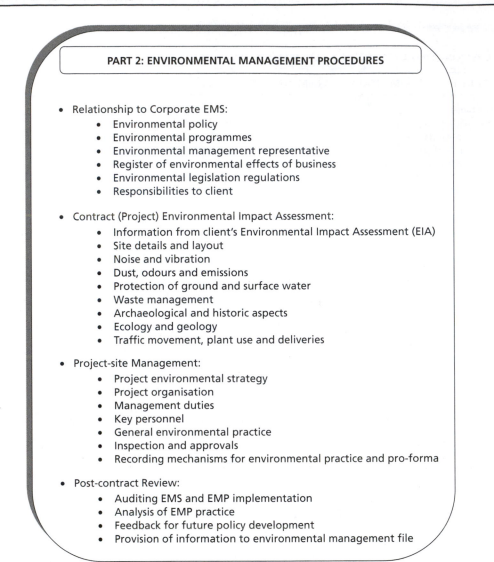

PART 2: ENVIRONMENTAL MANAGEMENT PROCEDURES

- Relationship to Corporate EMS:
    - Environmental policy
    - Environmental programmes
    - Environmental management representative
    - Register of environmental effects of business
    - Environmental legislation regulations
    - Responsibilities to client

- Contract (Project) Environmental Impact Assessment:
    - Information from client's Environmental Impact Assessment (EIA)
    - Site details and layout
    - Noise and vibration
    - Dust, odours and emissions
    - Protection of ground and surface water
    - Waste management
    - Archaeological and historic aspects
    - Ecology and geology
    - Traffic movement, plant use and deliveries

- Project-site Management:
    - Project environmental strategy
    - Project organisation
    - Management duties
    - Key personnel
    - General environmental practice
    - Inspection and approvals
    - Recording mechanisms for environmental practice and pro-forma

- Post-contract Review:
    - Auditing EMS and EMP implementation
    - Analysis of EMP practice
    - Feedback for future policy development
    - Provision of information to environmental management file

Figure 11.14 *Environmental Management Systems Manual development: key sections describing environmental management procedures*

execution of the works. These important stages provide systematic identification, evaluation and management of key potential environmental effects, for example: atmospheric emissions; discharges; spillages; noise; vibration; dust; fumes; waste; and hazardous substances" (Griffith, 1994; 1997; Griffith *et al.*, 2000).

Within each of the three stages identified, a number of tasks need to be performed:

Environmental Appraisal

- Undertake an environmental risk assessment as an integral part of pre-tender planning
- Consider the potential key environmental effects that will need to be managed on site
- Develop the EMP, following the principles set out in the EMS, where applicable
- Determine relationship between the company and site, with particular regard to auditing.

- Working Instructions:
  - Environmental induction, training and instruction in specific site tasks with environmental risk
  - Hazard identification
  - Risk assessment
  - Mitigation and control mechanisms
  - Site meetings on environmental issues
  - Environmental toolbox talks
  - Inspection routines
  - Working with hazardous materials
  - Record keeping
  - Reuse, recycling and disposal routines

Figure 11.15 *Environmental Management Systems Manual development: key sections describing environmental working instructions*

Environmental Familiarisation

- Brief managers and workforce on the environmental issues identified during project planning
- Conduct a site tour to familiarise the team with the site and environmental aspects
- Provide environmental awareness through induction
- Undertake training in special environmental procedures needed for the particular works.

Environmental Monitoring

- Provide detailed guides on all management procedures and working instructions
- Confirm potential actions to be taken should environmental effects manifest themselves
- Specify senior managers at company level who can be consulted easily and quickly should issues not be resolved by site management
- Implement self-audit and review guidelines for site use.

To achieve effective transfer of this information into application the EMP must clearly: (i) specify environmental requirements; (ii) assign management responsibilities; and (iii) implement control mechanisms.

Specify Environmental Requirements

In each area of site activity the principal contractor should ensure that:

- the potential for environmental effect is clearly identified
- the magnitude of potential is determined
- there is active monitoring to detect effects should they occur
- planned controls are in place to mitigate effects
- effects are systematically recorded together with a record of actions taken

- occurrences are investigated and reviewed
- experience is fed back into the EMP for both project and future use.

Assign Management Responsibilities

It is a key task of management to ensure that the above requirements are met by assigning responsibilities to supervisors and operatives which:

- ensure that daily checks are made in all vital areas of site activity
- demand the reporting of any breach of environmental integrity in the works
- review daily checking procedures and feed them into weekly site meetings
- collate weekly data for input to the monthly site progress meeting
- integrate information for long-term performance review.

Implement Control Mechanisms

To assist the practical implementation of the EMP, simple control mechanisms, based on good observation, recording and reporting, should be implemented. A series of task pro-forma documents (Griffith *et al.*, 2000) can be developed to assist in formally recording the following:

- occurrence of any environmental effect
- location of the incident
- reasons for the occurrence
- any action taken
- review of action to determine effectiveness
- further actions needed
- notifiable nature of specific incidents, for example a report to authorities in the event of a breach in regulations, or health and safety report where persons are involved
- feedback to project files and records for audit purposes.

The Project Environmental File

At the conclusion of the construction project the principal contractor should gather all the documentation and site records relevant to the environmental aspect and submit these to the project environmental file. The file assists the site team and the corporate organisation with project evaluation and review. Environmental issues experienced during the course of the project can help the principal contractor share best practice throughout the wider organisation. In addition, it provides useful information for tendering purposes on future contracts.

Case Study – Environmental Example

Project Quality and Environmental Plan (see pages 446 to 496)

This Case Study example illustrates the implementation of a project quality and environmental plan for a major civil engineering construction project. The comprehensiveness and level of detail needed to describe the work in terms of documentation are clearly evident.

Rather than utilise individual and separate project plans for quality and environment, this major construction organisation sought to utilise considerable proactiveness and forward thinking, and to integrate its project quality plans and project environmental plans into a singular dedicated plan. The result is an integrated, systematic, comprehensive and practical project plan document set.

The set of documents provides detailed information such as: the suite of documents used; project information; contract requirements; inspection and verification procedures; testing; audits, review and archiving; guidance for environmental management on site; and operational site evaluation. As a project management plan package, the documents embrace all aspects necessary to the successful management of quality and environment during the siteworks of a major construction project.

It was identified earlier in the chapter that a project environmental plan may well encompass: policy; management responsibilities; organisation charts; legislation; and method statements. These are not included in the example document but would be presented in a set of documents held by the company's head office as an intrinsic part of the company's management systems documentation. A complete framework for corporate systems and project management plans featured earlier in this chapter and is illustrated in Figure 11.10.

Environmental Inspection Monitoring (see pages 497 to 499)

This Case Study example describes environmental inspection monitoring for a major metropolitan infrastructure development project. The siteworks involved conventional construction activities, methods and sequences, and these were complicated by the requirements to undertake works adjacent to a river and to undertake considerable demolition works and landscape reinstatement.

Aspects of the project environmental plan need to be translated into systematic and practical procedures for inspecting and checking the works as they are undertaken. The documents presented relate to environmental management monitoring and control procedures used on the project. For the project described, a weekly environmental checklist was implemented. For key sitework activities a weekly inspection was formally undertaken and a tick-box record maintained. Where additional aspects were identified, action points were recorded for subsequent attention or revisit. The checklist is designed to be signed-off by the appointed inspector and countersigned by the site manager before being entered into the principal contractor's environmental site records. Information from these records was used to feed into the monthly site progress meetings and also, in a collated and summarised form, provided information for the project environmental file.

Summary

A carefully conceived and well-structured approach to environmental management from an all-embracing EMS to a project-specific EMP can provide an effective framework for environmental management by the principal contractor. The key to successful environmental management is the establishment of a clear policy and strategy on environmental matters, a structural management framework and good site environmental control practices. The systematic and effective management of environmental aspects during the siteworks will not only reduce the environmental effects of construction, but contribute significantly to more effective working by the principal contractor, leading to cost savings by eliminating remediation works. Moreover, a demonstrated commitment to the environment may lead to improved turnover of work with clients in both the public and private sectors who take environmental issues seriously and who pre-qualify principal contractors for contracts based on their environmental performance.

References

CIRIA (1993a) *Environmental Issues in Construction – A Review of Issues and Initiatives Relevant to the Building, Construction and Related Industries. Vol. 1: Overview Including Executive and Technical Summaries.* Special Publication SP93, Construction Industry Research and Information Association, London.

CIRIA (1993b) *Environmental Issues in Construction – A Review of Issues and Initiatives Relevant to the Building, Construction and Related Industries. Vol. 2: Technical Review.* Special Publication SP94, Construction Industry Research and Information Association, London.

CIRIA (1995) *Waste Minimisation and Recycling in Construction – A Review.* Construction Industry Research and Information Association, London.

CIRIA (1999a) *Environmental Good Practice on Site.* Construction Industry Research and Information Association, London.

CIRIA (1999b) *Environmental Issues in Construction – A Desk Study.* Project Report PR73, Construction Industry Research and Information Association, London.

CIRIA (1999c) *The Reclaimed and Recycled Construction Materials Handbook.* Construction Industry Research and Information Association, London.

CIRIA (2000) *Environmental Management in Construction.* Construction Industry Research and Information Association, London.

Department of the Environment (DoE) (1989a) *Environmental Assessment: Explanatory Leaflet.* HMSO, London.

Department of the Environment (DoE) (1989b) *Environmental Assessment: A Guide to the Procedures.* HMSO, London.

Department of the Environment, Transport and the Regions (DETR) (1998) *Less Waste More Value.* HMSO, London.

Environment Agency (EA) (1974) *The Control of Pollution Act 1974.* HMSO, London.

Environment Agency (EA) (1979) *The Ancient Monuments and Archaeological Areas Act 1979.* HMSO, London.

Environment Agency (EA) (1981) *The Wildlife and Countryside Act 1981.* HMSO, London.

Environment Agency (EA) (1990a) *The Environmental Protection Act 1990.* HMSO, London.

Environment Agency (EA) (1990b) *The Town and Country Planning Act 1990.* HMSO, London.

Environment Agency (EA) (1990c) *The Planning (Listed Buildings and Conservation Areas) Act 1990.* HMSO, London.

Environment Agency (EA) (1993) *The Clean Air Act 1993.* HMSO, London.

Environment Agency (EA) (1995) *The Environment Act 1995.* HMSO, London.

European Union (EU) (1993) *The Eco-Management and Audit Scheme (EMAS), EU Council Regulation 1836/93.* European Union, Brussels.

Goode D. (2000) The Best Job in London, *Landscape Design*, No. 294. Landscape Institute, London.

Griffith A. (1994) *Environmental Management in Construction.* Macmillan (now Palgrave), Basingstoke.

Griffith A. (1995) The Current Status of Environmental Management Systems in Construction, *Engineering, Construction and Architectural Management*, 2(1), 5–16.

Griffith A. (1997) *Environmental Management in the Construction Process.* Construction Paper 75, Chartered Institute of Building, Ascot.

Griffith A. (2001) *Environmental Management of Demolition Works: Effective Project Control.* Construction Papers 128, Chartered Institute of Building, Ascot.

Griffith A., Stephenson P. and Watson P. (2000) *Management Systems for Construction.* Addison Wesley Longman, Harlow.

Health and Safety Executive (HSE) (1974) *The Health and Safety at Work, etc. Act 1974.* HMSO, London.

Health and Safety Executive (HSE) (1989) *The Noise at Work Regulations 1989.* HMSO, London.

Health and Safety Executive (HSE) (1991) *The Environmental Protection Act (Prescribed Processes and Substances) Regulations 1991.* HMSO, London.

Health and Safety Executive (HSE) (1994) *The Construction (Design and Management) Regulations 1994.* HMSO, London.

Health and Safety Executive (HSE) (1996a) *Health and Safety in Construction.* HMSO, London.

Health and Safety Executive (HSE) (1996b) *The Construction (Health, Safety and Welfare) Regulations 1996.* HMSO, London.

Health and Safety Executive (HSE) (1996c) *The Control of Pollution (Special Waste) Regulations 1996.* HMSO, London.

Health and Safety Executive (HSE) (1996d) *The Carriage of Goods by Road Regulations 1996.* HMSO, London.

HM Government (1988a) *The Town and Country Planning (Assessment of Environmental Effects) Regulations.* HMSO, London.

HM Government (1988b) *The Environmental Assessment (Scotland) Regulations.* HMSO, London.

HM Government (1989) *The Planning (Assessment of Environmental Effects) Regulations (Northern Ireland)*. HMSO, London.

ISO (1996) *BS EN ISO 14001: Environmental Management Systems – Specification with Guidance for Use*. International Organisation for Standardisation, Geneva.

ISO (2000) *BS EN ISO 9000: 2000: Quality Systems – Specification with Guidance for Use*. International Organisation for Standardisation, Geneva.

Porritt J., Martin G. and Masero S. (1999) *Sustainable Construction, A Challenge for the Construction Industry*. CIOB Millennium Project: Sustainable Construction, Chartered Institute of Building, Ascot.

Principal Contractor's Project Quality and Environmental Plan

PROJECT QUALITY AND ENVIRONMENTAL PLAN

Project Name:	
Project No:	Client Reference Number:
Controlled Copy No:	

Purpose:

This Project Quality and Environmental Plan is a statement of intent to ensure that the management of the contract meets the requirements of the client and those of BS EN ISO 9001 and BS EN ISO 14001. It is a unique working document to provide guidance and direction for the effective management of this contract.

The Project Quality and Environmental Plan identifies the scope of the project, key personnel, areas of responsibility, the quality and environmental practices, resources, controls and procedures to be used on the contract as required and established by the   Co Ltd Management System to which reference must be made.

PROJECT QUALITY AND ENVIRONMENTAL PLAN APPROVAL	
Prepared By: .... (Name) .... (Signature) Senior Engineer............. .... (Title)	**Date:**
Approved By: .... (Name) .... (Signature) Contracts Manager .... (Title)	**Date:**
Client Approval: .... (Name) .... (Signature) .... (Title)	**Date:**

Project Name:	Section: Cover Sheet	Revision:
Project Number:	Page No: 1 of 1	Date of Issue:.

SECTION ONE – RECORD OF AMENDMENTS

Section	Date	Revision No	Description of Change	Approved By
All	April		FIRST DRAFT	

Project Name:	Section: One	Revision:
Project Number:	Page No: 1 of 1	Date of Issue:

SECTION TWO – CONTROLLED COPY DISTRIBUTION LIST			
Controlled Copy No	**Date of Issue**	**Title**	**Name**
01		Site Copy	File
02		Contracts Manager	
03		Senior Engineer	
04		Senior Engineer	
05		Resident Engineer	

Project Name:		**Section: Two**	**Revision:**
Project Number:		**Page No: 1 of 1**	**Date of Issue:**

SECTION THREE – CONTENTS LIST

Section No	Date of Issue	Name
One		Record of Amendments
Two	"	Controlled Copy Distribution List
Three	"	Contents List
Four	"	Project Information
	"	4.1 Company Information
	"	4.2 Client Information
	"	4.3 Additional Information
Five	"	Schedule of Project Documentation
	"	5.1 Contract Documentation
	"	5.2 Project Programme
	"	5.3 Project Method Statement Register
	"	5.4 Project Drawing Register
	"	5.5 Operational Environmental Evaluation Sheets
Six	"	Particular Contract Requirements
	"	6.1 Quality Requirements
	"	6.2 Environmental Requirements
Seven	"	Schedule of Significant Materials and Suppliers
Eight	"	Special Material and/or Equipment Storage Requirements
Nine	"	Schedule of Client Supplied or Free Issue Materials
Ten	"	Schedule of Sample and Trials Requirements
Eleven	"	Schedule of Management System Documents
	"	11.1 Procedure/Work Instruction
	"	11.2 Non–Standard or Client Required Forms
Twelve	"	Inspection, Testing and Verification Plan
Thirteen	"	Schedule of Measuring and Test Equipment
Fourteen	"	Emergency Information
Fifteen	"	Project Audits
Sixteen	"	Review of Project Quality and Environmental Plan
Seventeen	"	Archiving
Appendix A	"	Organisational Charts
Appendix B	"	Supporting Contract Documentation (as applicable)
Appendix C	"	Project Programme
Appendix D	"	Operational Environmental Evaluation Sheets
Appendix E	"	Applicable Consents, Licences and Authorisations
Appendix F	"	Quality and Environmental Policies

Project Name:	Section: Three	Revision:
Project Number:	Page No: 1 of 1	Date of Issue:

SECTION FOUR – PROJECT INFORMATION

4.1 Company Information (Attach Organisation Chart at Appendix A)

Company Address:	Company Site Address:
Tel No:	Tel No: TBA
Fax No:	Fax No: TBA

Key Company Personnel (Head Office and Site Based)			
Name	Appointment	Sample Signatures (if required)	Contact Number
	Contract Director		
	Contracts Manager		
	Project Manager		
	Site Agent		
	Quantity Surveyor		
	General Foreman		
	H & S Manager		
	Quality Manager		
	Environmental Officer		
	Site Safety Coordinator		
	Temporary Works Coordinator		
	Environmental Matters		

	Section: Four	Revision:
Project Name:		
Project Number:	Page No: 1 of 5	Date of Issue:

SECTION FOUR – PROJECT INFORMATION

4.2   Client Information (Attach Organisation Chart At Appendix A, if appropriate)

Client Address: **Tel No:** **Fax No:**	 **Tel No:** **Fax No:**

Key Client Personnel (Head Office and Site Based)		
Name	**Appointment**	**Contact Number**
	Project Sponsor	

Note: Highlight with asterisk name of client representative to whom contract correspondence should be addressed.

Project Name:	**Section: Four**	**Revision:**	
Project Number:	**Page No: 2 of 5**	**Date of Issue:**	

SECTION FOUR−PROJECT INFORMATION

4.3   Additional Information

4.3.1 CDM Information

	Telephone No:
Name of Principal Contractor: **(for purposes of CDM Regs)**	**Telephone No:**

4.3.2   Consultant(s) Information (Attach Organisation Chart At Appendix A, if appropriate)

Name: Engineer	**Telephone No:**
Name: **Resident Engineer**	**Telephone No: TBA**
Name:	**Address:** **Telephone No:**

Project Name:	**Section: Four**	**Revision:**
Project Number:	**Page No: 3 of 5**	**Date of Issue:**

SECTION FOUR – PROJECT INFORMATION

4.3   Additional Information

4.3.3   Subcontractor Information (Attach Organisation Chart at Appendix A, if appropriate)

Subcontractor Name: Service/Product Supplied: Fencing	Subcontractor Address: Telephone No:	Name of On-Site Representative: Telephone No:
Subcontractor Name: Service/Product Supplied: Spoil Haulage	Subcontractor Address: Telephone No:	Name of On-Site Representative: Telephone No:
Subcontractor Name: _____ Service/Product Supplied: _____	Subcontractor Address: Telephone No:	Name of On-Site Representative: Telephone No:
Subcontractor Name: _____ Service/Product Supplied: _____	Subcontractor Address: Telephone No:	Name of On-Site Representative: Telephone No:
Subcontractor Name: _____ Service/Product Supplied: _____	Subcontractor Address: Telephone No:	Name of On-Site Representative: Telephone No:

Project Name: _____ Project Number:	Section: Four Page No: 4 of 5	Revision: Date of Issue:

SECTION FOUR – PROJECT INFORMATION

4.3   Additional Information

4.3.4   Regulatory Authority Information

Environmental Agency Name:	Environmental Agency Address: Telephone No:	Name of Representative: Telephone No:
Local Authority Name:	Local Authority Address: Telephone No:	Name of Representative: Telephone No:
Sewerage Undertaker Name:	Sewerage Undertaker Address: Telephone No:	Name of Representative: Telephone No:
Heritage Organisation Name:	Heritage Organisation Address: Telephone No:	Name of Representative: Telephone No:
Nature Conservation Organisation Name:	Nature Conservation Organisation Address: Telephone No:	Name of Representative: Telephone No:

Project Name:	Section: Four	Revision:
Project Number:	Page No: 5 of 5	Date of Issue:

SECTION FIVE – SCHEDULE OF PROJECT DOCUMENTATION

5.1 Contract Documentation
(Copy of information to be included at Appendix B as appropriate, e.g. Client List of Delegated Authorities)

Originator	Date	Reference	Title
	Oct 2001	Volume 1	Instructions to Tenderers
		Volume 2	Form of Tender Conditions of Contract Specification Bill of Quantities
		Volume 3	Drawings (2 Folders)
		Volume 4	Pre-Tender Health and Safety Plan
		Volume 5	Additional Information (3 Folders)

5.2 Project Programme(s) (Latest versions to be included at Appendix C)

Reference	Date	Revision	Title
		0	Clause 14 Programme (for approval)
		1	Clause 14 Programme (for approval)

	Section: Five	Revision:
Project Name:		
	Page No: 1 of 4	Date of Issue:
Project Number:		

SECTION FIVE — SCHEDULE OF PROJECT DOCUMENTATION

5.3   Project Method Statement Register

Method Statement Reference	Date	Revision	Title
GEEMS 1		03	Emergency Guide
GEEMS 2		03	Spillage to Land
GEEMS 3		03	Spillage to Foul Water Drain
GEEMS 4		03	Spillage to Surface Water
GEEMS 5		03	Who to Contact
GEMS 1		01	Habitat Protection
GEMS 2		01	Excavation
GEMS 3		01	Product Handling
GEMS 4		01	Dust Blow
GEMS 5		01	Powered Machinery Usage
GEMS 6		01	Noise and Vibration
GEMS 7		01	Waste Storage and Handling
GEMS 8		01	General Waste Disposal
GEMS 9		01	Special Waste Disposal
GEMS 10		01	Contaminated land
GEMS 11		01	Discharges to Water and Work Over, In or Near Water
GEMS 12		01	Carriage of Dangerous Goods
GEMS 13		01	Work In or Near Areas of Disease Infectious to Farm Animals
MS/01			Trial Excavations
MS/02			Site Establishment
MS/03			Confined Spaces Access and Egress
MS/04			Shaft Sinking Access Shaft 002
MS/05			Shaft Sinking Access Shaft 001
MS/06			Shaft Sinking Upstream Portal
MS/07			Groundwater and Sewage Discharge
MS/08			Admiralty Road Access
MS/09			Installation of 1500 mm dia. Pipes (MH001–MH003)
MS/10			Installing and burying TBM
MS/11			Construction of 2100 mm internal diameter tunnel (ch 920–2556)
MS/12			Construction of 2100 mm internal diameter tunnel (ch 281–920)
MS/13			Installation of 1500 mm dia. pipe under Ferry Toll Road
MS/14			Installation of 1500 mm dia. pipes (MH004A–Throttle Chamber–Existing Sewer)
MS/15			Installation of the 600 mm diameter throttle pipe

MS/16			Railway Crossing of the 600 mm diameter throttle pipe
MS/17			Burn Crossing for the 1500 mm dia. concrete pipe
MS/18			Construction of the Drawpits and Cable ducts in the WWTW
MS/19			Construction of Throttle Chamber
MS/20			Construction of Manholes
MS/21			Temporary and Permanent Reinstatement of roads, footpaths and grasslands
MS/22			Installation of 1200 mm dia. concrete pipes (MH037− MH001− MH012)
MS/23			Construction of the Flow Separation Chamber

Project Name:	Section: Five	Revision:
Project Number:	Page No: 2 of 4	Date of Issue:.

SECTION FIVE – SCHEDULE OF PROJECT DOCUMENTATION

5.4   Project Drawing Register

Drawing Reference	Date	Revision	Title
56200/GLA/001		A	Site Location & Layout Plan
56200/GLA/002		A	Upstream Connection Area Arrangement
56200/GLA/003		A	Downstream Connection Area Arrangement
56200/GLA/004		A	Access Shaft 1 Area Plan
56200/GLA/005		A	Access Shaft 2 Area Plan
56200/GLA/006		0	Standard Pipe Bedding Details
56200/GLA/007		0	Longitudinal Section Sheet 1 of 2
56200/GLA/008		0	Longitudinal Section Sheet 2 of 2
56200/GLA/009		0	Access Shafts Standard Details
56200/GLA/010		0	Typical Manhole Details
56200/GLA/011		0	Upstream Connection Details
56200/GLA/012		0	Upstream Flow Control Chamber
56200/GLA/015		0	Throttle Chamber General Arrangement
56200/GLA/016		0	Downstream Connection General Arrangement of Connection at WWTW
56200/GLA/017		A	Miscellaneous Details
56200/GLA/019		0	Tunnel Lining General Arrangement
56200/GLA/021		A	Existing Services Layout Plan (Sheet 1 of 5)
56200/GLA/022		A	Existing Services Layout Plan (Sheet 2 of 5)
56200/GLA/023		A	Existing Services Layout Plan (Sheet 3 of 5)
56200/GLA/024		A	Existing Services Layout Plan (Sheet 4 of 5)
56200/GLA/025		A	Existing Services Layout Plan (Sheet 5 of 5)
56200/GLA/026		0	Manhole Schedule
56200/GLA/030		0	Tunnel Lining Reinforced Concrete Details
56200/GLA/031		0	Monitoring Points
56200/GLA/032		0	Details of Drawpits and Cable Ducts
56200/GLA/033		0	Portal Manholes Standard Details

		Section: Five	Revision:
Project Name:			
		Page No: 3 of 4	Date of Issue:
Project Number:			

SECTION FIVE – SCHEDULE OF PROJECT DOCUMENTATION

5.5 Operational Environmental Evaluation Sheets
(Copy of relevant Environmental Evaluation Sheets to be held in Appendix D)

Environmental Aspect	Environmental Impact	Last Reviewed
Work on site using powered machinery	Air pollution from exhaust emissions	
Disposal to landfill of waste materials	Contamination of land, depletion of resources and release of landfill gases	
Storage & handling of materials on site	Contamination of land, controlled waters or sewers	
Excavation of site	Removal or disturbance of invasive species	
Dust blow from exposed ground, stored materials or unpaved access roads	Pollution of controlled waters or sewers from run off	
Work on a site that contains protected structures, flora or fauna, trees and hedges, archaeological remains or nesting birds	Environmental damage caused by the work to any of the protected items	
Noise caused by site work	Noise or vibration resulting from the site work that is liable to cause nuisance	
Work on site using powered machinery	Noise or vibration resulting from the use of plant that is liable to cause nuisance	
Work on site using powered machinery	Contamination of land, controlled waters or sewers through leaks and drips while in operation or during refuelling	
Work on site using powered machinery	Dust or mud problems resulting from the use of plant that is liable to be a nuisance	
Storage & handling of materials on site	Dust blow from storage of powdered materials that is liable to be a nuisance	
Excavation of site	Dust blow from soil exposed by excavations or from soil stock piled as a result of excavations	
Excavation of site	Pollution of controlled waters or sewers by run off from excavated materials	
Dust blow from exposed ground, stored materials or unpaved access roads	Nuisance caused to sensitive receptors such as local residents or crops	
Site work involving discharges to water or work over, in or near water	Contamination of controlled waters or sewers by discharges from site or by virtue of the works themselves	
Carriage of dangerous or hazardous goods to or from site	Contamination of land, controlled waters or sewers by leakage during transport of dangerous or hazardous materials	
Dust caused by cutting, grinding, blasting or operations involving the breaking out of materials	Nuisance caused to sensitive receptors such as local residents or crops	
Debris from cutting, grinding, blasting or operations involving the breaking out of materials	Contamination of land, controlled waters or sewers	

Project Name:	Section: Five	Revision:
Project Number:	Page No: 4 of 4	Date of Issue:.

SECTION SIX – PARTICULAR CONTRACT REQUIREMENTS

6.1 Quality Requirements

1. The Works are divided into four sections with the main activities being:

    a, Upstream Works – The works include connecting into the existing sewer, construction of a flow separation chamber and the laying of 280 m of 1500 mm id pipes c/w three manholes.

    b, Tunnelling Works – The activities at this section include the construction of 2.24 km of a 2100 mm internal diameter tunnel and the sinking of two manholes at the u/s and d/s tunnel portals.

    c, Access Shaft Works – The works include the construction of two access shafts and the provision of access to the access shafts.

    d, Downstream Works – The works include laying of 30 m of 1500 mm pipe, the construction of a new throttle chamber, the laying of 370 m of 600 mm GRP pipe and connections into the   sewer and the WWTW.

2. The defects date is 52 weeks for works not involving permanent reinstatement of roads and footpaths and 104 weeks for permanent reinstatement of roads and footpaths. It is 156 weeks for permanent reinstatement of roads, car parks and footpaths if the excavation is greater than 1.5 m deep.

3. The Specifications for the contract are:

    'Civil Engineering Specification for the Water Industry, 5th Edition' augmented by Supplementary Clauses.

    'Specification for Tunnelling, Thomas Telford, London, 2000' augmented by Supplementary Clauses.

    'Water Industry Mechanical and Electrical Specification' augmented by Supplementary Clauses.

4. In so far as any Supplementary Clause may conflict, or be inconsistent, the Supplementary Clause shall always prevail.

Project Name:	Section: Six	Revision:
Project Number:	Page No: 1 of 2	Date of Issue:

SECTION SIX – PARTICULAR CONTRACT REQUIREMENTS

6.2 Environmental Requirements
(Copy of any consents, licences, authorisations, etc. to be included at Appendix E)

Environmental Issue	Contract Requirement
Site set-up, e.g. • **Lighting** • **Visual impact** • **Signage** • **Siting of plant**	• Lighting not to adversely affect the surrounding areas • Signage to be erected as per the Client's instructions • Plant will be situated away from watercourses. And where possible away from residential areas • Vehicle wheel washing facilities to be provided where appropriate
Habitat protection **Wildlife protection** **Tree protection** **Hedge protection** **Plant protection** **Protected structures**	Trees and roots to be protected from damage and excess loads not to be placed over tree roots. No equipment to be attached to trees. All roots to be protected against frost and air.
Powered machinery, e.g. • **Dust** • **Noise** • **Drip trays**	• Wheel washing facility to be established. Roads to be kept clear of mud • Compressors used on site shall be silenced • Dewatering plant, if run 24 hours, shall be silenced • Ancillary pneumatic tools shall be fitted with silencers • Noise level requirements to be obtained from the EHO prior to work commencing Working hours within the site: Mon–Fri 0800 to 1800 / Sat 0800 to 1300 only (24 hour working below ground).
Storage, e.g. • **Plant** • **Materials** • **Excavations** • **Topsoil/Turf**	• All tanks/drums to be bunded in accordance with Oil Storage Regulations 2001 • All tanks/drums shall be secured when not in use to prevent unauthorised use • Static plant shall be surrounded by an impermeable bund
Waste, e.g. • **Storage** • **Transfer** • **Disposal** • **Recycling**	• RE/Client to be notified of any potentially contaminated material required to be disposed of off-site • All waste will be disposed at a suitably licensed tip • SEPA to be notified where appropriate
Contaminated land, e.g. • **Ground investigation techniques** • **Sampling** • **Disposal**	• Where suspected potentially contaminated material is encountered, and Council Planning and Building Control department, as well as the Engineer shall be notified as soon as practical • All waste will be disposed at a suitably licensed tip
Water, e.g. • **Discharges** • **Abstraction** • **Use of chemicals** • **Alterations to watercourse profile or floodplain**	All discharges/abstractions to be approved by the Environment Protection Agency.
Communications, e.g. • **Meetings** • **Induction** • **Toolbox talks** • **Noticeboards**	• Signage to be erected as per the Client's instructions • Consultation with interested parties • Letter drops and the appointment of a member of staff as a point of contact

Project Name:	Section: Six	Revision:
Project Number:	Page No: 2 of 2	Date of Issue:

SECTION SEVEN – SCHEDULE OF SIGNIFICANT MATERIALS AND SUPPLIERS

Material	Supplier
BITUMINOUS MATERIAL	
CEMENT	
CONCRETE	
CONCRETE PIPES	
DOWELS AND BOLTS	
DUCTILE IRON PIPES	
GEOTEXTILES	
GRP PIPEWORK AND CHAMBERS	
HYDROPHILIC GASKET	
MANHOLE COVERS	
MISCELLANEOUS STEELWORK	

PIPE BEDDING	
PRECAST CONCRETE MANHOLE RINGS	
REINFORCEMENT	
SAND	
SHAFT LININGS	
TUNNEL LININGS	
TYPE 1	
UPVC DUCTS	

Project Name:	Section: Seven	Revision:
Project Number:	Page No: 1 of 1	Date of Issue:.

SECTION EIGHT – SPECIAL MATERIAL AND/OR EQUIPMENT STORAGE REQUIREMENTS

Material/Equipment	Risk to be Protected From	Method of Storage/ Protection
CEMENT	MOISTURE	– MOISTURE RESISTANT BAGS – STORED IN DRY ENVIRONMENT
CONCRETE	ENVIRONMENTAL PROTECTION	– DISPOSAL AS PER SEPA'S GUIDELINES – WASHOUT OF MIXERS AS PER SEPA'S GUIDELINES
DIESEL	ENVIRONMENTAL PROTECTION	– TO BE STORED IN A BUNDED DOUBLE SKINNED CONTAINER – DRIP TRAYS TO BE PLACED UNDER EQUIPMENT WHERE APPROPRIATE
GROUT	ENVIRONMENTAL PROTECTION	– DISPOSAL AS PER SEPA'S GUIDELINES
HIGH TENSILE STEEL BARS	MOISTURE	– STORED IN DRY ENVIRONMENT
HYDROPHILLIC GASKET	MOISTURE	– STORED IN DRY ENVIRONMENT
REINFORCEMENT	CONTAMINATION	– STORED ON RACKS OR SUPPORTS SUFFICIENT TO PREVENT CONTAMINATION FROM THE GROUND – FOR LONG-TERM STORAGE , THE REINFORCEMENT WILL BE COVERED

Project Name:	Section: Eight	Revision: A
Project Number:	Page No: 1 of 1	Date of Issue:

SECTION NINE – SCHEDULE OF CLIENT SUPPLIED OR FREE ISSUE MATERIALS	

Material	Acceptance Criteria/Special Storage Requirements etc.
NONE	NOT APPLICABLE

Project Name:	Section: Nine	Revision:
Project Number:	Page No: 1 of 1	Date of Issue:.

SECTION TEN – SCHEDULE OF SAMPLE AND TRIALS REQUIREMENTS

Item to be Sampled or Trialed	Sample or Trial Requirements	Frequency
Atmospheric	Levels of Oxygen, Carbon Monoxide, Hydrogen Sulphide and Explosive gases	Continually during tunnel occupation
Bituminous Painting	DFT of 0.78 mm	Locations specified by Engineer
Concrete	Slump measurement, Crushing Strength	Spec. Cl. 4.3
Ground Settlement	As per Drawing 56200/GLA/031	As per Drawing 56200/GLA/031
Grout	Crushing Strength	Spec. Cl. 4.3 & 353
Noise	Specification Clause 1.21	As required
Pipes	Visual Inspection for Cracks	Before use
Pipework	Specification Clause 7	Once
Road Make-up	CBR	Locations specified by Engineer
Shaft Lining	Dimensional Survey	As required
Structures	Specification Clause 7	Once
Tunnel Lining	Dimensional Survey	As required

Project Name:	Section: Ten	Revision:
Project Number:	Page No: 1 of 1	Date of Issue:.

SECTION ELEVEN – SCHEDULE OF MANAGEMENT SYSTEM DOCUMENTS
APPLICABLE TO THE PROJECT

11.1 Procedures/Work Instructions

Procedure No	Title	Site Specific Y/N	Issue Status (site specific only)
PR05	Hold Start-Up Meeting	N	
PR08	Review & Plan Labour Requirements & Prepare Histogram	N	
PR09	Develop Project Execution Plan	N	
PR10	Allocate Labour & Mobilise Site	N	
PR12	Manage Project and Changes	N	
PR13	Establish Material Requirements & Raise Requisition	N	
PR14	Establish Plant Requirements & Raise Requisition	N	
PR15	Identify Subcontract Requirements	N	
PR16	Mobilise Subcontractor & Manage on Site	N	
PR18	Recruit Labour	N	
PR18a	Manage Labour	N	
PR19	Manage Material on Site	N	
PR20	Check Plant Received, Manage Plant & Register	N	
PR21	Manage Commissioning & Handover	N	
PR22	Manage Defects Period	N	
PR23	Review Project Performance & Feedback	N	
Q03	Project Quality Planning	N	
Q04	Internal Auditing	N	
Q06	Non-Conformance & Customer Complaint Reporting	N	
Q08	Records	N	
Q10	Document and Data Control	N	
C09	Identify Changes, Monitor & Agree Actual Project Costs	N	
C11	Prepare CVR Pack	N	
C13	Carry Out Client Valuation	N	
C15	Agree Final Account	N	
C17	Archiving	N	
CP09	Develop & Maintain Project Risk Register	N	
CP12	Training and Development	N	
CP13	Recruitment	N	

Project Name:	Section: Eleven	Revision:
Project Number:	Page No: 1 of 3	Date of Issue:

┌───┐
│ SECTION ELEVEN – SCHEDULE OF MANAGEMENT SYSTEM DOCUMENTS   │
│ APPLICABLE TO THE PROJECT   │
└───┘

11.1 Procedures/Work Instructions (continued)

Procedure No	Title	Site Specific Y/N	Issue Status (site specific only)
D02	Develop Design & Technical Options	N	
D05	Finalise Temporary Works Design	N	
D06	Review, Develop & Finalise Design	N	
D07	Complete As Built Drawings & Technical Docs	N	
D08	Calculations and Weight Take-Offs	N	
D09	Design Checking System	N	
D10	Design Changes	N	
D11	Design Procedure	N	
D13	Prepare and Maintain Plant & Equipment	N	
D14	Calibration of Inspection, Measuring & Test Equipment	N	
HS06	Develop, Maintain & Handover Health & Safety File	N	
HS07	Develop & Maintain Health & Safety Plan	N	
HS08	Site Environmental Control	N	
HS14	Waste Management	N	
HS15	Control of Environmental Emergencies	N	
HS17	General Waste Transfer	N	
HS18	Special Waste Transfer	N	
HS20	Accident, Incident & Near Miss Reporting	N	
HS21	Risk & COSHH Assessment	N	
HS22	Carriage of Dangerous Goods	N	
PC02	Supplier & Subcontractor Assessment	N	
PC03	Send & Receive Enquiries	N	
PC07	Procure Subcontractor	N	
PC08	Procure Plant	N	
PC09	Procure Material	N	

Project Name:	Section: Eleven	Revision:
Project Number:	Page No: 2 of 3	Date of Issue:

SECTION ELEVEN – SCHEDULE OF MANAGEMENT SYSTEM DOCUMENTS
APPLICABLE TO THE PROJECT

11.2 Non-Standard or Client Required Forms

Form Reference	Title	Issue Status	Date of Issue

Project Name:	Section: Eleven	Revision:
Project Number:	Page No: 3 of 3	Date of Issue:

SECTION TWELVE – INSPECTION/TESTING AND VERIFICATION PLAN

INSPECTION/TESTING/VERIFICATION PLAN (This document forms part of the Project Quality Plan. It can however be used as a separate document if so desired.)

CLIENT NAME:

CLIENT ADDRESS:

CONTRACT TITLE:

CLIENT REFERENCE NO:

CONTRACT REFERENCE NO:

Reference Documents (Continue on Separate Sheet if required)
Contract Drawings (see section 5.1), Specification (see section 6.1)

INSPECTION CODE

H	Hold Point
W	Witness (Routine Inspection)
S	Surveillance (On-going Random Observation)
FI	Final Inspection
A	Approval
D	Document Review
NR	Not Required
NA	Not Applicable
S10	Sample Testing 10%
S25	Sample Testing 25%
S100	Sample Testing 100%

A	B	C	1	2	3	4	5	6	7

PREPARED BY:

DATE:

APPROVED BY:

DATE:

Section: Twelve	**Revision:**
Page No: 1 of 3	**Date of Issue:**

Project Name:

Project Number:

SECTION TWELVE – INSPECTION/TESTING AND VERIFICATION PLAN

Activity No	Activity	Controlling Document/Instruction Reference	Characteristic to be Verified	Acceptance Criteria	Verifying Document	Inspection Requirements		
							Client	Other
1	Establish Setting-out Controls	Contract Drawings	Drawings tie in with existing features/ structures	Compliance	Instrument file Setting-out file	H	D	
2	Pipe Installation	Contract Drawings Specification Manufacturer's Rec.	– Formation suitability – Pipe bedding – Placement of selected fill – Pressure testing	– Contractor/Engineer acceptance – Compliance with specification	Checklist testing	H, FI T10, T100 S FI	W, FI W S W, FI	
3	Manhole Construction	Contract Drawings Specification Manufacturer's Rec.	– Materials in good condition	No damage	Checklist	H, FI	W, FI	
4	Shaft Construction	Contract Drawings Specification Manufacturer's Rec.	– Dimensional survey – Visual inspection when grouting – Visual inspection for visible leakage	– Within tolerances – No voids – No leakage	Survey Sheets Checklists	H, FI H, FI H, FI	W, FI W, FI W, FI	
5	Tunnel Construction	Contract Drawings Specification Manufacturer's Rec.	– Dimensional survey – Visual inspection when grouting – Visual inspection for visible leakage – Ground settlement	– Within tolerances – No voids – No leakage	Survey Sheets Checklists	H, FI H, FI H, FI	W, FI W, FI W, FI	

Section: Twelve	Revision:
Page No: 2 of 3	Date of Issue:

Project Name:

Project Number:

SECTION TWELVE – INSPECTION/TESTING AND VERIFICATION PLAN

Activity No	Activity	Controlling Document/Instruction Reference	Characteristic to be Verified	Acceptance Criteria	Verifying Document	Inspection Requirements		
							Client	Other
6	Structures	Contract Drawings Specification	– Dimensional Survey – Water test/Visual inspection for leakage – Penstocks working	– Within tolerances – No leakage – Penstocks working	Survey Sheets Checklists	H, FI H, FI H, FI	W, FI W, FI W, FI	
7	Backfilling	Contract Drawings Specification	Compaction	Compacting Method	Checklists	S	S	

Project Name:	Section: Twelve	Revision:
Project Number:	Page No: 3 of 3	Date of Issue:

SECTION THIRTEEN – SCHEDULE OF MEASURING AND TEST EQUIPMENT

Item	Reference No	Manner of Calibration	Date of Last Service/Calibration	Calibration Interval
Automatic Level		Service/Check Calibration		Weekly Yearly
Gas Detector		Service/Check Calibration		Weekly 6 Monthly
Laser Level		Service/Check Calibration		Weekly Yearly
Pipe Laser		Service/Check Calibration		Weekly Yearly
Rescue Sets		Inspect/Test Service/Check Calibration		Daily Monthly 6 Monthly
Theodolite		Service/Check Calibration		Monthly Yearly
Total Station		Service/Check Calibration		Monthly Yearly

Project Name:	**Section: Thirteen**	**Revision:**
Project Number:	**Page No: 1 of 1**	**Date of Issue:**

SECTION FOURTEEN – EMERGENCY INFORMATION

Site Address: Site Office   Tel: TBA

Emergency Access: TBA

Local Hospital: Tel:

Normal Site Access: TBA

Spillkit Equipment:

Booms ✔   Socks ☐   Cushions ✔   Mats ✔   Spilldry ☐   Dammit Clay ☐

Other (please specify)...

Location of Spillkit Equipment: ... STORES..

External Environmental Emergency Contacts:

		Day	Night
Pollution –	Environmental Agency –	Tel:	
		Tel:	
Protected Species/Habitats –	Nature Conservation Organisation –	Tel:	
Protected Structures –	Local Authority –	Tel:	

Project Name:	Section: Fourteen	Revision:
Project Number:	Page No: 1 of 1	Date of Issue:

SECTION FIFTEEN – PROJECT AUDITS

Audit No	J	F	M	A	M	J	J	A	S	O	N	D
1						/						
2							/					
3								/				
4									/			
5										/		
6											/	
7												/
8	/											
9		/										
10			/									
11				/								
12					/							
13						/						
14							/					

AUDIT PLANNED			/							Audit Legend	
AUDIT UNDERTAKEN				◣							
AUDIT CLOSED OUT						■					

Project Name:	Section: Fifteen	Revision
Project Number:	Page No: 1 of 1	Date of Issue:

SECTION SIXTEEN – REVIEW OF PROJECT QUALITY AND ENVIRONMENTAL PLAN

Review Date	Reviewed By	Signature

Note:

This Project Quality and Environmental Plan is to be formally reviewed every six months throughout the duration of the project.

Where the duration of the project is less than six months the Project Quality and Environmental Plan shall be reviewed at least once during the life of the project.

	Section: Sixteen	Revision:
Project Name:		
	Page No: 1 of 1	**Date of Issue:**
Project Number:		

SECTION SEVENTEEN – ARCHIVING

Project Completion:	Date:
Project Quality and Environmental Plan Archived:	Date:
Signature:	(Site Manager)

This Project Quality and Environmental Plan plus its supporting documentation are to be returned to the Contract Administration Department for archiving in accordance with the company archive procedures.

GUIDANCE for ENVIRONMENTAL EMERGENCY MANAGEMENT on SITE
(GEEMS) 2
SPILLAGE TO LAND

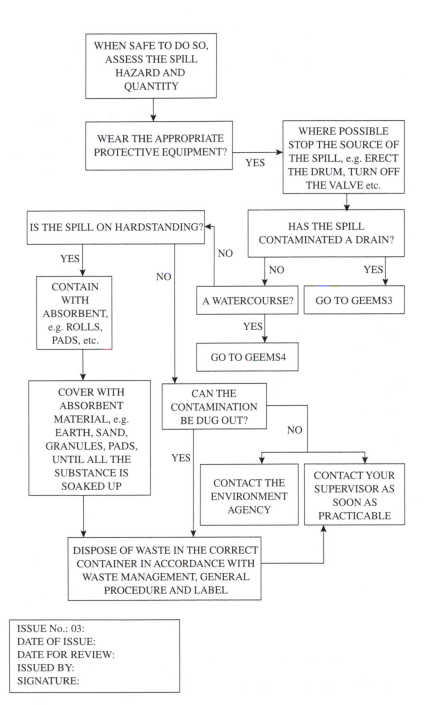

GUIDANCE for ENVIRONMENTAL MANAGEMENT on SITE
(GEMS) 1 – HABITAT PROTECTION

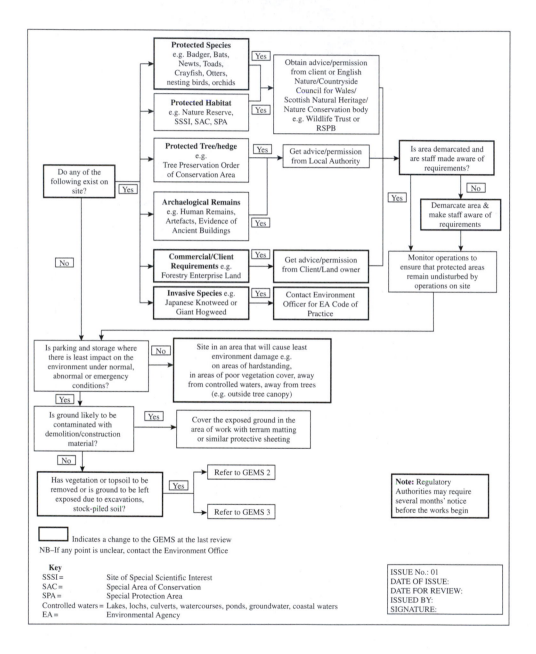

GUIDANCE for ENVIRONMENTAL MANAGEMENT on SITE (GEMS) 2 – EXCAVATION

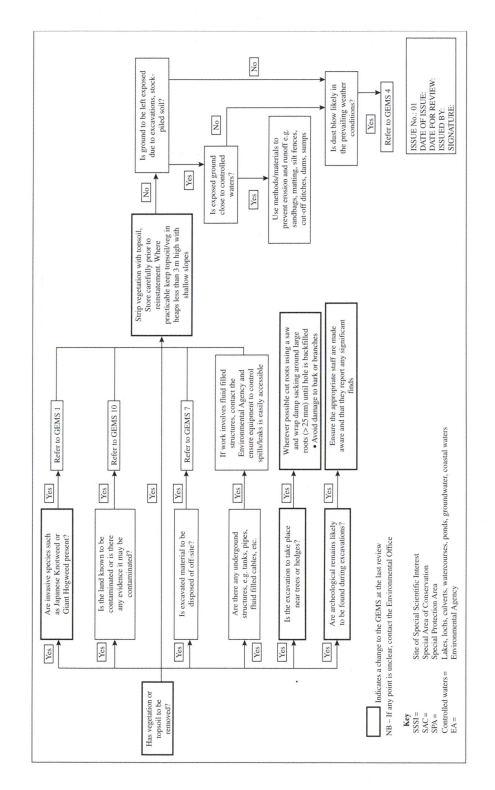

Has vegetation or topsoil to be removed?

Are invasive species such as Japanese Knotweed or Giant Hogweed present? — Yes → Refer to GEMS 1

Is the land known to be contaminated or is there any evidence it may be contaminated? — Yes → Refer to GEMS 10

Is excavated material to be disposed of off-site? — Yes → Refer to GEMS 7

Are there any underground structures, e.g. tanks, pipes, fluid filled cables, etc. — Yes → If work involves fluid filled structures, contact the Environmental Agency and ensure equipment to control spills/leaks is easily accessible

Is the excavation to take place near trees or hedges? — Yes → Wherever possible cut roots using a saw and wrap damp sacking around large roots (>25 mm) until hole is backfilled • Avoid damage to bark or branches

Are archeological remains likely to be found during excavations? — Yes → Ensure the appropriate staff are made aware and that they report any significant finds

Strip vegetation with topsoil, Store carefully prior to reinstatement. Where practicable keep topsoil/veg in heaps less than 3 m high with shallow slopes

Is ground to be left exposed due to excavations, stock-piled soil? — No →

Is exposed ground close to controlled waters? — No / Yes → Use methods/materials to prevent erosion and runoff e.g. sandbags, matting, silt fences, cut-off ditches, dams, sumps

Is dust blow likely in the prevailing weather conditions? — Yes → Refer to GEMS 4

ISSUE No.: 01
DATE OF ISSUE:
DATE FOR REVIEW:
ISSUED BY:
SIGNATURE:

☐ Indicates a change to the GEMS at the last review

NB – If any point is unclear, contact the Environmental Office

Key
SSSI = Site of Special Scientific Interest
SAC = Special Area of Conservation
SPA = Special Protection Area
Controlled waters = Lakes, lochs, culverts, watercourses, ponds, groundwater, coastal waters
EA = Environmental Agency

GUIDANCE for ENVIRONMENTAL MANAGEMENT on SITE (GEMS) 7 – WASTE STORAGE AND HANDLING

Waste is anything a person produces or possesses which they intend or are required to discard (includes scrap metal and products that are past their expiry date)

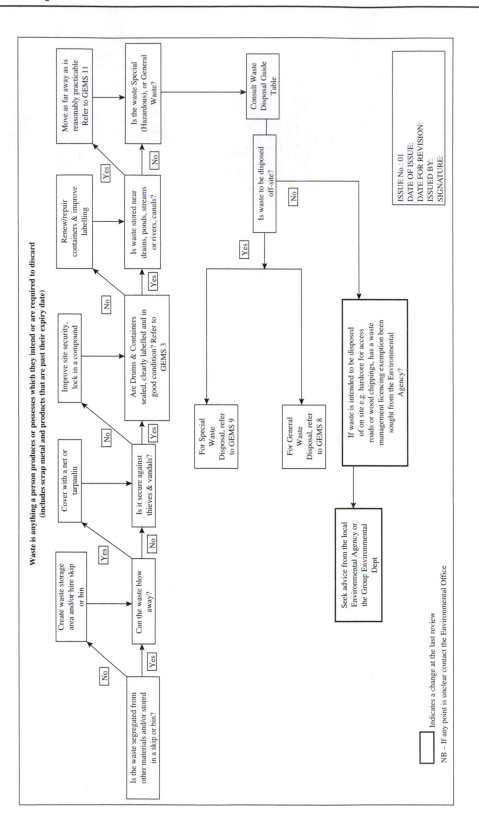

ISSUE No.: 01
DATE OF ISSUE:
DATE FOR REVISION:
ISSUED BY:
SIGNATURE:

Indicates a change at the last review

NB – If any point is unclear contact the Environmental Office

GUIDANCE for ENVIRONMENTAL MANAGEMENT on SITE (GEMS) 11 – DISCHARGES TO WATER AND WORK OVER, IN OR NEAR WATER

When working next to water/drains ensure:
- materials are stored as far as possible away from controlled waters/drains
- plant is stored/operated as far as possible away from controlled waters/drains
- pollution control equipment is accessible, e.g. mats, rolls, booms, sandbags and where an event is forseeable put in preventative measures e.g. filters or sandbags across drains or booms across a river
- refuelling occurs away from controlled waters and drains

When working over water ensure:
- platforms and scaffolding are properly sealed or tarpauling is used to prevent construction debris polluting the water
- If bridge cleaning, etc. that physical cleaning methods are used in preference to chemical cleaning methods

When working in water ensure:
- oversized pumps are not used and are raised above the river bed to prevent disturbance to bed sediments
- disturbance to the watercourse bed is minimised e.g. by use of temporary bridges
- measures are in place to reduce the movement of any disturbed sediments downstream e.g. coffer dams, filters, sandbags, sediment capture mats, etc.
- measures are taken to ensure any obstructions to flow introduced by the project works will not increase the chances of flooding under high water level conditions (consult Environmental Agencies)

When discharging nuisance water from excavations or controlling run-off from site roads, exposed ground, etc. use the following methods to reduce pollution:

Silt
- use cut-off dams on ditches to act as a settlement ponds or use settlement tanks
- use silt fences at the toe of exposed soil (the ends should turn up the slope to prevent flow around the ends)
- use grassed areas, geotextile filters, sediment capture mats or flocculants to remove sediments (if using flocculants consult the Environmental Agency)

Oil
- Place a skirted or oil absorbent boom across a tank, setting pond or around discharge point

GEMS 3

GEMS 5

Monitor to ensure consent conditions are met

ISSUE No.: 01
DATE OF ISSUE:
DATE FOR REVISION:
ISSUED BY:
SIGNATURE:

Note: Regulatory authorities may require several months' notice prior to works commencing

Ensure appropriate materials for dealing with oil, fuel or chemical spills are available on site and if working near a large watercourse, that a boom(s) is available

Apply to Environment Agency/Scottish Environmental Protection Agency for advice/consent

Apply to Local Water Undertaker for advice/discharge consent

Are there any controlled waters, e.g. river, canal, pond, coastal area, ground water, or drains associated with the site?

Yes

Is work over water, in water, adjacent to water or drains, or are there any discharges to controlled waters/drains?

Yes

Are there any discharges to controlled waters?
OR
Will there be any alterations to the profile of the watercourse or to the floodplain e.g. due to river diversions, culvert relining works, temporary bridge or large deposits of soil on the floodplain?
OR
Will the site need to abstract water from controlled waters?
OR
Are sensitive operations such as herbicide spraying or drainage work to take place adjacent to controlled waters or wetland areas?
OR
Is work next to, over or in controlled waters controlled by a byelaw?

Yes

No

Are there any discharges to drains?

Yes

Indicates a change to the GEMS at the last review
NB – If any point is unclear, contact the Environmental Office

Operational Sites Evaluation Sheets

Aspect
All work on site involving the use of any type of powered machinery

Reference Number	Impact	Controlling Legislation	Score under Worst Case Conditions	Significance under Worst Case Conditions
	Air pollution from exhaust emissions.	None Applicable	4.3	S3 Significant

Worst Case Scenario
All equipment/engines running at the same time.

Company Objective — **Score after Control Measures** 2.3 — **Significance after Control Measures** R2 Not Significant

Action taken to Deal with Objective — Maintain present control measures.

Directive Liable to be Breached under Worst Case Scenario

Legislation Liable to be Breached under Worst Case Scenario
None Applicable.

Control Measures
Guidance for Environmental Management on Site (GEMS) 5 provides information and assistance in working with equipment and plant on site

Aspect
Disposal to landfill of all types of waste materials produced during the course of the work.

Reference Number	Impact	Controlling Legislation	Score under Worst Case Conditions	Significance under Worst Case Conditions
	Contamination of land. Depletion of resources. Release of landfill gases to the atmosphere, e.g. methane and carbon dioxide.	Special Waste Regulations 1996, as amended. Environmental Protection (Duty of Care) Regs 1991, as amended.	4.3	S3 Significant

Worst Case Scenario
All waste sent to landfill.

Company Objective — **Score after Control Measures** 4.3 — **Significance after Control Measures** R3 Significant

Action taken to Deal with Objective — Added to programme of Objectives & Targets.

Directive Liable to be Breached under Worst Case Scenario

Legislation Liable to be Breached under Worst Case Scenario
Control of Pollution Act 1974.

Controlled Waste (Registration of Carriers and Seizure of Vehicles) Regulations 1991, as amended.

Controlled Waste Regulations 1992, as amended.

Control Measures
Where site work is controlled by a main contractor, it is usual to use his facilities and procedures for disposal of waste materials.

Waste minimisation programmes and recycling initiatives are being undertaken to attempt to reduce the quantity of waste sent to landfill.

Guidance for Environment Managemental on Site (GEMS) 7, 8 & 9 provide information and assistance in dealing with the keeping, treating and disposal of waste materials generated on site.

Aspect
Storage of raw materials and waste on site and handling of these items.

Reference Number	
Impact	Contamination of land, controlled waters or sewers.
Worst Case Scenario	All available fluid enters a sewer, drain or watercourse
Controlling Legislation	The Control of Pollution (Oil Storage) (England) Regulations 2001.
Directive Liable to be Breached under Worst Case Scenario	
Legislation Liable to be Breached under Worst Case Scenario	Environmental Protection (Duty of Care) Regulations 1991, as amended. Environmental Protection Act 1990. Finance Act 1996. Landfill Tax (Contaminated Land) Order 1996. Landfill Tax (Qualifying Material) Order 1996.
Significance under Worst Case Conditions	S2 Not Significant
Score under Worst Case Conditions	3.3
Significance after Control Measures	R2 Not Significant
Score after Control Measures	2.3
Company Objective	
Action taken to Deal with Objective	Maintain present control measures.
Control Measures	Guidance for Environmental Emergencies on Site (GEEMS) 2, 3 & 4 provide information and assistance in dealing with and reporting environmental accidents and incidents on site. Guidance for Environmental Management on Site (GEMS) 3 provides information and assistance in dealing with material handling and storage on site. Anti-Pollution Works Regulations 1999. Control of Pollution Act 1974. Environmental Protection (Duty of Care) Regulations 1991, as amended. Groundwater Regulations 1998. Public Health Act 1936, Section 259. Salmon and Freshwater Fisheries Act 1975, Section 4.

Aspect
Dust blow from exposed ground, stored materials or unpaved access roads.

Reference Number	Impact	Controlling Legislation	Score under Worst Case Conditions	Significance under Worst Case Conditions
	Pollution of controlled waters or sewers from run-off.	None Applicable	4.3	S3 Significant
			Score after Control Measures 2.3	Significance after Control Measures R2 Not Significant

Worst Case Scenario

	Company Objective	Action taken to Deal with Objective	
All run-off discharges to a drain, sewer or watercourse during torrential or prolonged rainfall.		Maintain existing control measures.	

Directive Liable to be Breached under Worst Case Scenario

Legislation Liable to be Breached under Worst Case Scenario

Anti-Pollution Works Regulations 1999.

Control of Pollution (Applications, Appeals and Registers) Regulations, 1996
Control of Pollution Act 1974.
Groundwater Regulations 1998.
Public Health Act 1936, Section 259.

Control Measures

Guidance for Environmental Emergencies on Site (GEEMS) 2, 3 & 4 provide information and assistance in dealing with and reporting environmental accidents and incidents on site.

Guidance for Environmental Management on Site (GEMS) 4 provides information and assistance in dealing with conditions liable to result in dust blow and run-off.

Aspect
Work on a site that contains protected structures, flora and fauna, trees and hedges, archaeological remains, listed buildings or nesting birds.

Reference Number	Impact	Controlling Legislation	Score under Worst Case Conditions	Significance under Worst Case Conditions
	Environmental damage caused by the work to any of these protected items.	Planning (Listed Buildings and Conservation Areas) (Scotland) Act 1997. Wildlife and Countryside Act 1981.	3.5	S4 Significant
			Score after Control Measures 1.5	Significance after Control Measures R2 Not Significant

Worst Case Scenario

	Company Objective	Action taken to Deal with Objective	
Protected building, habitat or site is irretrievably damaged		Maintain existing control measures.	

Directive Liable to be Breached under Worst Case Scenario

Legislation Liable to be Breached under Worst Case Scenario

Conservation (Natural Habitats, etc.) Regulations 1994, as amended.

Environment Act 1995.

Hedgerow Regulations 1997.

Planning (Listed Buildings and Conservation Areas) (Scotland) Act 1997.

Protection of Badgers Act 1992.

Town and Country Planning (Trees) Regulations 1999.

Wildlife and Countryside Act 1981.

Control Measures

Guidance for Environmental Management on Site (GEMS) 1 provides information and assistance in dealing with protected sites.

Aspect
Noise caused by site work.

Reference Number	*Impact*	*Controlling Legislation*	*Score under Worst Case Conditions*	*Significance under Worst Case Conditions*
	Noise or vibration resulting from the site work, including use of plant, that is liable to be a nuisance.	Noise and Statutory Nuisance Act 1993.	3.4	S4 Significant

	Worst Case Scenario	*Company Objective*	*Action taken to Deal with Objective*	*Score after Control Measures*	*Significance after Control Measures*
	Plant and equipment operated in close proximity to local housing with no control measures.		Maintain existing control measures.	2.3	R2 Not Significant

Directive Liable to be Breached under Worst Case Scenario

Legislation Liable to be Breached under Worst Case Scenario

Control of Noise (Appeals) (Scotland) Regulations 1983.

Control of Noise (Codes of Practice for Construction and Open Sites) (Scotland) Order 1985.

Environmental Protection Act 1990.

Noise and Statutory Nuisance Act 1993.

Noise at Work Regulations 1989.

Statutory Nuisance (Appeals) (Scotland) Regulations 1996.

Control Measures

Guidance for Environmental Management on Site (GEMS) 6 provides information and assistance in dealing with noise issues and calculating noise levels from individual and combinations of plant and equipment.

Aspect
All work on site involving the use of any type of powered machinery.

Reference Number	Impact	Controlling Legislation	Score under Worst Case Conditions	Significance under Worst Case Conditions
	Contamination of land, controlled waters or sewers through leaks and drips while in operation or during refuelling.	None Applicable	2.5	S4 Significant

Worst Case Scenario	Company Objective	Action taken to Deal with Objective	Score after Control Measures	Significance after Control Measures
Fuel tank bursts and entire contents escape to drain or sewer.		Maintain existing control measures.	2.3	R2 Not Significant

Directive Liable to be Breached under Worst Case Scenario

Legislation Liable to be Breached under Worst Case Scenario
Anti-Pollution Works Regulations 1999.

Control of Pollution Act 1974.

Environmental Protection (Duty of Care) Regulations 1991, as amended

Groundwater Regulations 1998.

Public Health Act 1936, Section 259.

Salmon and Freshwater Fisheries Act 1975, Section 4.

Sewerage (Scotland) Act 1968.

Trade Effluent (Prescribed Processes and Substances) Regulations 1989, as amended.

Control Measures
Guidance for Environmental Emergencies on Site (GEEMS) 2, 3 & 4 provide information and assistance in dealing with and reporting environmental accidents and incidents on site.

Guidance for Environmental Management on Site (GEMS) 5 provides information and assistance when working with powered machinery on site.

Aspect
All work on site involving the use of any type of powered machinery.

Reference Number	Impact	Controlling Legislation	Score under Worst Case Conditions	Significance under Worst Case Conditions
	Dust or mud problems resulting from the use of plant that is liable to be a nuisance.	None Applicable	4.4	S4 Significant

Worst Case Scenario	Company Objective	Action taken to Deal with Objective	Score after Control Measures	Significance after Control Measures
Excessive mud deposits left on road causing a traffic accident.		Maintain existing control measures.	3.3	R2 Not Significant

Directive Liable to be Breached under Worst Case Scenario

Legislation Liable to be Breached under Worst Case Scenario
Environmental Protection Act 1990.

Statutory Nuisance (Appeals) (Scotland) Regulations 1996.

Control Measures
Guidance for Environmental Management on Site (GEMS) 5 provides information and assistance when working with powered machinery on site.

Guidance for Environmental Management on Site (GEMS) 4 provides guidance in dealing with mud on roads and dust blow.

Aspect
Excavation of site.

Reference Number	Impact	Worst Case Scenario	Controlling Legislation	Score under Worst Case Conditions	Significance under Worst Case Conditions	Score after Control Measures	Significance after Control Measures
	Disturbance of underground structures leading to leakage of oils or chemicals and pollution of land or controlled waters.	Puncture of underground structure leading to escape of all fluid contained.	None Applicable	5.5	S4 Significant	3.3	R2 Not Significant

Company Objective

Action taken to Deal with Objective Maintain existing control measures.

Directive Liable to be Breached under Worst Case Scenario

Legislation Liable to be Breached under Worst Case Scenario
Anti-Pollution Works Regulations 1999.

Environmental Protection (Duty of Care) Regulations 1991, as amended.

Public Health Act 1936, Section 259.

Salmon and Freshwater Fisheries Act 1975, Section 4.

Control of Pollution Act 1974.

Control Measures
Guidance for Environmental Emergencies on Site (GEEMS) 2, 3 & 4 provide information and assistance in dealing with and reporting environmental accidents and incidents on site.

Guidance for Environmental Management on Site (GEMS) 2 provides information and assistance in carrying out excavations on site.

Aspect
Excavation of site.

Reference Number	
Impact	Pollution of controlled waters or sewers by run-off from excavated materials
Worst Case Scenario	Run-off during torrential or prolonged rainfall.
Controlling Legislation	None Applicable
Score under Worst Case Conditions	3.3
Significance under Worst Case Conditions	S2 Not Significant
Score after Control Measures	1.3
Significance after Control Measures	R2 Not Significant
Company Objective	
Action taken to Deal with Objective	Maintain present control measures.

Directive Liable to be Breached under Worst Case Scenario

Legislation Liable to be Breached under Worst Case Scenario
Anti-Pollution Works Regulations 1999.

Control of Pollution Act 1974.

Public Health Act 1936, Section 259.

Salmon and Freshwater Fisheries Act 1975, Section 4.

Sewerage (Scotland) Act 1968.

Trade Effluent (Prescribed Processes and Substances) Regulations 1989, as amended.

Control Measures
Guidance for Environmental Management on Site (GEMS) 2 provides information and assistance in carrying out excavations on site.

Guidance for Environmental Emergencies on Site (GEEMS) 2, 3 & 4 provide information and assistance in dealing with and reporting environmental accidents and incidents on site.

Aspect
Dust blow from exposed ground, stored materials or unpaved access roads.

Reference Number	
Impact	Nuisance caused to sensitive receptors such as local residents or crops.
Worst Case Scenario	High winds blowing loose unprotected materials.
Controlling Legislation	None Applicable
Score under Worst Case Conditions	3.3
Significance under Worst Case Conditions	S2 Not Significant
Score after Control Measures	2.3
Significance after Control Measures	R2 Not Significant
Company Objective	
Action taken to Deal with Objective	Maintain present control measures.

Directive Liable to be Breached under Worst Case Scenario

Legislation Liable to be Breached under Worst Case Scenario
Environmental Protection Act 1990.

Statutory Nuisance (Appeals) (Scotland) Regulations 1996.

Statutory Nuisance (Appeals) Regulations 1995.

Control Measures
Guidance for Environmental Management on Site (GEMS) 4 provides information and assistance in dealing with materials likely to be affected by dust blow.

Aspect
Site work involving discharges to water or work over, in or near water.

Reference Number

Impact
Contamination of controlled waters or sewers by discharges from site or by virtue of the works themselves.

Worst Case Scenario
Uncontrolled contaminated discharge to drains or sewers.

Controlling Legislation
None Applicable

Company Objective

Action taken to Deal with Objective
Maintain existing control measures.

Score under Worst Case Conditions
3.4

Score after Control Measures
2.3

Significance under Worst Case Conditions
S3 Significant

Significance after Control Measures
R2 Not Significant

Directive Liable to be Breached under Worst Case Scenario

Legislation Liable to be Breached under Worst Case Scenario
Anti-Pollution Works Regulations 1999.

Control of Pollution (Applications, Appeals and Registers) Regulations, 1996.

Control of Pollution Act 1974.

Groundwater Regulations 1998.

Land Drainage Act 1991.

Public Health Act 1936, Section 259.

Salmon and Freshwater Fisheries Act 1975, Section 4

Control Measures
Guidance for Environmental Management on Site (GEMS) 11 provides information and assistance for making discharges to controlled waters or sewers and also for working over, in or near water.

Guidance for Environmental Emergencies on Site (GEEMS) 3 & 4 provide information and assistance in dealing with and reporting environmental accidents and incidents on site.

Aspect
Carriage of dangerous or hazardous goods to or from site.

Reference Number	Impact	Controlling Legislation	Significance under Worst Case Conditions
	Contamination of land, controlled waters or sewers by leakage during the transport of dangerous or hazardous materials.	None Applicable	S3 Significant
			Significance after Control Measures R2 Not Significant

Worst Case Scenario
All materials being transported escape and enter a drain or sewer.

Score under Worst Case Conditions
3.4

Score after Control Measures
1.4

Directive Liable to be Breached under Worst Case Scenario

Legislation Liable to be Breached under Worst Case Scenario
Anti-Pollution Works Regulations 1999.

Company Objective   **Action taken to Deal with Objective**   Maintain existing control measures.

Control Measures
Guidance for Environmental Management on Site (GEMS) 12 provides information and assistance when transporting dangerous goods.

Guidance for Environmental Emergencies on Site (GEEMS) 2, 3 & 4 provide information and assistance in dealing with and reporting environmental accidents and incidents on site.

Information regarding dangerous goods transported by the company together with dangerous goods cards are available on this database as menu option.

Carriage of Dangerous Goods (Amendment) Regulations 1998

Carriage of Dangerous Goods (Classification, Packaging and Labelling) and Use of Transportable Pressure Receptacles Regulations 1996.

Carriage of Dangerous Goods (Driver Training) Regulations 1996.

Carriage of Dangerous Goods by Road Regulations 1996.

Environmental Protection (Duty of Care) Regulations 1991, as amended.

Public Health Act 1936, Section 259.

Salmon and Freshwater Fisheries Act 1975, Section 4.

Aspect

Dust caused by cutting, grinding, blasting, drilling or operations involving the breaking out/excavation of material.

Reference Number	Impact	Controlling Legislation		Score under Worst Case Conditions	Significance under Worst Case Conditions

Impact
Nuisance caused to sensitive receptors such as local residents or crops.

Controlling Legislation
None Applicable

Score under Worst Case Conditions
3.3

Significance under Worst Case Conditions
S2 Not Significant

Score after Control Measures
2.3

Significance after Control Measures
R2 Not Significant

Worst Case Scenario
High winds blow dust from operations towards a sensitive receptor.

Company Objective   **Action taken to Deal with Objective**
Maintain present control measures

Directive Liable to be Breached under Worst Case Scenario

Control Measures
Guidance for Environmental Management on Site (GEMS) 4 provides information and assistance in dealing with materials likely to be affected by dust blow.

Legislation Liable to be Breached under Worst Case Scenario
Environmental Protection Act 1990.

Statutory Nuisance (Appeals) (Scotland) Regulations 1996.

Aspect
Debris from cutting, grinding, blasting, drilling or operations involving the breaking out of material.

Reference Number

Impact
Contamination of land, controlled waters or sewers.

Controlling Legislation
None Applicable

Score under Worst Case Conditions
3.4

Significance under Worst Case Conditions
S3 Significant

Score after Control Measures
1.4

Significance after Control Measures
S2 Not Significant

Worst Case Scenario
All debris enters controlled waters or a sewer.

Company Objective   **Action taken to Deal with Objective**
Maintain existing control measures

Directive Liable to be Breached under Worst Case Scenario

Control Measures
Guidance for Environmental Management on Site (GEMS) 1 provides information and assistance in minimising the effect on habitats during construction work.

Guidance for Environmental Management on Site (GEMS) 4 provides information and assistance in dealing with materials likely to be affected by dust blow.

Guidance for Environmental Management on Site (GEMS) 11 provides information and assistance for making for working over, in or near water/sewers.

Legislation Liable to be Breached under Worst Case Scenario
Anti-Pollution Works Regulations 1999.

Control of Pollution Act 1974.

Environmental Protection (Duty of Care) Regulations 1991, as amended.

Groundwater Regulations 1998.

Public Health Act 1936, Section 259.

Salmon and Freshwater Fisheries Act 1975, Section 4.

Sewerage (Scotland) Act 1968.

Trade Effluent (Prescribed Processes and Substances) Regulations 1989, as amended.

Aspect
Waste-water tanks for site cabin waste water overflow onto the ground.

Reference Number	Impact	Controlling Legislation		Score under Worst Case Conditions 2.3	Significance under Worst Case Conditions S2 Not Significant

Reference Number

Impact
Contamination of land, controlled waters or sewers.

Controlling Legislation
None Applicable

Score under Worst Case Conditions
2.3

Significance under Worst Case Conditions
S2 Not Significant

Score after Control Measures
1.3

Significance after Control Measures
S2 Not Significant

Worst Case Scenario
Waste water allowed to flow freely on to land that drains to a watercourse.

Company Objective

Action taken to Deal with Objective
Maintain present control measures.

Directive Liable to be Breached under Worst Case Scenario

Legislation Liable to be Breached under Worst Case Scenario

Anti-Pollution Works Regulations 1999.
Control of Pollution Act 1974.

Environmental Protection (Duty of Care) Regulations 1991, as amended.

Groundwater Regulations 1998.

Salmon and Freshwater Fisheries Act 1975, Section 4.

Sewerage (Scotland) Act 1968.

Control Measures
Guidance for Environmental Management on Site (GEMS) 7, 8 & 9 provide information and assistance in dealing with the keeping, treating and disposal of waste materials generated on site.

Aspect
Spillage of fuel, oil or chemicals during the draining, pumping out, purging or decanting of a tank, drum or other container.

Reference Number

Impact
Contamination of land, controlled waters or sewers.

Controlling Legislation
None Applicable

Score under Worst Case Conditions
2.5

Significance under Worst Case Conditions
S3 Significant

Score after Control Measures
1.3

Significance after Control Measures
R2 Not Significant

Worst Case Scenario
All the contents of the tank enter controlled waters.

Company Objective

Action taken to Deal with Objective
Maintain existing control measures.

Directive Liable to be Breached under Worst Case Scenario

Legislation Liable to be Breached under Worst Case Scenario

Anti-Pollution Works Regulations 1999. Control of Pollution Act 1974.

Environmental Protection (Duty of Care) Regulations 1991, as amended.

Groundwater Regulations 1998.

Salmon and Freshwater Fisheries Act 1975, Section 4.

Sewerage (Scotland) Act 1968.

Control Measures

Guidance for Environmental Management on Site (GEMS) 3 provides information and assistance when storing and handling materials on site.

Aspect
Removal of vegetation, trees and hedges.

Reference Number	*Impact*	*Worst Case Scenario*	*Controlling Legislation*	*Company Objective*	*Action taken to Deal with Objective*	*Score under Worst Case Conditions*	*Score after Control Measures*	*Significance under Worst Case Conditions*	*Significance after Control Measures*
	Degradation or loss of habitat (including impact on visual amenity).	Damage to a protected habitat or species	None Applicable		Maintain existing control measures.	2.4	1.2	S4 Significant	R1 Not Significant

Directive Liable to be Breached under Worst Case Scenario

Legislation Liable to be Breached under Worst Case Scenario

Conservation (Natural Habitats, etc.) Regulations 1994, as amended.

Hedgerow Regulations 1997.

Protection of Badgers Act 1992.

Town and Country Planning (Trees) Regulations 1999.

Wildlife and Countryside Act 1981.

Control Measures

Guidance for Environmental Management on Site (GEMS) 1 provides information and assistance for habitat protection.

Aspect
Positioning of temporary lighting in sensitive locations (e.g. close to residential properties or protected sites/species).

Reference Number		
Impact	Nuisance or disturbance caused to local residents or protected species.	**Controlling Legislation** None Applicable
		Score under Worst Case Conditions 3.3
		Significance under Worst Case Conditions S2 Not Significant
Worst Case Scenario Lighting shines through bedroom window at night causing loss of sleep.	**Company Objective**	
	Action taken to Deal with Objective	**Score after Control Measures** 1.2
		Significance after Control Measures R1 Not Significant

Directive Liable to be Breached under Worst Case Scenario

Legislation Liable to be Breached under Worst Case Scenario
Environmental Protection Act 1990.
Statutory Nuisance (Appeals) (Scotland) Regulations 1996.

Control Measures

Environmental Inspection Monitoring Procedures

WEEKLY ENVIRONMENTAL CHECKLIST

Date ....

Tick appropriate box on inspection	Yes No N/A	Comments/action to be taken

1. WASTE DISPOSAL

a) Are Controlled Waste Transfer notes in place for all skips on site?

b) Are all skips secure?

c) Are all skips leaving site covered/netted?

d) Do all skips in use contain only the designated material?

e) Is the site clear of waste materials?

2. **MATERIALS STORAGE**

a) Are all materials stored in accordance with the manufacturer's instructions?

b) Are all storage areas/containers secured?

c) Are all non-compliant materials marked/segregated?

d) Is river floodplain clear of stored materials and waste?

3. **FUEL STORAGE**

a) Are all fuel storage areas properly bunded?

b) Are all items of plant being refuelled in accordance with Project Method Statement?

c) Are bund areas free of rainwater/spillage to maintain capacity?

4. **NOISE**

a) Have the required noise measurements been taken?

b) Are there any unusually noisy operations in progress?

c) Have affected person(s) been notified/informed?

5. **PLANT**

a) Have any items of plant been serviced in the period?

b) Are any items of plant in use overdue for a service?

c) Are there any items of plant which require servicing (on visual inspection)?

d) Are service records available and up-to-date for all company-owned plant on site?

6. **MATERIALS USE/WASTAGE**

a) Are all material reconciliations required up-to-date?

b) Have target wastage figures been reviewed against actual?

c) Are unused materials being returned to storage/ saved for further use?

d) Are all water supply pipes free of leaks?

7. **DUST**

a) Is dust under control?

b) Are there any uncontrolled dusty operations in progress?

c) Are roads clear of mud?

8. **TRAFFIC MANAGEMENT**

a) Are all traffic and pedestrian routes properly signed?

9. **HABITAT/RIVER**

a) Are all hedges, trees and plant protected as required?

b) Is the river protected from pollution by rainfall run-off?

c) Is the river free from pollution?

d) Are the wash-out concrete mixers/trucks properly controlled?

10. **TRAINING**

a) Have operatives/staff on site had induction training?

b) Have operatives been trained in general site environment management?

c) Have all operatives been trained in the activity method statement for their current activity?

d) Is any further training required?

Signature of person making inspection ....

Job Title .... Date ....

Countersigned by Contracts/Project Manager .... Date ....

12 Post-Contract Review

Introduction

This short concluding chapter outlines the concept and practice of post-contract review. Construction projects are often subject to two distinct types of post-contract review. The first is the client's post-contract evaluation, focusing on the client's perceptions of performance by those parties employed by the client on the project – such as design consultants, the principal contractor and nominated sub-contractors. The second type of post-contract review is the principal contractor's internal, or intra-organisational, evaluation of its own performance throughout the construction project. This focuses on the efficiency and effectiveness of corporate and project management systems, procedures and site practices. The purpose of both types of review is to provide the contractual parties with an opportunity to reflect on their practices with a view to importing continuous improvement to the management of the construction process.

Client's Post-Contract Evaluation

From the client's perspective, it is essential that a post-contract evaluation is conducted to determine the real success, or otherwise, of the project which the client procured. The focus of the evaluation is to assess the effectiveness of procurement used, the form of contract administered, the general control of the contract, including the siteworks, and the quality of the finished product, together with an appraisal of the contribution and performance of the various contractual parties involved. This may include the project's consultants, nominated subcontractors and nominated suppliers in addition to the principal contractor. This evaluation can be used to inform and guide the client organisation when considering the procurement and administration of subsequent construction projects.

Contract performance of the principal contractor, for example, can be carried out by measuring actual performance on the project against pre-tender expectations and known benchmarks and performance indicators. The outcome of such evaluation may be used to consider the principal contractor's suitability for future works to be let by the client. It is therefore of utmost interest to the principal contractor to have performed well on the project and for this to be recognised and acknowledged by the client.

Post-Contract Evaluation Meetings

It is not uncommon for the client to hold a post-contract evaluation meeting to further inform the judgements made of project performance. Such meetings are attended by the representatives from each of the contractual parties – for example client/client's lead consultant, other specialist consultants, client's financial consultant/quantity surveyor, planning supervisor (health and safety), key nominated subcontractors, principal contractor's contracts manager/site manager, site quantity surveyor, key sub-contractors, health and safety officer/manager environmental manager and a personal assistant/secretary (to take minutes of the meeting). This formal meeting will seek to identify aspects of good practice and matters which raised concern during the contract, with a view to discussion and reflection focusing primarily on encouraging improvements to be sought from project experience to inform and enhance future projects undertaken by the participants. As clients frequently work with the same consultants and principal contractors from project to project, in particular when selective tendering and partnering arrangements are used, such reflection allows the opportunity not only to improve processes and procedures but to reinforce inter-organisational and inter-personal relationships.

Assessing the Principal Contractor's Performance

Post-contract evaluation of the principal contractor's performance by the client organisation will embrace specific aspects. In addition to evaluating, in overall terms, the effectiveness of procurement form used, the client may seek to examine the principal contractor's:

- understanding of the project requirements and objectives
- ability to maintain effective dialogue and communication
- timeliness in carrying out the works
- co-ordination and scheduling of contractors (subcontractors)
- timeliness in handling contract variations
- effectiveness in keeping to project programme/budget
- arrangements for maintaining supervision and control
- maintenance of the site and its environs (environment)
- effectiveness of construction safety
- quality of delivery (workmanship)
- ability to complete the project
- position regarding defects remediation.

To facilitate the client's evaluation, a series of questions related to the above points can be rated on a 'Likert' scale, from not-acceptable performance to excellent performance. These can then be summarised to present a satisfaction score as a quantitative, although judgemental, measure of performance. Qualitative information may also be gathered in the form of evaluator's comments which assist in contextualising the rating score determined. Such a mechanism is shown in Figure 12.1. Similarly, the client may undertake an evaluation of other inputs to the project, an example being a consultant post-contract evaluation as shown in Figure 12.2.

Principal Contractor's Post-Contract Review

Post-contract review presents an opportunity for the principal contractor to reflect on the siteworks and learn from its experiences. In addition to focusing on its own site construction activities, review will take in an evaluation of its relationships with the client organisation, consultants, subcontractors, suppliers, regulatory bodies and other parties with whom interaction has taken place. Contract documentation will also be reviewed, together with an evaluation of general site administration and control (Figure 12.3).

Systems and Procedures Improvement

As many aspects of managing the project centre on the implementation of management systems at both corporate level and project level, such systems will be evaluated comprehensively. An intrinsic part of maintaining and checking management systems during the project focuses on periodic and systematic auditing. Such information can be gathered, collated and synthesised to provide a profile of systems effectiveness throughout the project. A key element for scrutiny is the suite of project management plans – quality, safety and environment used in translating the corporate management systems into management procedures implemented on site. Attention will be given to carefully evaluating the efficiency and effectiveness of management procedures and marking practices, as these embrace all the

---

CLIENT'S POST-CONTRACT REVIEW
PRINCIPAL CONTRACTOR'S EVALUATION FORM

Project No. .... Project Title ....

Name of Principal Contractor ....

Project Location .... Start Date .... Finish Date ....

EVALUATOR .... RATING SCALE .... LEGEND:

N/A: not applicable; 1: not acceptable; 2: poor; 3: average; 4: good; 5: excellent

---

		RATING
1	General understanding of project requirements and objectives	
2	Ability to maintain effective dialogue and communication	
3	Timeliness in carrying out the works	
4	Co-ordination and scheduling of contractors	
5	Timeliness in handling variations	
6	Effectiveness in maintaining programmes/budgets	
7	Arrangements for maintaining supervision and control	
8	Maintenance of the site and its environs	
9	Effectiveness of construction safety measures	
10	Ability to complete the project	
11	Effectiveness/ability in carrying out defects remediation	

EVALUATOR'S COMMENTS

....

....

....

....

....

....

....

| | OVERALL RATING | |

Figure 12.1 *Client's post-contract review: principal contractor's evaluation form*

```
┌─────────────────────────────────────────────────────────────────┐
│                  CLIENT'S POST-CONTRACT REVIEW                    │
│                  CONSULTANT EVALUATION FORM                       │
└─────────────────────────────────────────────────────────────────┘
```

Project No. .... Project Title ....

Name of Consultant ....

Project Location .... Start Date .... Finish Date ....

EVALUATOR .... RATING SCALE .... LEGEND:

N/A: not applicable; 1: not acceptable; 2: poor; 3: average; 4: good; 5: excellent

		RATING
1	Design reflected an understanding of the project elements	
2	Design reflected a comprehensive investigation of site conditions	
3	Design addressed functional space relationship	
4	Design met building regulations	
5	Relationship between client organisation and consultant	
6	Design solution reflected innovative construction options	
7	Consultant's perception of project variables (time/cost/quality)	
8	Ability to meet time/cost programmes/budgets	
9	Aesthetics of the building/structure	
10	Ability to complete/distribute drawings/specifications	
11	Degree of co-ordination among areas of work (development/construction)	
12	Degree of definition of specifications	
13	Timeliness of flow of project communication	
14	Effectiveness of project control (meetings/administration/site checks)	
15	Promptness in checking/handling payments	
16	Promptness in handling project variations	
17	Ability to handle requests for further information	
18	Responsiveness to handling problems/difficulties	
19	Effectiveness of overall site/project control	
20	Ability to complete final project phase and handle defects	

EVALUATOR'S COMMENTS

....

....

....

| | OVERALL RATING | |

Figure 12.2 *Client's post-contract review: consultant evaluation form*

PRINCIPAL CONTRACTOR'S
POST-CONTRACT EVALUATION FORM

Project No. .... Project Title ....

Client .... Lead Consultant ....

Project Location .... Procurement Form ....

Contract Form .... Dates/Duration ....

EVALUATOR .... RATING SCALE .... LEGEND:

N/A: not applicable; 1: not acceptable; 2: poor; 3: average; 4: good; 5: excellent

		RATING
1	Effectiveness of construction methods adopted	
2	Efficiency of resource provision (labour)	
3	Efficiency of resource provision (plant)	
4	Efficiency of resource provision (materials)	
5	Effectiveness of site management and control	
6	Effectiveness of supervision on site	
7	Appropriateness of health and safety procedures	
8	Appropriateness of environmental procedures	
9	Maintenance of quality/workmanship	
10	Effectiveness of contract administration	
11	Relationship with client organisation	
12	Relationship/liaison with lead consultant	
13	Relationship/liaison with contractors	
14	Relationship/liaison with suppliers	

COMMENTS

....

....

....

....

....

OVERALL RATING

Figure 12.3 *Principal contractor's post-contract review*

project oversight management and workplace supervisory activities. Management systems, commonly based on recognised international standards, will need to be reviewed in relation to those standards to determine compliance and any non-compliance.

It is essential that all generic and project-specific elements of management plans be considered. While project-specific aspects will, obviously, vary from project to project, the generic aspects will commonly be carried forward from project to project. It would not be in the principal contractor's interest to be carrying forward deficient procedures to subsequent projects, as repeated misuse would only compound system/plan ineffectiveness. Procedures should be reviewed to identify difficulties and also effective practice such that difficulties can be resolved and improvements made. Such management systems improvement will provide invaluable evidence when the principal contractor seeks to revalidate its systems under standards-based certification schemes.

Summary

Post-contract review is an important and concluding phase in the principal contractor's management of the project. Feedback from the client's post-contract evaluation of service inputs to the project, together with the principal contractor's own post-contract review, provides much invaluable information on organisation performance. Moreover, it can provide information useful to the betterment of corporate and project organisational systems and project management planning for subsequent construction contracts. In the wider perspective, post-control review can also contribute usefully and significantly to the implementation, upkeep and improvement of management systems. This is needed for these systems to maintain their effectiveness, currency and validity within management standards assessment, or certification, schemes.

Select Bibliography

Many books have been written over the years on almost every aspect of construction management. The topics contained in each chapter of this book are an introduction to key management concepts, principles, systems and practices. For detailed information on each aspect and other aspects of construction management, the reader is directed to key texts listed in this Select Bibliography.

Ashford J.L. (1990) *The Management of Quality in Construction*. Spon, London.

Fryer B. (1997) *The Practice of Construction Management*. Blackwell, Oxford.

Griffith A. (1994) *Environmental Management in Construction*. Macmillan (now Palgrave), Basingstoke.

Griffith A. and Howarth T. (2001) *Construction Health and Safety Management*. Longman, Harlow.

Griffith A., Stephenson P. and Watson P. (2000) *Management Systems for Construction*. Longman, Harlow.

Harris F. and McCaffer R. (1991) *Management of Construction Equipment*. Macmillan (now Palgrave), Basingstoke.

Harris F. and McCaffer R. (1995) *Modern Construction Management*. Blackwell, Oxford.

Illingworth J. (2000) *Construction Methods and Planning*. Spon, London.

Lavender S. (1996) *Management for the Construction Industry*. Longman, Harlow.

Moore D.R. and Hague D.J. (1999) *Building Production Management Techniques*. Longman, Harlow.

Naoum S. (2001) *People and Organisational Management in Construction*. Thomas Telford, London.

Oxley R. and Poskitt J. (1996) *Management Techniques Applied to the Construction Industry*. Blackwell, Oxford.

Pilcher R. (1992) *Principles of Construction Management*. McGraw-Hill, London.

Index